AF332977

IET ENERGY ENGINEERING SERIES 90

Clean Energy Microgrids

Other volumes in this series:

Clean Energy Microgrids

Edited by
Shin'ya Obara and Jorge Morel

The Institution of Engineering and Technology

Published by The Institution of Engineering and Technology, London, United Kingdom

The Institution of Engineering and Technology is registered as a Charity in England & Wales (no. 211014) and Scotland (no. SC038698).

First published 2017

The Institution of Engineering and Technology
Michael Faraday House
Six Hills Way, Stevenage
Herts, SG1 2AY, United Kingdom

www.theiet.org

British Library Cataloguing in Publication Data
A catalogue record for this product is available from the British Library

ISBN 978-1-78561-097-4 (hardback)
ISBN 978-1-78561-098-1 (PDF)

Typeset in India by MPS Limited
Printed in the UK by CPI Group (UK) Ltd, Croydon

Contents

7   Microgrids in Japan   **233**
Jorge Morel

8   Microgrids in Europe   **259**
Sergio Rivera and Tomas Valencia

9   Microgrids in the United States   283
Sergio Rivera and Miguel Leon

Preface

We are blessed to be part of the ongoing energy revolution and to have the great chance to contribute to the construction of a sustainable society for future generations. The drive for major changes to power systems configuration and operational approaches comes from the urgent need to reduce pollution and greenhouse gas emissions responsible for global warming and largely caused by the burning of conventional fossil fuels in centralised conventional power plants. Another reason is that fossil fuels are depleting, and new forms of energy generation are needed, and that the increasing world population will need more a more energy in the future. Thus, in order to continue the extraordinary progress humanity is experiencing in key aspects of our lives, the energy generation and consumption have to be done in a sustainable way.

One way to forge a sustainable energy sector is the use of renewable generation with zero or lower greenhouse gas emissions. The main issues with most renewables, such as wind and solar, are the inherent uncertainty and variability of their outputs. Storage systems appear then, as the key to solve these issues, but unfortunately at an additional cost. In order to successfully integrate these variable renewable-generation units into existing power grids and also to improve reliability and energy security in a given region, technologies such as those applied to the smart grid and microgrids are now under continuous development.

Microgrids possess special characteristics different from those of large centralised conventional power systems, and their technology is evolving rapidly. The motivation for the book came from the need for updated material that presents the latest microgrid projects around the world, together with the main aspects of this emerging technology. In this book, the authors of each chapter present a timely and relevant material with the latest technology, pilot projects, and other important aspects of microgrids. We expect that the book can enrich and contribute to the objectives and aspirations of the reader in the knowledge of microgrids and, indirectly, smart grids.

Acknowledgements

We would like to thank The Institution of Engineering and Technology, and particularly Christoph von Friedeburg and Paul Deards, for extending us the invitation to edit a book on microgrids and for the great opportunity to publish in one of the most prestigious publishers of the world. Also, we would like to thank Jennifer Grace and Olivia Wilkins who supported us in all the process of book preparation and submission.

We would like to thank very specially the authors of each chapter for accepting the challenge of contributing to the book.

Finally, we are very grateful to Evans Chogumaira, currently with Transpower New Zealand, for reviewing the book and giving us his invaluable feedback and ideas on content and style.

Chapter 1

Origin of clean energy systems

Shin'ya Obara

1.1 Introduction: origin of clean energy systems

Microgrids (small-scale power grids), which are distributed energy systems, emerged due to the need of clean and sustainable generation to reduce CO_2 emissions and to improve energy security in a given region. In these grids, clean renewable energy or 'green energy', which output is typically unstable, can be utilised effectively by introducing a battery.

The design and operation planning of a compound energy system forming a microgrid with renewable energy sources is a dynamic, multivariate and non-linear problem. This chapter describes the case of optimisation of the design of a microgrid using neural networks (NNs) and genetic algorithms (GAs).

The first section describes the production-of-electricity prediction algorithm (PEPA) for a solar cell to inject power into a microgrid. In the PEPA, a layered NN is trained on past weather data; and the operation plan of the compound system of a solar cell and other energy systems is examined using this prediction algorithm. A dynamic operational-scheduling algorithm is developed using an NN, and a GA to provide predictions for solar cell power output. A study case of a microgrid with photovoltaics supplying electricity to nine residences in the city of Sapporo in Japan is described. In this work, the relationship between the accuracy of output prediction of the solar cell and the operation plan of the microgrid was determined. GA provides a facile method for solving optimised problem and can be easily adapted to complicated energy systems. Conventional GA requires a long runtime when the microgrid contains numerous energy sources and the solution must be highly accurate.

The second section introduces a preliminary step in which experimental design techniques, namely, an orthogonal array experiment and a factorial-effect chart, are used to find an operating point that is close to the optimal solution for the energy system. The optimal operation solution is determined by using the operation method obtained from the orthogonal array experiment as the initial generation of chromosomes for the conventional GA. This proposed method does not find a strictly mathematical optimal solution, but the quasi-optimum solution is far more accurate than that from conventional GAs and can be used industrially. The characteristics of the output power sources are found to strongly affect the analytic accuracy for the microgrid example.

1.2   Dynamic operational scheduling for a microgrid with renewable energy

1.2.1   Introduction

Microgrid technology with the capacity for sustainable energy operation has been widely discussed recently from the point of view reducing the environmental impact of society [1–3]. In this setup, the operation optimisation programme installed in the controller of a combined system is the most important aspect of the technology for determining the performance of the system [4]. However, as an output prediction for the green energy contribution to the system is required, the dynamic operation plan of a system that combines conventional energy equipment (for example a diesel engine, a gas engine, a fuel cell etc.) and green-energy equipment can be very difficult to design. In this work, we use a NN to obtain output predictions for a solar cell. Weather data from the past 14 years (amount of solar radiation and ambient temperature) is fed into the learning process of the NN. This NN PEPA was developed by the author and is described in [5].

Power fluctuations are known to occur in systems that utilise green energy on an independent microgrid and that experience large or rapid changes in load [6]. Given this, power storage equipment must be introduced to improve the dynamic characteristics of the microgrid. The cost of a storage-of-electricity system falls by commercial production of electric vehicles or a hybrid cars, on the other hand, the performance is rising [7]. With this in mind, this chapter investigates algorithms for the operation planning of a microgrid that combines conventional type energy equipment, a solar cell and a battery. As a microgrid is typically built up of two or more energy systems, we have to solve a non-linear problem with many variables. In this chapter, the operation condition of the generating equipment is expressed with chromosome codes; these are fed into a GA that determines the optimal operation plan.

1.2.2   Independent microgrid with renewable energy and battery

1.2.2.1   System configuration

Figure 1.1 shows the case of an independent microgrid that distributes power from a small collection of generating equipment (here we use gas engine generators as an example, but our analysis is not limited to this case). The microgrid is controlled by a controller in which we have installed a dynamic operation planning software. In this section, we investigate the power supply-and-demand characteristics of the microgrid.

1.2.2.2   Dynamic operation planning

Operation planning based on a solar cell output power model
Figure 1.2 shows the operation method of a microgrid with a solar cell. Dynamic operation of the microgrid is planned based on the output power model of a solar cell. In this section, the solar cell output model is based on PEPA predictive data and past average weather data [8,9]. The PEPA was designed specifically for the solar cell [5]. Figure 1.2(a) shows the method of planned dynamic operation based on the output power model of the solar cell described in the introduction.

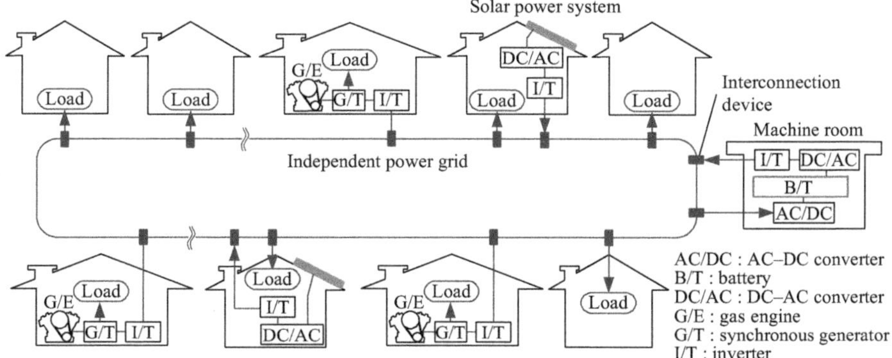

Figure 1.1   Independent microgrid system using gas engines

From time t_0 to time t_2, there is no output from the solar cell, and so in this interval, more electric energy is stored in the battery than is required by the load. In light of this, generating equipment is operated during this period [Generator output (1)]. Each piece of generating equipment is operated at fixed load near the maximum efficiency point. The supply of electric power in the microgrid is adjusted by controlling the number of pieces of power equipment in operation. When the supply of electric power exceeds the load in the period from t_0 to t_2, the amount of surplus charge is moved into the battery. There is an output of the solar cell in the period t_3 to t_6. In the period t_2 to t_4, the system responds to the load by discharging the battery and supplying charge from output of the solar cell. As the output of the solar cell exceeds the load in the period t_5 to t_4, surplus power can again be stored in the battery. If the output characteristics of the solar cell and the characteristic of power load are able to be predicted in the early morning, one can minimise the number of hours of operation of equipment and the number of charge and discharge cycles of the battery throughout the day. It is known, however, that load following with operation of the generator is disadvantageous, in terms of installed capacity and the hours of operation required of the power equipment, when compared with the output predictive model of a solar cell [5].

Error of the output predictive model of a solar cell, and its influence on an operation plan

Figure 1.2(b) shows the operation plan when there is a difference between the output model of the solar cell as shown in Figure 1.2(a) and an actual solar cell output. This system starts out (during t_0 to t_2) by following the operation plan shown by the command of the controller in Figure 1.2(a). However, as shown in Figure 1.2(b), as there is so little power being produced in the solar cell, the period of t_4 to t_5 shown in Figure 1.2(b) requires additional operation of power equipment [Generator output (2)]. The system is continuously controlled by the controller to follow the operation shown in Figure 1.2(a). However, there is little storage of electricity from the solar cell in Figure 1.2(b) compared with Figure 1.2(a). For this reason, when the battery is discharged in the period t_7 to t_8, the system will change

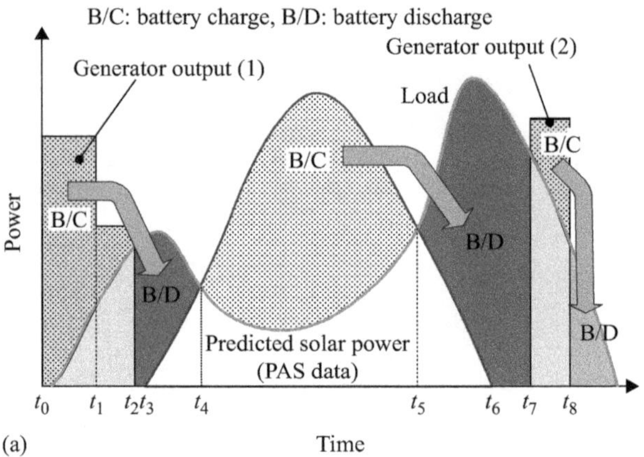

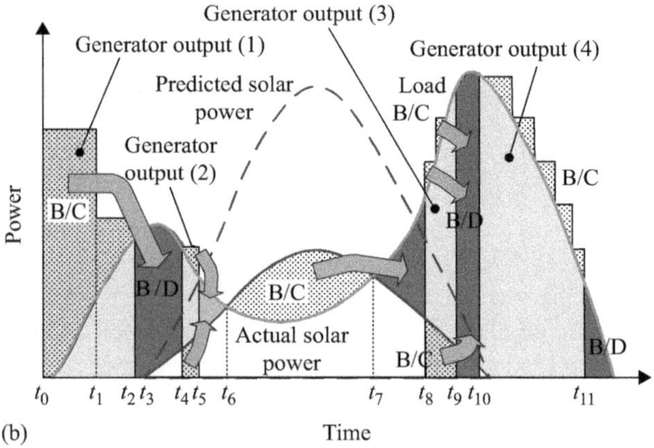

*Figure 1.2   Operation planning of a microgrid with a solar power: (a) operation
planning based on the PEPA predictive data and (b) operation
planning when taking into consideration the error between PEPA
predictive data and actual solar power*

to load following operation [Generator outputs (3) and (4)]. As a result, the number
of hours of operation of power equipment increases, and the additional capacity of
the battery cannot be fully utilised.

1.2.2.3   Solar cell system

The area of power-generation of the solar cell connected to the microgrid will
be called S_{sol}, and we are assuming a polycrystalline silicon type solar cell. The
production of electricity of the solar cell P_{sol} is calculated using (1.1). The power-
generation efficiency changes as the temperature of the solar cell T_{ref} changes (the
efficiency falls as the temperature increases). The temperature coefficient in this case
is called R_T. T_∞ is a reference temperature, and η_{sol} is the power-generation efficiency
of the solar cell at T_∞. H_D in (1.1) expresses the solar radiation intensity for direct

delivery (the intensity of radiation which enters into the acceptance surface). Moreover, H_S in (1.1) expresses the solar radiation intensity of the dispersion component.

$$P_{\text{sol}} = S_{\text{sol}} \cdot \eta_{\text{sol}} \cdot (H_D + H_S) \cdot \left\{ 1 - (T_{\text{ref}} - T_\infty) \cdot \left(\frac{R_T}{100} \right) \right\} \tag{1.1}$$

1.2.3 Power balance and objective function

1.2.3.1 Power balance

Equation (1.2) expresses the power balance equation in the proposed microgrid. The left-hand side of the equation is the power output from the equipment that makes up the system, and the right-hand side expresses the power consumed by the microgrid. $E_{\text{gen},i,t}$, $E_{bt,t}$ and $E_{\text{sol},t}$ of the left-hand side express the output of the generating equipment, battery and solar cell between time t and $t+1$, respectively. Moreover, N_{eng} is the number of units of generating equipment connected to the microgrid. $E_{\text{need},j,t}$ is the power demand at time t of the house j. N_{house} is the number of the houses connected to the microgrid. The last term on the right-hand side of (1.2) ($\Delta E_{\text{loss},t}$) expresses the power loss in the system. The charge-and-discharge efficiency of a battery, power transmission loss etc. are included in this term. In the analysis of this section, the charge-and-discharge efficiency of a battery is the only effect included in this term. The sampling interval for each piece of equipment, such as the generating equipment, a solar cell or a battery, is set as 1 h in this section.

$$\sum_{i=1}^{N_{\text{eng}}} E_{\text{gen},i,t} + E_{bt,t} + E_{\text{sol},t} = \sum_{j=1}^{N_{\text{house}}} E_{\text{need},j,t} + \Delta E_{\text{loss},t} \tag{1.2}$$

1.2.3.2 Objective function

The number of hours of operation of the generating equipment EOT_t between time t to $t+1$ is obtained by calculating (1.3). However, $n_{\text{gen},i,t}$ expresses the operational status of the generating equipment i in the time between t and $t+1$ ('1' indicates operation, '0' indicates idle). N_{gen} is the number of units of generating equipment installed in the microgrid. The total number of hours of operation for all units of generating equipment that were in operation during the periods $t = 1, 2, \ldots, P_{\text{sys}}$ is calculated by (1.4). The optimisation of the dynamic operation plan of the microgrid is examined using a GA. In the GA, the objective function shown by (1.4) is defined as an adaptive value. The solution that closely satisfies (1.4) is described as having a 'large adaptive value'.

$$\text{EOT}_t = \sum_{i=1}^{N_{\text{gen}}} n_{\text{gen},i,t} \tag{1.3}$$

$$\text{Engine operation time} = \text{minimize} \left(\sum_{t=1}^{P_{\text{sys}}} \text{EOT}_t \right) \tag{1.4}$$

1.2.4   Analysis method

1.2.4.1   Production-of-electricity prediction algorithm of solar cell (PEPA) [5]

A layered NN is introduced, and the production of electricity of a solar cell is predicted according to the following procedures.

Input-and-output data of NN used by PEPA

Figure 1.3 expresses the input-and-output data of the NN used for PEPA. *dw* expresses the present date, and *t* expresses the present time. In the learning and analysis process of the NN, the average amount of solar radiation and average outdoor air temperature are inputs for each time of the present date. The input data described in the introduction is fed into (1.1), and the production of electricity P_{sol} of the solar cell is obtained. This P_{sol} is used as the teaching data in the learning process of the NN.

Input data

The data entered into the NN learning process includes the average amount of solar radiation and average outdoor air temperature for each time of the day (from time 0 to *t*). On the other hand, the average amount of solar radiation and average outdoor air temperature, from $t = 1$ to 24 for a given day, give the values measured on the same time and the same day of the previous year, as obtained from 'the standard weather and the solar radiation database on weather from the government office and AMEDAS (1990–2003)' [8] and 'NEDO technical information data base (METPV-3)' [9]. The data entered into the NN during the analysis process is the same as that entered during the learning process. The average amount of solar radiation and average outdoor air temperature from $t = 1$ to 24 for a given day give the data is measured at same time the previous day.

Structure of the NN and output data

Figure 1.4 shows the structure of the NN used in PEPA. Figure 1.4(a) and (b), respectively, shows the learning process and the analysis process. The NN used in this proposal has three layers. In the learning process, the input data (x_1 to x_{48}) (described in Section 1.2) is fed into the first layer (the power-input layer), and the teaching data (y_1 to y_{24}) is fed into the third layer (the output layer). The weight of each network connection between neurons is decided using back propagation [10] so that the relationship between each input data point and each teaching data point

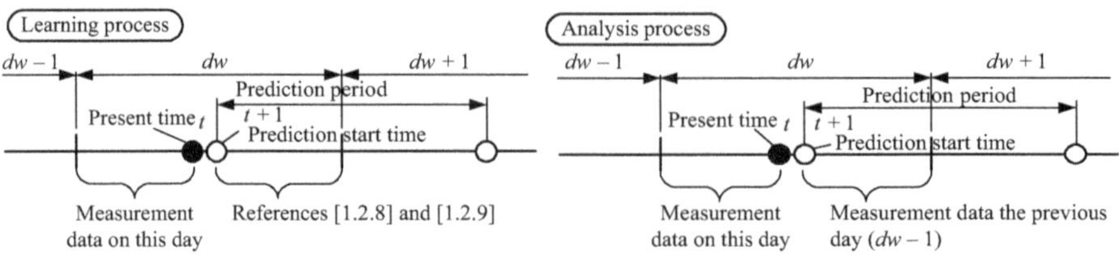

Figure 1.3   Input data for the NN

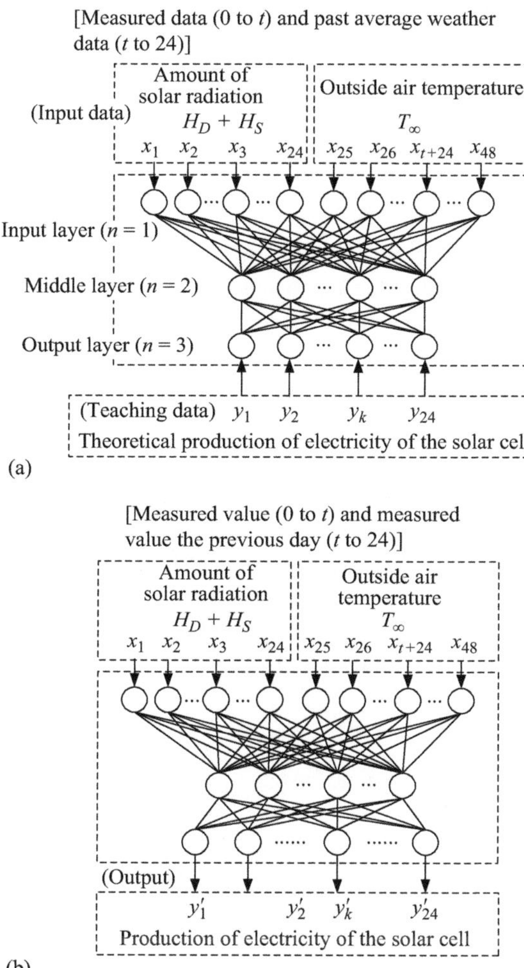

Figure 1.4   Layered neural network of the proposed system: (a) learning process and (b) analysis process

may be realised. Input data can be given to a learned NN (in the analysis process), and the solar cell output power (y'_1 to y'_{24}) in each time of dw can be obtained from the output layer.

1.2.4.2   Optimisation of dynamic operation using a genetic algorithm (GA)

Chromosome model
Dynamic operation planning of the microgrid is optimised using a GA based on the solar cell output power model (the PEPA predictive data and the past average

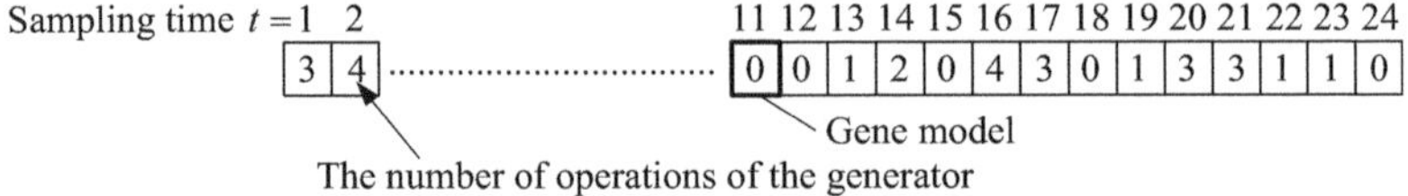

Figure 1.5   Chromosome code used for genetic algorithm

weather data). Figure 1.5 expresses the chromosome code used by the proposed GA. One chromosome (individual) consists of 24 genes. Each gene shows the operational status of the various units of generating equipment that are on the grid during each time step in the date dw. With the number of the units of generating equipment connected to the grid and available for generation, the value of each gene is an integer and takes a value between 0 and N_{gen}. The genes of the initial generation's chromosome are decided using random numbers.

Multiplication and selection

By decoding the genes in a chromosome, the operating condition $n_{gen,i,t}$ of the generating equipment in time t can be obtained. By giving $n_{gen,i,t}$ to (1.3), EOT_t is calculable. Furthermore, the value of the objective function of each chromosome is calculated from (1.4). The software implementation is such that the number of individuals of high adaptive value (obtained by evaluating the objective function) may be multiplied at a fixed rate.

Crossover

Change of generation is repeated, adding the genetic manipulation of crossover to chromosomes to maintain diversity (an evaluative process). In the chromosome code, when arriving at the last generation, a fitness value decides the individual most suited to be the optimal operating method. In the crossover calculation, two parent chromosomes are selected by the crossover probability P_{cros} given beforehand, and the crossover position common to both of parent chromosomes is decided at random. The genes of both of the parent chromosomes on one side of the crossover position are swapped (crossed over), and the chromosome with a new gene is generated.

1.2.4.3   Analysis flow of operation planning

Analysis of the proposed GA optimisation of the operation plan for the microgrid is shown in Figure 1.6. In the Calculation part (a) of the figure, an electricity demand pattern, a solar cell output power model (PEPA or output power pattern of the solar cell based on the past average weather data) and the parameter of GA are entered. The initial generation's chromosome group is generated in Calculation part (b). In Calculation part (c), the fitness value of all the chromosomes is calculated and the order of chromosomes is decided at Calculation part (d) according to the magnitude of their fitness values. Chromosomes of large fitness value are made to increase in number at a fixed rate, while chromosomes of small fitness value are deleted. In Calculation part (e), a parent chromosome is

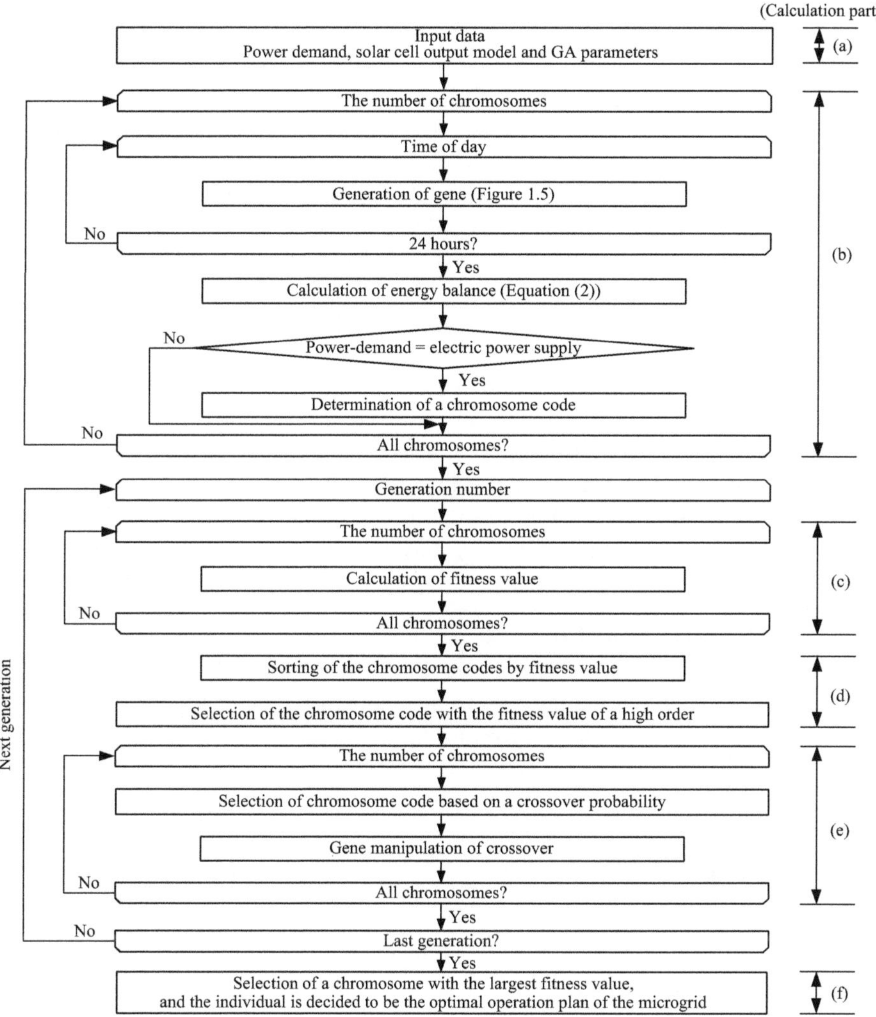

Figure 1.6   Calculation flow of the optimal operation plan using GA

chosen at random under crossover probability, a crossover position is decided at random, and genes are exchanged. In Calculation part (c)–(e), the calculation of the fitness value, the operation of arranging chromosomes in order of fitness value and exchange of the gene by crossover is done. These operations are calculated and repeated until we reach the final generation number. In the last generation's chromosomes, an individual with the highest fitness value is taken to be the optimal solution. Decoding the chromosome code of the optimal solution provides the operational plan of the generating equipment for each time step of day.

1.2.5   Case analysis

1.2.5.1   Analysis system

In this case analysis, we assume a microgrid connected to nine houses in Sapporo. In this section, the difference between the operation plan is calculated using the proposed method and that with the past average weather data is investigated. The generating equipment connected to the microgrid is arranged in five sets, and the power output of each piece of generating equipment set is 1 kW. As described in Section 1.2.5.2, the power load in this case analysis has a maximum of about 7.5 kW (Figure 1.7).

The solar cell is a flat plate type of polycrystalline silicon and covers an area of 150 m^2 facing south and sloping by 30 degrees. As the area of the solar cell installed in an average house is 20–40 m^2, the capacity of the solar cell linked to the proposed microgrid is equivalent to four to seven houses. The battery connected to the system is a nickel hydrogen type. The performance and specification were obtained from [10]. The analysed conditions of the solar cell, battery, each converter used in the simulation and an inverter are shown in Table 1.1. The loss of charge efficiency, discharge efficiency and natural discharge are all included in the table.

1.2.5.2   Analysis conditions

Power load

The average over time of the electricity demand pattern of nine houses in the representative day of the every month in Sapporo is shown in Figure 1.7 [11].

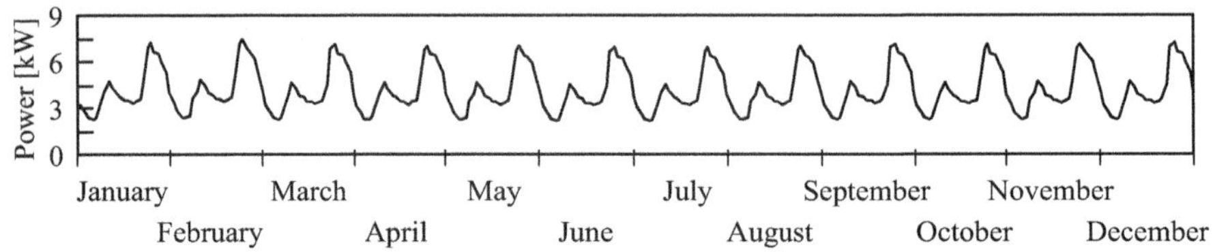

Figure 1.7   Power demand of the nine-houses microgrid in Sapporo

Table 1.1   Specifications of equipment

Solar cell type	Multicrystalline silicon
Generation efficiency of the solar cell	14%
Temperature coefficient of the solar cell	0.4%/K
Battery type	Nickel–hydrogen
Battery efficiency	90%
Efficiency of AC–DC converter	95%
Efficiency of DC–AC converter	95%
Efficiency of DC–DC converter	95%

Air-conditioning load is not included in the electricity demand pattern of Figure 1.7, and it is assumed that there are four residents per house on average. Space heating load is supplied by engine exhaust heat, and cooling load is unnecessary. Although the actual electricity demand pattern changes sharply on short time scales, a gradually varying time-averaged value is used throughout this chapter.

Amount of solar radiation and outdoor air temperature
In this case analysis, we will investigate about 6 days, from the second day to seventh day in each of March, June, September and December (the first day of every month is used to check the prediction of the PEPA analysis). Figures 1.8 and 1.9 show the measured amount of solar radiation and outside temperature for seven days in every month from 1990 to 2004 in Sapporo [8,9]. The daily fluctuations in the amount of solar radiation are typically large when compared to the daily fluctuations of the outdoor air temperature in every month.

GA parameters
In this case analysis, the number of an initial generation's chromosome codes is 8,000. In the genetic manipulation of multiplication and selection, 20 chromosome codes (that is 5% of the population) with the largest fitness values are made to increase their number in the next generation. The crossover probability is set to 0.2. The generation number was set to ten because of the large total number of chromosome codes. These parameters of GA were arrived at by a trial-and-error method, and the final values decided using analysis accuracy as a reference.

1.2.6   Analysis results

1.2.6.1   Prediction of solar cell output power via PEPA
Relationship between prediction start time and analysis accuracy
Figure 1.10 shows the predicted amount of solar radiation on 6 June 2007 in a south-facing set of 30 distinct angular orientations, is calculated using PEPA. In Figure 1.10, it will be the present time in 5, 8 and 11 o'clock, and each prediction start time will be at 6, 9 and 12 o'clock. Actual data refers to the weather data measured on 6 June 2007. We see that the difference between past average weather data [8,9] and actual data is larger than the difference between the actual data and the predicted results. Moreover, the data that came from using a start time of 9 or 12 o'clock is closer to the actual data than that which came from using a start time of 6 o'clock. In the result of Figure 1.10, if there is a large power input when using the weather data measured on that day (accordingly, the prediction start time is late), the analysis accuracy will have improved. In the analysis using PEPA later, prediction start time is 6 o'clock.

Prediction result in every month
The result based on the prediction of solar cell output power and the actual output power value, and past average weather data [8,9] obtained using PEPA in representative days every month is shown in Figure 1.11. One can see from Figure 1.11 that in many cases, one gets closer to the actual value by using the PEPA prediction than one would get by using past average weather data.

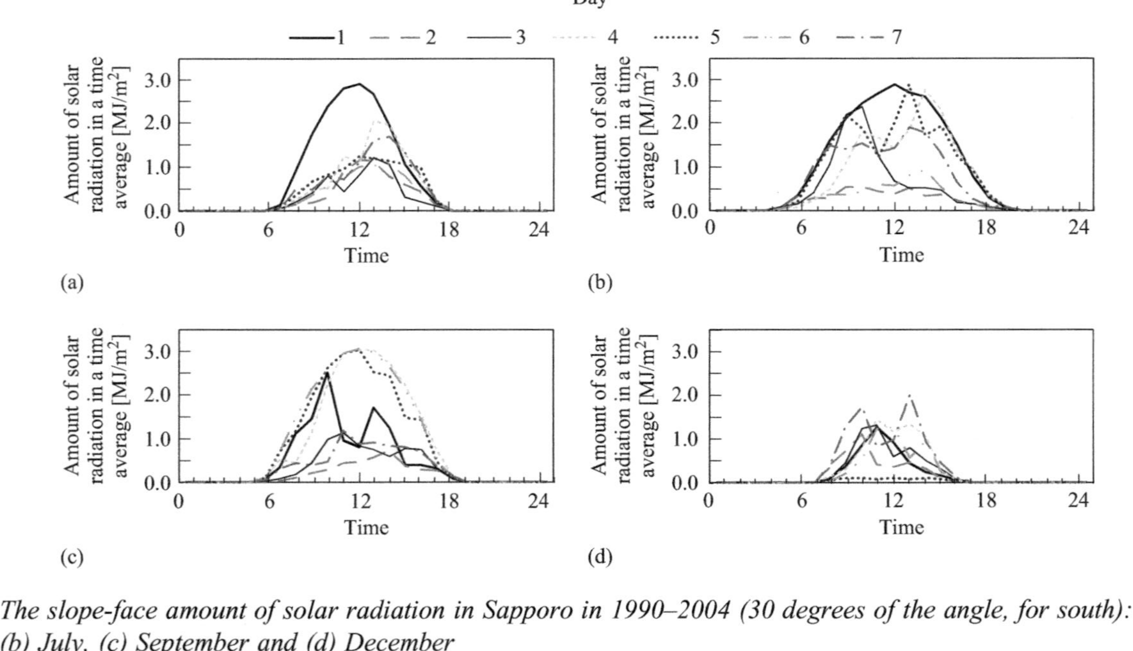

Figure 1.8   The slope-face amount of solar radiation in Sapporo in 1990–2004 (30 degrees of the angle, for south): (a) March, (b) July, (c) September and (d) December

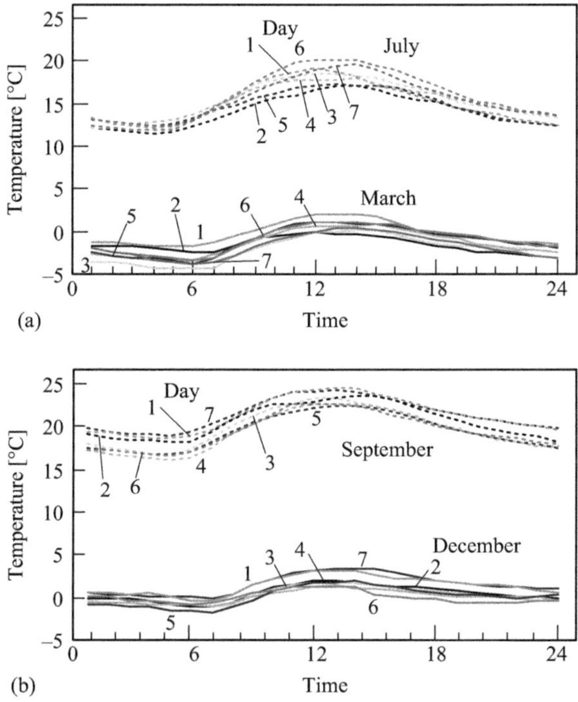

Figure 1.9 *Outside temperature data in Sapporo in 1990–2004: (a) March and July and (b) September and December*

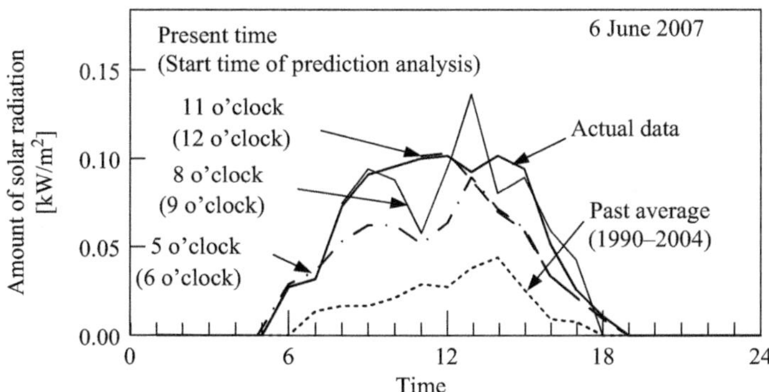

Figure 1.10 *The predicted amount of solar radiation on 6 June 2007 in a south-facing set of 30 distinct angular orientations*

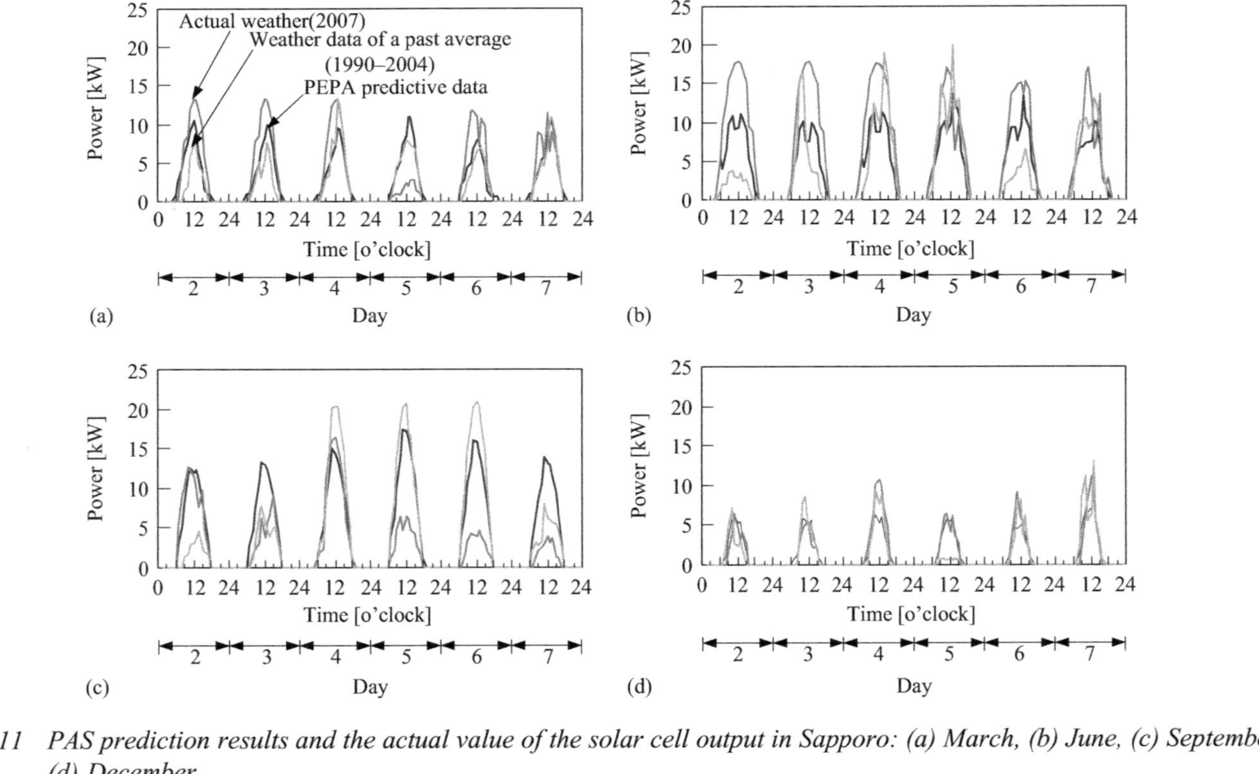

Figure 1.11   PAS prediction results and the actual value of the solar cell output in Sapporo: (a) March, (b) June, (c) September and (d) December

1.2.6.2   Prediction error of PEPA and operation method of generating equipment

Operation planning of generating equipment

Figure 1.12 shows the result of the dynamic operations analysis on 2 March 2007. To investigate the influence of operation planning on battery efficiency, we here set the battery efficiency to 100%. The battery efficiency in Table 1.1 means charge-and-discharge efficiency. In analyses other than Figure 1.12, the battery efficiency shown in Table 1.1 is used. Accordingly, the loss based on battery efficiency is not taken into consideration in Figure 1.12. Figure 1.12 shows the calculation result except charge-and-discharge loss of the battery, to clarify the power balance of consumption, the solar cell and the generating equipment. Figure 1.12(a) shows the result of operation planning of generating equipment using the solar cell output power prediction generated by PEPA. On the other hand, Figure 1.12(b) shows the result of operation planning of generating equipment using the actual amount of solar radiation and outdoor air temperature. If the PEPA-generated solar cell output power is the same as that coming from actual data, then the optimal operation planning method is shown in Figure 1.12(a). However, one must also note that errors are introduced in the PEPA predictions and actual operation may more closely resemble the result of Figure 1.12(c). Accordingly, the output power of the generating equipment as controlled by PEPA and that controlled by actual weather conditions are compared, and the system is operated in such a way as to accommodate the worst conditions (this is a typical way of doing things in similar systems).

Operation planning of battery

Figure 1.12(d) shows the results that come from the operation planning of a battery when using either PEPA predictions or actual data to calculate the solar cell output power. One sees that the operating characteristics of the battery greatly influence the method and duration of operation of the generating equipment. Looking at Figure 1.12(d), one sees that, in at least two different places, there is a significant difference in the battery capacity calculated in the two different schemes.

Output characteristics of the solar cell and operation planning

Figure 1.12(e) shows the operation planning of the generating equipment when using either past average weather data or actual weather data. Figure 1.12(f) shows the resultant battery capacity planning under the same conditions. Figure 1.12(g) shows the result of operation planning of generating equipment. When Figure 1.12(c) is compared with Figure 1.12(g), one sees that distinct methods of planning for generating equipment begin to deliver noticeably divergent results after about time 15. This happens as the output power models of the solar cell introduced into operation planning begin to produce diverging results. Accordingly, the accuracy of the output power prediction generated by PEPA has a large influence on operation planning of the generating equipment and the capacity planning of the battery.

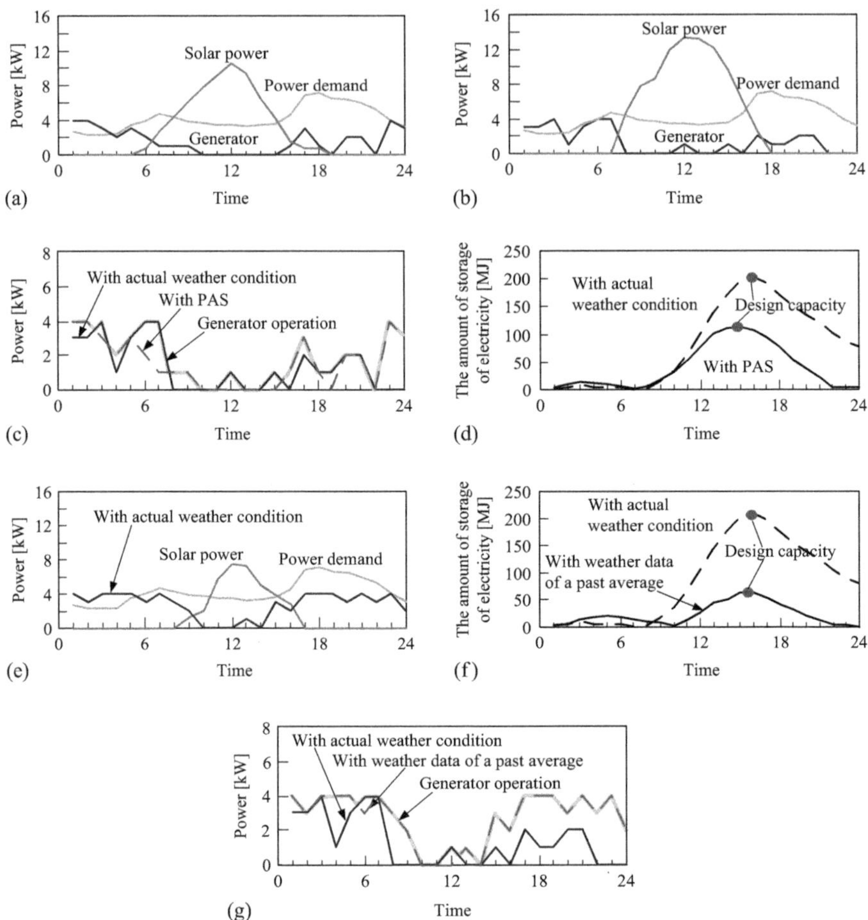

Figure 1.12   Result of the dynamic operation plan of the microgrid system (2 March 2007): (a) operation plan based on the PAS predictive data, (b) operation plan using actual weather data, (c) dynamic operation planning of the generator using the PAS predictive data and the actual weather, (d) operation plan of the amount of storage of electricity in the battery, (e) operation plan based on the weather data of a past average, (f) operation plan of the amount of storage of electricity in the battery based on the weather data of a past average and (g) dynamic operation planning of the generator using the weather data of a past average and the actual weather

1.2.6.3   Result of dynamic operation planning

Operation planning for a microgrid based on PEPA and the past average weather data

Figure 1.13 shows the result of dynamic operation planning of the system (including generating equipment and batteries) when using either past average weather data or PEPA prediction data as the solar cell output power model. The error in the solar cell output power in Figure 1.13 is the difference between the output determined by a given simulation and that determined from actual weather conditions. In operation of an actual microgrid, the difference is in the solar cell output power model described in the introduction and the solar cell output power operating under actual weather conditions. The planning of dynamic operations using the solar cell output power model as described in Section 1.2.6.2, an additional operation of generating equipment is expected in an unfavourable condition. For example we see, in Figure 1.13 (b), that the partial output power of the generating equipment on December 5 (calculated in the past average weather data scheme) exceeds 5 kW, the maximum power of the proposed system. We can make sense of this as the smallest amount of solar radiation comes in the month of December. Because of this, the solar cell output power grows small and the discrepancy with past average weather data grows large. Thus, if the error between the model output power and the actual output power is large, the working time of the generating equipment is expected to increase. In comparison, for operations planning using the PEPA predictive data (Figure 1.13), the generating equipment is never asked to exceed 5 kW in the entire month of December. We see here that operations planning by PEPA has an advantage over planning with past average weather data.

Actual system operation using the PEPA

If a difference occurs in the solar cell output power based on PEPA predictive value and actual weather conditions, the operation of generating equipment will follow the method described below. Power balance (see (1.2)) of the microgrid in the sampling time t is calculated, and when electricity demand exceeds supply, the generating equipment starts operation immediately. The amount of units of generating equipment sent into operation at this time is decided to be the minimum number plausible to avoid the case where the power supply exceeds the demand. However, this is not a limitation when large electricity demand is predicted at later time from the analysis of dynamic operation planning in that case. Accordingly, when demand is expected to exceed the maximum power supply (5 kW), we can increase the number of pieces of equipment in operation beforehand. Figure 1.14 shows the operation of the generating equipment and battery as a result of adding the modification (the operation method described above) to the prediction error of operation planning using PEPA shown in Figure 1.13. Figure 1.14 shows operation planning of the generating equipment using PEPA predictive data, as well as operation planning with adjustment of PEPA prediction error (here called 'Adjustment value'). In the actual operation of a microgrid, operation planning with PEPA and 'operation planning of the generating equipment by PEPA predictive data' and 'operation planning by adjustment value' in each sampling time are compared, during high power output.

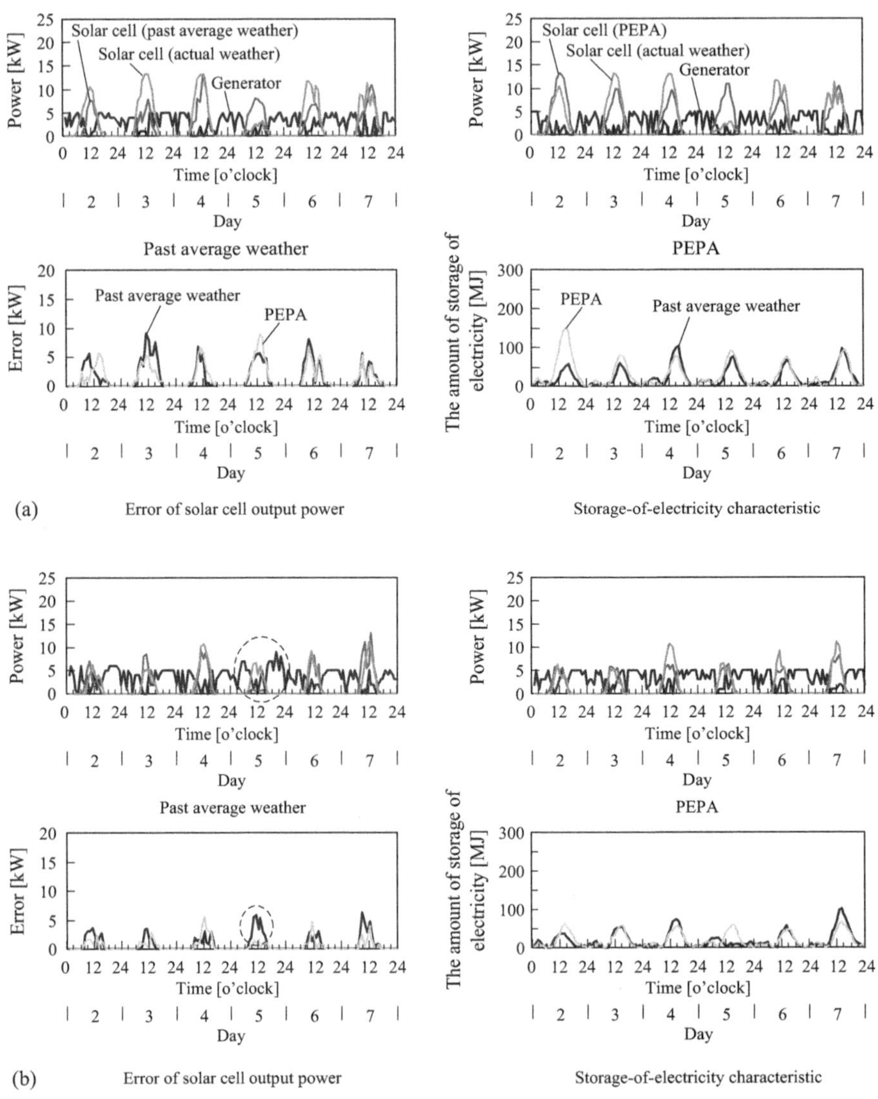

*Figure 1.13   Dynamic operation planning based on the past average weather
data and the PEPA predicted data. In these analyses, the influence
on operation planning by the error of 'the past average weather
data' and 'the PEPA predictive data' to actual weather is not
taken into consideration: (a) March and (b) December*

Hours of operation of generating equipment

Figure 1.15 shows the result of operating the generating equipment in the microgrid
according to the method described in the 'Actual system operation using the PEPA'
section. When the hours of operation using past average weather data and PEPA
prediction data are compared, there are fewer hours of operation of the generating

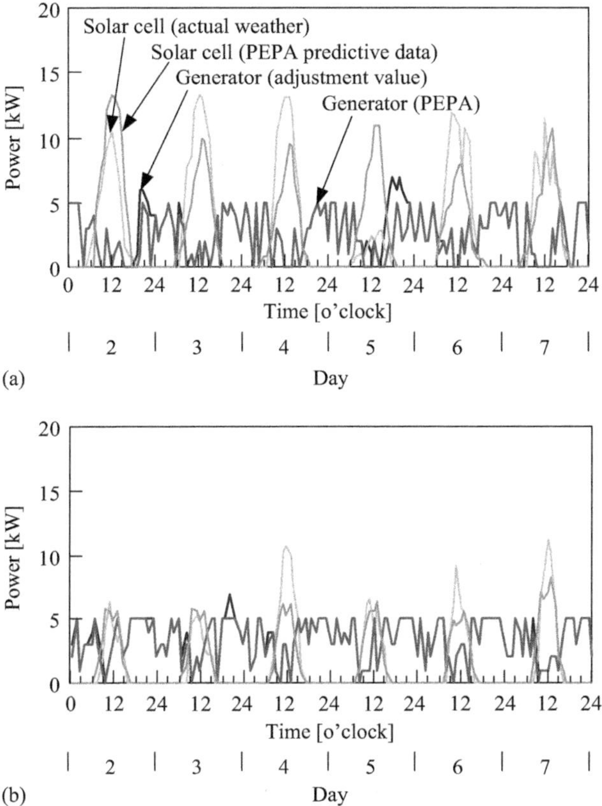

Figure 1.14   *Results of the dynamic operation planning of the power generator with PEPA prediction. The error of the PEPA predictive data to the actual weather is taking into consideration: (a) March and (b) December*

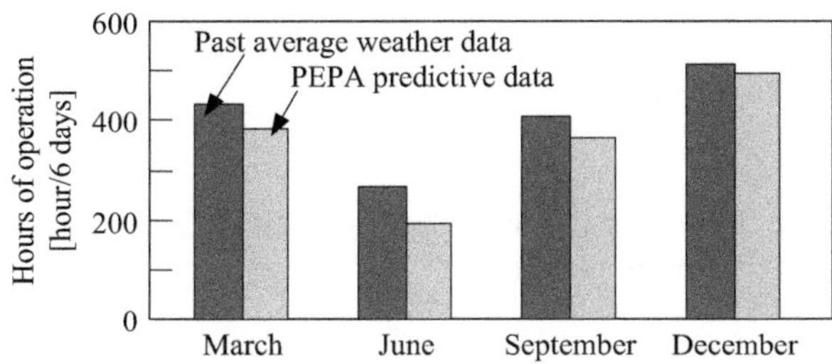

Figure 1.15   *Result of the generator hours of operation showing comparison between the past average weather data and the PEPA predictive data*

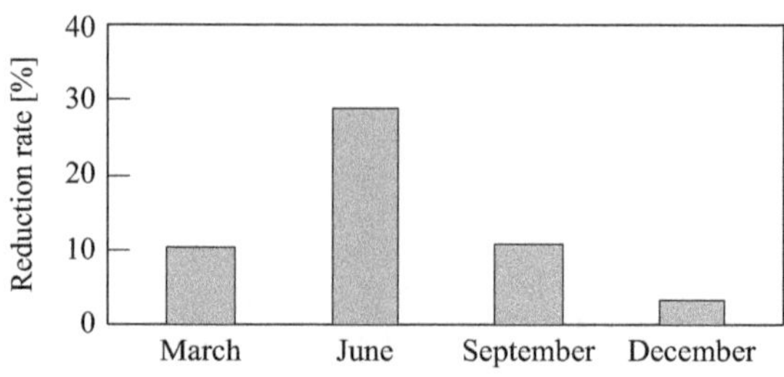

Figure 1.16   Reduction of the rate of power generator operating hours with PEPA predictive data compared to the past average weather data

equipment with PEPA for a whole month. This is the advantage of this operation planning method over the operation planning method using the past average weather data. Moreover, Figure 1.16 shows the reduction in the total equipment working hours from using the PEPA prediction data rather than using the past average weather data. By introducing PEPA into dynamic operation planning of the microgrid, the working time of generating equipment is reduced, as compared to using past average weather data, from 30% to 3%.

1.2.7   Conclusions

The operations planning of a microgrid was investigated here using a proposed algorithm. Operation of the proposed microgrid was analysed using actual weather data (amount of solar radiation and outside temperature) collected from the first to the seventh in the months March, June, September and December of 2007 in Sapporo. However, it is thought that the following results change by the region where the microgrid is. The following conclusions have been obtained:

1.   If the PEPA predictive value is introduced as the predictive value of solar cell output power compared with the past average weather data, the working time of the generating equipment can be reduced from 30% to 3%. However, working time of generating equipment is influenced by battery capacity.
2.   However, there is the possibility that in seasons with only a small amount of solar radiation, if the prediction error of PEPA is large, the original operation planning will change greatly. In this case, storage capacity over and above the capability of the generating equipment connected to the microgrid is needed.

1.3   Operation analysis of microgrids using an orthogonal array-GA hybrid method

1.3.1   Introduction

Decentralising a vehicular energy system can decrease power transmission losses, effectively use the heat from exhaust and promote the use of renewable energy.

However, renewable energy systems must be stabilised by combining two or more power sources. Distributed energy microgrids have been studied widely: energy management strategies have been reported for a renewable energy-based residential microgrid [12]; an actual microgrid has been optimised in Iceland [13]; and a smart grid has been reported, which was accompanied by various value-added features [14–16]. The optimal operation of a compound energy system must integrate different energy networks and various power sources. In general, dynamic operation optimisation is required because renewable energy systems must be able to store electricity and heat. In addition, the input–output characteristics of energy devices are often non-linear, so the operation planning of a compound energy system is a dynamic, non-linear and multivariate problem. Analysis methods have recently been developed to improve the operation optimisation of energy systems; such methods include the conjugate gradient method, integer programming and GAs. Nazar and Haghifam [17] reported a multi-objective electric distribution system using a hybrid energy hub; GA has been used to optimise the control of hydrogen-fuelled engines [18], and non-linear programming theory was used to optimise the calibration of a hydrogen-fuelled engine [19]. Even though mixed-integer programming is a powerful analysis method, its application to the dynamic planning of energy storage requires complicated modifications.

In contrast, GA analysis is a facile method for solving dynamic, non-linear, multivariate problems and can be easily adapted to a complex energy system [20]. When an analysis must incorporate a large number of energy sources and a high level of analytical accuracy, very long runtimes are required; this results from an increase in the number of chromosome models and the length of the genes contained therein. The objective of this chapter is to develop a computer algorithm with industrially relevant accuracy that can control the operation of compound energy systems with many design parameters. That is, quasi-optimum solutions are obtained with a sufficient degree of analytical accuracy that can be used for short times. First, the proposed algorithm introduces an orthogonal array and factorial-effect chart [21], which are experimental design techniques, and determines an operation method that is close to the optimal solution of the compound energy system. The design of an energy system using experimental design techniques has only been reported previously [22–24]. The near-optimum operation method is entered into the chromosome models of the GA as initial values, and the conventional GA searches for the optimal operation parameters for the microgrid. The conventional GA searches the operation area near the optimal solution; when the initial value is chosen well, the optimal operation solution is obtained more efficiently than if the GA analysis method was used alone [25]. This study also examines the effects of using various kinds of power sources, which exhibit different output characteristics, like the ratio of electric power to heat. The operation planning of the microgrid is solved for these different scenarios. The aim is to determine how differing power output characteristics affect the operation method and its analytical accuracy.

1.3.2   Analysis methods

The basic principle of the conventional GA and the orthogonal array-GA hybrid methods described below is the analysis method developed in the past by Obara and Watanabe [25,26].

1.3.2.1   Genetic algorithm

Chromosome model

The relationship between the input and output values of electric power and heat is shown in (1.5) and (1.6) for the microgrid. The left-hand side of both equations gives the supply terms (for electric power or heat) at sampling time t, and the right-hand side comprises the consumption terms. The left-hand side of (1.5) accounts for electric power output via generators, such as an engine generator, a fuel cell, or a solar cell; the total number of electric generators is given by M. The left-hand side of (1.6) accounts for heat outputs, such as boilers, heat pumps, engine generators and fuel cells, as well as the thermal power of heating equipment, such as solar heat. The number of heat outputs is given by N_h. The first term on the right-hand side of (1.5) and (1.6) is the demand; the second term is the total consumption of electric power and heat by the sum of all pieces of equipment I and J, respectively.

The energy storage terms, such as accumulation of electricity and heat storage, are included in the second term on the right-hand side of each equation and the energy supply terms, such as the electric discharge of a battery and the heat output from a heat storage tank, are included in the left-hand side.

$$\sum_{m=1}^{N_p} E_{m,t} + E_{bt,btd,t} = E_{\text{needs},t} + E_{bt,btc,t} + \sum_{i=1}^{I} \Delta p_{i,t} \tag{1.5}$$

$$\sum_{n=1}^{N_h} H_{n,t} + H_{st,\text{out},t} = H_{\text{needs},t} + H_{st,\text{in},t} + \sum_{j=1}^{J} \Delta h_{j,t} \tag{1.6}$$

The power demand $E_{\text{needs},t}$ and heat demand $H_{\text{needs},t}$ are fixed for a representative day, and $E_{m,t}$ and $H_{n,t}$ in (1.5) and (1.6) are determined by the chromosome model of the GA, as shown in Figure 1.17. The chromosome model n_{cr} in Figure 1.17 consists of the outputs $E_{n_{cr},m,t}$ and $H_{n_{cr},n,t}$ from the generator m and the heat source n, respectively, at sampling time t. Here, n_{cr} is the number of chromosomes, and m and n are the numbers of generators and heat sources, respectively. The region enclosed by a broken line in Figure 1.17 is the genetic information included in one chromosome. The gene is written for the data $E_{n_{cr},m,t}$ or $H_{n_{cr},n,t}$, and the information on one chromosome represents the operation method for a single sampling time during a representative day of the target energy system.

The operation method of the battery and the heat storage tank at sampling time t is influenced by the past $(t - 1, t - 2, \ldots)$ operation methods. When the past operation methods of the system influence the future operation method, it is defined as the dynamic operation planning. The analysis example of this chapter is a dynamic operation plan, because the operation period is shifted using a battery and a heat

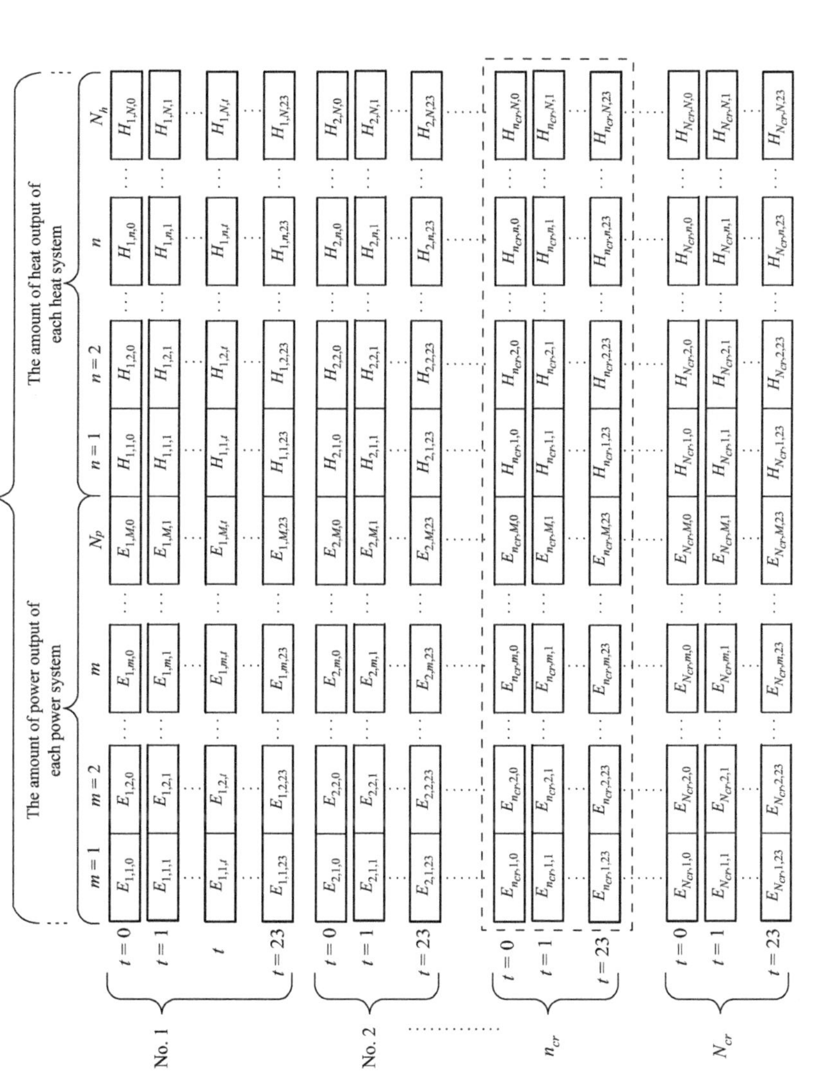

Figure 1.17 Chromosome model using GA

storage tank. As for the operation optimisation by the GA in this chapter, equipment operations at each sampling time is determined using the chromosome model of the GA. As for the proposed analysis method, the operation method of each equipment is decided using the predicted values of the load and renewable energy outputs.

Objective function (adaptive value)

A standard objective function for an energy system is given in (1.7). The cost of equipment, fuel expenses, environmental impact and maintenance costs are all taken into consideration by the terms in (1.7). α_1, α_2, α_3 and α_4 are weighting factors; λ is an operating period (year); and CT_m, CT_n and CT_k are the costs of generators, heat sources and maintenance of the system for one year, respectively. θ_p and θ_h are the unit fuel prices of generators and heat equipment, respectively; f_m and f_n are the fuel consumptions of generators and heat equipment, respectively. On the right-hand side of (1.7), the second set of brackets in the second term denotes the environmental impact cost of the system. ϕ_p and ϕ_h are the costs accompanying the discharge of greenhouse gases by generators and heat equipment.

$$
\begin{aligned}
\text{FOB} = \alpha_1 \cdot & \left(\sum_{m=1}^{N_p} CT_m + \sum_{n=1}^{N_h} CT_n \right) + \sum_{mh=1}^{12} \left\{ \alpha_2 \cdot \lambda \cdot \sum_{t=1}^{24} \left(\theta_p \cdot \sum_{m=1}^{N_p} f_{m,mh,t} + \theta_h \cdot \sum_{n=1}^{N_h} f_{n,mh,t} \right) \right. \\
& \left. + \alpha_3 \cdot \lambda \cdot \sum_{t=1}^{24} \left(\phi_p \cdot \sum_{m=1}^{N_p} f_{m,mh,t} + \phi_h \cdot \sum_{n=1}^{N_h} f_{n,mh,t} \right) \right\} + \alpha_4 \cdot \lambda \cdot CT_k
\end{aligned}
\tag{1.7}
$$

Calculation of the payback period λ_{pb}, as shown in (1.8), is generally used to evaluate an energy system. θ_{conv} and f_{conv} are the unit price and quantity of fuel consumption for the system. Equation (1.7) was integrated into the objective function to retain flexibility in setting the values of θ_{conv} and f_{conv}.

$$
\lambda_{pb} = \frac{\left(\sum_{m=1}^{N_p} CT_m + \sum_{n=1}^{N_h} CT_n \right)}{\sum_{mh=1}^{12} \left[\sum_{t=1}^{24} \left\{ \theta_{\text{conv}} \cdot f_{\text{conv},mh,t} - \left(\theta_p \cdot \sum_{m=1}^{N_p} f_{m,mh,t} + \theta_h \cdot \sum_{n=1}^{N_h} f_{n,mh,t} \right) \right\} \right]}
\tag{1.8}
$$

Analysis flow

The flow of the analysis that optimises the operation of the energy system via GA is shown in Figure 1.18. First, many chromosome models of N_{cr}, described in the 'Chromosome model' section, are randomly generated by a computer [Figure 1.18(a)]. The electric power and heat outputs for each sampling time are determined by decoding these chromosome models [Figure 1.18(b)]. Quantifying the output characteristics of each piece of equipment enables the analysis to account for the load factor, the amount of exhaust heat, the power-generation efficiency, and the efficiency of thermal power [Figure 1.18(c)]. The amount of electric-power storage or output to a battery is calculated by introducing these values into the power balance of (1.5) [Figure 1.18(d) and (e)].

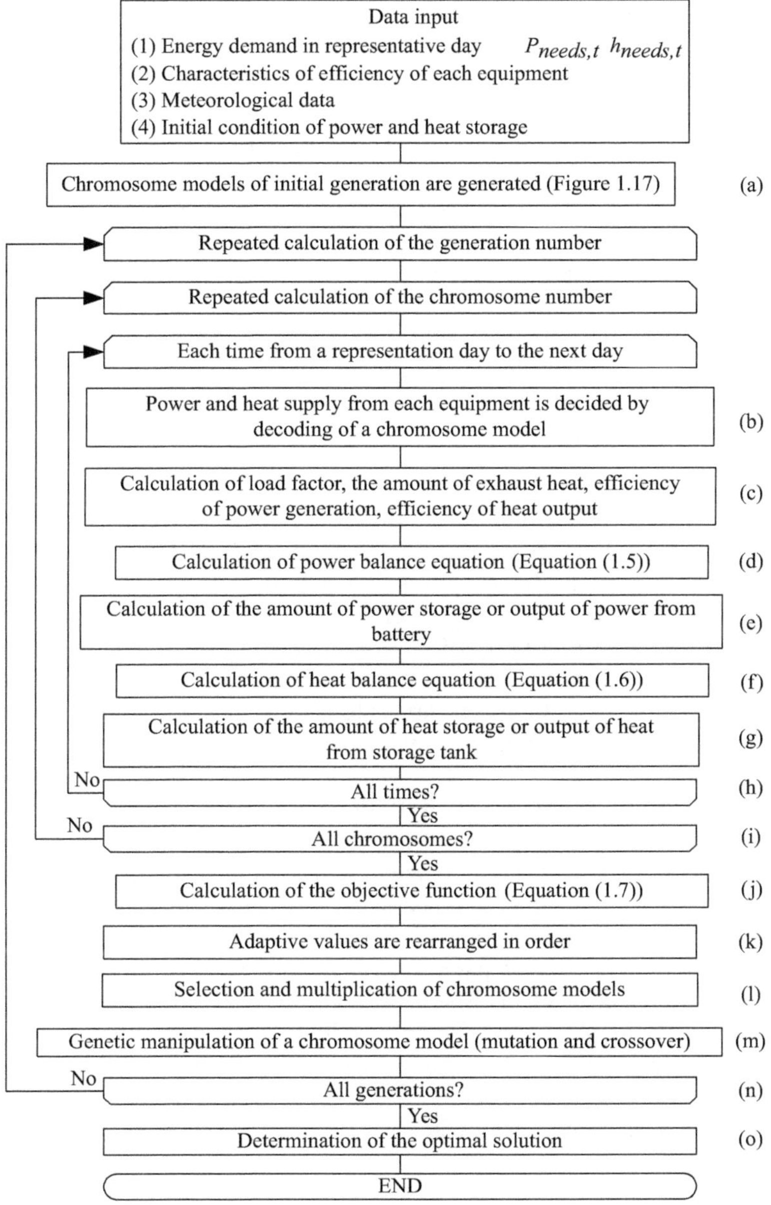

Figure 1.18   *Analysis flow using GA for a representative day*

The heat energy describes the total amount of thermal power for each piece of heat equipment and is determined by decoding the chromosome model of the GA. The amount of exhaust heat from the generators is added to the heat balance in (1.6) [Figure 1.18(f)] to calculate the total amount of heat that must be stored and the thermal capacity of the heat storage tank [Figure 1.18(g)]. The operation method is determined by repeating the calculations from Figure 1.18(b)–(g) for every sampling time during a representative day [Figure 1.18(h)]. Moreover, the calculations from Figure 1.18(b)–(h) are repeated for all of the chromosome models [Figure 1.18(i)], and the adaptive value of (1.7) is calculated for each [Figure 1.18(j)]. The chromosome models are then ordered based on their adaptive values [Figure 1.18(k)]; chromosome models with low adaptive values are screened, and chromosome models with high adaptive values are increased [Figure 1.18(l)].

Chromosome models are randomly selected based on the prior probabilities of parent genes, which are manipulated by cross-over and mutation operations [Figure 1.18(m)] to diversify the chromosome models. The calculations shown in Figure 1.18(b)–(n) repeat for each generation, and an individual with the highest adaptive value among the final generation's chromosome models is chosen as the optimal solution [Figure 1.18(o)].

GA analysis method
Although GAs can solve non-linear, multivariate problems, the analysis time becomes very long when the number of genes is increased. The increased runtime is necessary to accommodate an increase in the number of design parameters or to improve the analytical accuracy of the solution. Therefore, an orthogonal array and factorial-effect chart are used to select values of each design parameter that produce an operation scenario that is close to the optimum for the system. Then, the values of these design parameters are entered into a conventional GA, which searches for the optimal operation parameters.

1.3.2.2   Orthogonal array-GA hybrid analysis
Reducing the number of trials by experimental design
Experimental design aims to increase the efficiency of an experiment; the effect of each design parameter can be examined using an orthogonal array, without testing each combination of the design parameters. An example of the L_{18} orthogonal array is shown in Figure 1.19; design parameters in the same line of the orthogonal array are independent of each other. The orthogonal array is arranged specifically to ensure this independence so that the effect of each design parameter can be evaluated independently when each line of the orthogonal array is totalled. Orthogonal arrays greatly reduce the number of design-parameter combinations. In the L_{18} array, one design parameter can take either of two values (the level value) and seven design parameters can take any of three values. The total number of round-robin trials would be $2^1 \times 3^7 = 4374$, but an L_{18} orthogonal array requires only 18 ($e1$ to $e18$) trials of a design parameter [Figure 1.19(A)–(H)].

Level values in the orthogonal array-GA hybrid analysis
Level values of 1–3 for each design parameter are shown in the orthogonal array in Figure 1.19. The level values x for each of the three levels of Parameters 1–8 are listed in Table 1.2. The level value table must be defined before the analysis is performed. For example, the level values of Parameter 2 are $x_{p2,1}$, $x_{p2,2}$, and $x_{p2,3}$.

		(A)	(B)	(C)	(D)	(E)	(F)	(G)	(H)	$t1$	$t2$	$\cdots\cdots$	$t24$
	$e1$	1	1	1	1	1	1	1	1	$f_{e1,t1}$	$f_{e1,t2}$	$\cdots\cdots$	$f_{e1,t24}$
	$e2$	1	1	2	2	2	2	2	2	$f_{e2,t1}$	$f_{e2,t2}$	$\cdots\cdots$	$f_{e2,t24}$
	$e3$	1	1	3	3	3	3	3	3	$f_{e3,t1}$	$f_{e3,t2}$	$\cdots\cdots$	$f_{e3,t24}$
	$e4$	1	2	1	1	2	2	3	3	$f_{e4,t1}$	$f_{e4,t2}$	$\cdots\cdots$	$f_{e4,t24}$
	$e5$	1	2	2	2	3	3	1	1	$f_{e5,t1}$	$f_{e5,t2}$	$\cdots\cdots$	$f_{e5,t24}$
	$e6$	1	2	3	3	1	1	2	2	$f_{e6,t1}$	$f_{e6,t2}$	$\cdots\cdots$	$f_{e6,t24}$
	$e7$	1	3	1	2	1	3	2	3	$f_{e7,t1}$	$f_{e7,t2}$	$\cdots\cdots$	$f_{e7,t24}$
Experiment number	$e8$	1	3	2	3	2	1	3	1	$f_{e8,t1}$	$f_{e8,t2}$	$\cdots\cdots$	$f_{e8,t24}$
	$e9$	1	3	3	1	3	2	1	2	$f_{e9,t1}$	$f_{e9,t2}$	$\cdots\cdots$	$f_{e9,t24}$
	$e10$	2	1	1	3	3	2	2	1	$f_{e10,t1}$	$f_{e10,t2}$	$\cdots\cdots$	$f_{e10,t24}$
	$e11$	2	1	2	1	1	3	3	2	$f_{e11,t1}$	$f_{e11,t2}$	$\cdots\cdots$	$f_{e11,t24}$
	$e12$	2	1	3	2	2	1	1	3	$f_{e12,t1}$	$f_{e12,t2}$	$\cdots\cdots$	$f_{e12,t24}$
	$e13$	2	2	1	2	3	1	3	2	$f_{e13,t1}$	$f_{e13,t2}$	$\cdots\cdots$	$f_{e13,t24}$
	$e14$	2	2	2	3	1	2	1	3	$f_{e14,t1}$	$f_{e14,t2}$	$\cdots\cdots$	$f_{e14,t24}$
	$e15$	2	2	3	1	2	3	2	1	$f_{e15,t1}$	$f_{e15,t2}$	$\cdots\cdots$	$f_{e15,t24}$
	$e16$	2	3	1	3	2	3	1	2	$f_{e16,t1}$	$f_{e16,t2}$	$\cdots\cdots$	$f_{e16,t24}$
	$e17$	2	3	2	1	3	1	2	3	$f_{e17,t1}$	$f_{e17,t2}$	$\cdots\cdots$	$f_{e17,t24}$
	$e18$	2	3	3	2	1	2	3	1	$f_{e18,t1}$	$f_{e18,t2}$	$\cdots\cdots$	$f_{e18,t24}$

Column groups: "Row number (design parameter)" spans (A)–(H); "Calculation results of system fuel consumption" spans $t1$, $t2$, $\cdots\cdots$, $t24$.

Figure 1.19   Orthogonal L_{18} array

Table 1.2   Level of each design parameter

Design parameters	First level	Second level	Third level
(A) Parameter 1	$x_{p1,1}$	$x_{p1,2}$	
(B) Parameter 2	$x_{p2,1}$	$x_{p2,2}$	$x_{p2,3}$
(C) Parameter 3	$x_{p3,1}$	$x_{p3,2}$	$x_{p3,3}$
(D) Parameter 4	$x_{p4,1}$	$x_{p4,2}$	$x_{p4,3}$
(E) Parameter 5	$x_{p5,1}$	$x_{p5,2}$	$x_{p5,3}$
(F) Parameter 6	$x_{p6,1}$	$x_{p6,2}$	$x_{p6,3}$
(G) Parameter 7	$x_{p7,1}$	$x_{p7,2}$	$x_{p7,3}$
(H) Parameter 8	$x_{p8,1}$	$x_{p8,2}$	$x_{p8,3}$

The mean value of Parameter 2 is chosen to be $x_{p2,2}$, and the minimum and maximum values are $x_{p2,1}$ and $x_{p2,3}$, respectively. Next, the evaluation value f_{ek} ($k = 1, 2, \ldots, 18$) is calculated from (5) for experiments $e1$ to $e18$ using the level values of Table 1.2 for each design parameter in the orthogonal array [Figure 1.19 (A)–(H)].

$$f_{ek} = \sum_{t=1}^{24} f_{ek,t} = \sum_{t=1}^{24} \left(\sum_{m=1}^{N_p} f_{m,t} + \sum_{n=1}^{N_h} f_{n,t} \right) \tag{1.9}$$

Determination of the initial values of the GA using a factorial-effect chart
The average evaluation value for the given level value ls of a design-parameter pl is $\overline{f_{pr,ls}}$. For example, the average evaluation value $\overline{f_{(B),\,2}}$ of the second level of design parameter (B) is the average of the evaluation values from $e4$, $e5$, $e6$, $e13$, $e14$ and $e15$ for the experiment numbers f_{e4}, f_{e5}, f_{e6}, f_{e13}, f_{e14} and f_{e15} in Figure 1.19. The factorial-effect chart shown in Figure 1.20 is an example taken from this result. Suppose $\overline{f_{pr,ls}}$ yields a small value for an objective function, the level that is closest to an optimal solution [for design parameters (A)–(H)] is set to the level values of the white round head in Figure 1.20. Level values that are close to the optimal solution are described as high-level values.

The design parameter (B) in Figure 1.20 appears to have an optimal solution in the second level (namely, near the average value) in Table 1.3. (A), (C), (D), (F) and (G) have optimal solutions in the third level (namely, near the maximum values). It appears that (E) and (H) have optimal solutions in the first level (namely, near the minimum values). In this study, the high level value of each high design parameter is used as the initial value in the chromosome models of the GA described in Section 1.3.2.1. The analysis efficiency is expected to increase significantly because the search range of the GA is concentrated near optimum values.

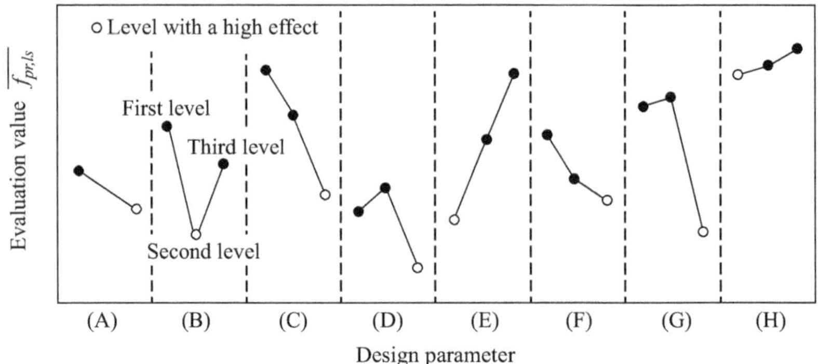

Figure 1.20   Factorial effect chart

Table 1.3   Efficiency of each equipment

Solar cell (with power conditioner) $\varphi_{cd,t}$	0.15
Heat storage tank $\varphi_{st,in,t}$, $\varphi_{st,out,t}$	0.8
Battery (efficiency of charge and discharge) φ_{btc}, φ_{btd}	0.9
Power conditioner using fuel cell (SOFC and PEFC) and G/E generator (included to $E_{fc,t}$, $E_{ge,t}$)	0.9
Power transmission of power grid	1.0
Heat supply to heat grid	1.0

1.3.3   Case study

1.3.3.1   System outline

An interconnected microgrid with two or more energy sources is shown in Figure 1.21. The power supply for the energy system exhibits non-linear characteristics. The system in Figure 1.21 supplies electric power and heat to 30 residences using various energy sources; the operation of the microgrid is optimised using both conventional GA and the proposed algorithm to draw a comparison. The microgrid in Figure 1.21 consists of an electric-power grid and a heat grid (hot water). The system comprises a fuel cell [solid oxide fuel cell (SOFC) or a proton-exchange membrane fuel cell (PEFC)] or a gas-engine generator (G/E generator), photovoltaics, power conditioners (1) and (2), a heat pump, a battery and a heat storage tank. The type of generator that is used as Equipment A in Figure 1.21 depends on the power conditioner (1); the equipment related to each power supply is shown in Figure 1.22.

1.3.3.2   Electric power supply system

When natural gas is supplied to a fuel cell or a G/E generator, alternating-current electric power (at 200 V and 50 Hz) will be input to the electric-power grid from power conditioner (1). Moreover, the output of the fuel cell or the G/E generator is controlled by adjusting the supply of natural gas. The power-generation efficiency of this equipment depends on the load factor; the relationship between the load factor and the efficiency of each generator (Figure 1.23) is used to determine the output of electric power and heat [27–30]. When the output characteristics shown in Figure 1.23(a)–(c) are used, the output ratio of electric power and heat can be obtained from the load factor of each power source. Idling operation is assumed as corresponding to of the rapid increase in load of a fuel cell. However, the fuel consumed by this operation mode is not taken into consideration. This analysis does not account for the effects of scaling each power supply. When photovoltaics are used to supply power to the grid via power conditioner (2) in Figure 1.21, excess electric power can be stored in the battery.

1.3.3.3   Heat supply system

The heat sources in the system are exhausted heat from the power sources, like the fuel cell, and an air-source heat pump. When exhaust heat is stored, it can be supplied to the demand side with a time-shift operation.

1.3.3.4   Energy flow of the system

Energy balance equation

Equations (1.10)–(1.15) are the balances for electric power and heat in the microgrid shown in Figure 1.21. The left-hand sides and the right-hand sides of each equation are input and output terms, respectively. The right-hand sides $E_{need,t}$ and $H_{need,t}$ of each equation are the demands of electric power and heat; the load pattern is used for the average residence in Sapporo, Japan during February (winter) (see Figure 1.24) [30]. The electric power and heat demands in February are much larger

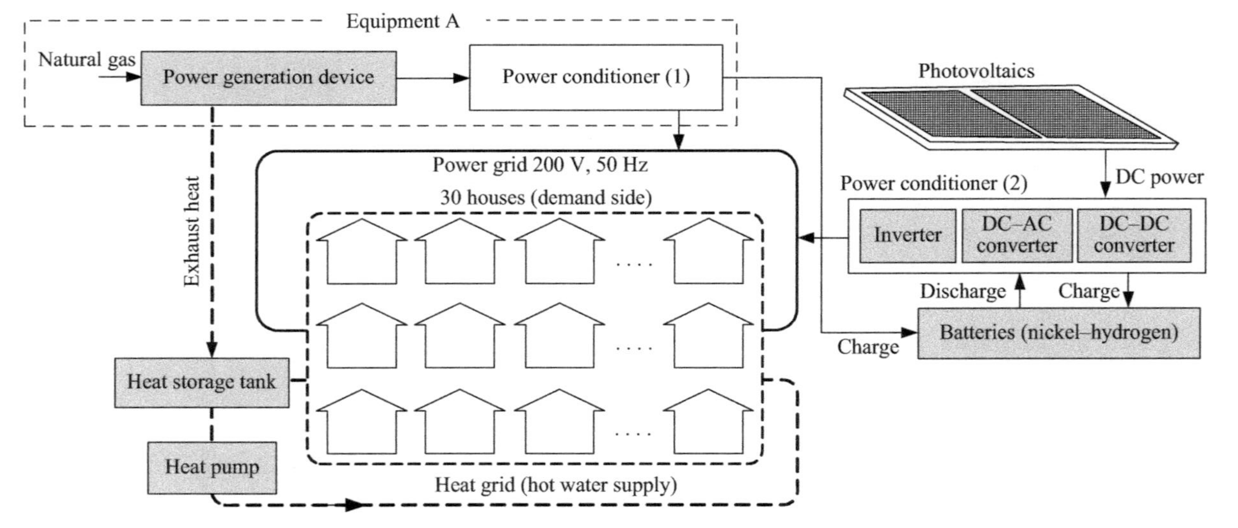

Figure 1.21   Independent microgrid system scheme

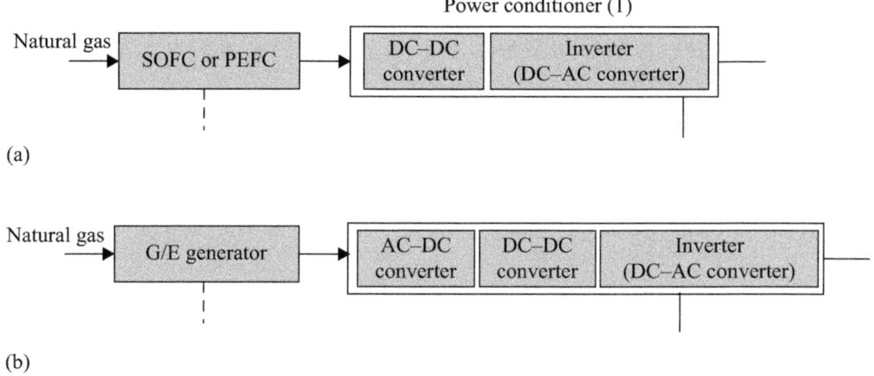

Figure 1.22   Equipment A: (a) fuel cell and (b) G/E generator

than during the summer season. Electric lamps and household appliances are encompassed by the power load; water heaters and space heaters are encompassed by the heating load. Losses from power conditioner (1) are included in the output of the power source $E_{fc,t}$ or $E_{ge,t}$ for sampling time t. $H_{\mathrm{rad},t}$ is the heat loss (heat radiation) from a heat storage tank. The heat storage loss sets the efficiency of thermal storage at 95% (mixed tank type heat storage), taking 5% of loss into consideration.

SOFC:

$$E_{\mathrm{SOFC},t} + E_{pv,t}\varphi_{cd,t} + E_{btd,t}\varphi_{btd} = E_{\mathrm{need},t} + E_{hp,t} + E_{btc,t}\varphi_{btc} \tag{1.10}$$

$$H_{\mathrm{SOFC},t} + H_{hp,t} + H_{st,\mathrm{out},t}\varphi_{st,\mathrm{out}} = H_{\mathrm{need},t} + H_{st,\mathrm{in},t}\varphi_{st,in} + H_{\mathrm{rad},t} \tag{1.11}$$

PEFC:

$$E_{\mathrm{PEFC},t} + E_{pv,t}\varphi_{cd,t} + E_{btd,t}\varphi_{btd} = E_{\mathrm{need},t} + E_{hp,t} + E_{btc,t}\varphi_{btc} \tag{1.12}$$

$$H_{\mathrm{PEFC},t} + H_{hp,t} + H_{st,\mathrm{out},t}\varphi_{st,\mathrm{out}} = H_{\mathrm{need},t} + H_{st,\mathrm{in},t}\varphi_{st,in} + H_{\mathrm{rad},t} \tag{1.13}$$

Gas engine generator:

$$E_{ge,t} + E_{pv,t}\varphi_{cd,t} + E_{btd,t}\varphi_{btd} = E_{\mathrm{need},t} + E_{hp,t} + E_{btc,t}\varphi_{btc} \tag{1.14}$$

$$H_{ge,t} + H_{hp,t} + H_{st,\mathrm{out},t}\varphi_{st,\mathrm{out}} = H_{\mathrm{need},t} + H_{st,\mathrm{in},t}\varphi_{st,\mathrm{in}} + H_{\mathrm{rad},t} \tag{1.15}$$

Output characteristics of the generators
The load factor for a power source is given by (1.16); (1.17)–(1.20) give the power-generation efficiency and fuel consumption of each power source. When the load factor $\eta_{fc,t}$ (or $\eta_{ge,t}$) obtained from (1.18) is substituted into (1.17)–(1.19), the power-generation efficiency $\varphi_{fc,\eta_{fc,t}}$ or $\varphi_{ge,\eta_{ge,t}}$ is obtained. C_{fc} (or C_{ge}) is the rated

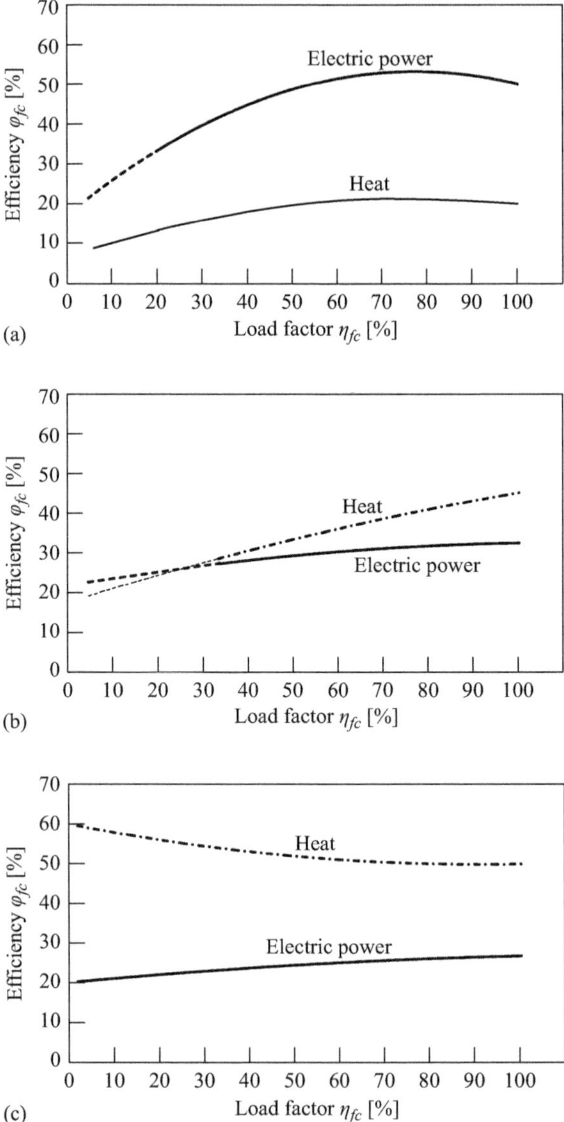

Figure 1.23   *Performance of (a) SOFC [9], (b) PEFC [27] and (c) G/E generator [28]*

capacity of the power source. The value of each coefficient in (1.17)–(1.19) is obtained from an approximate expression of the output characteristics of each power source shown in Figure 1.23. Approximated curves (broken lines) show the power-generation efficiency $\varphi_{fc,\eta_{fc,t}}$ of load factors below the lower limit of the fuel cell as shown in Figure 1.23(a) and (b). The fuel consumption $F_{fc,t}$ (or $F_{ge,t}$) of each

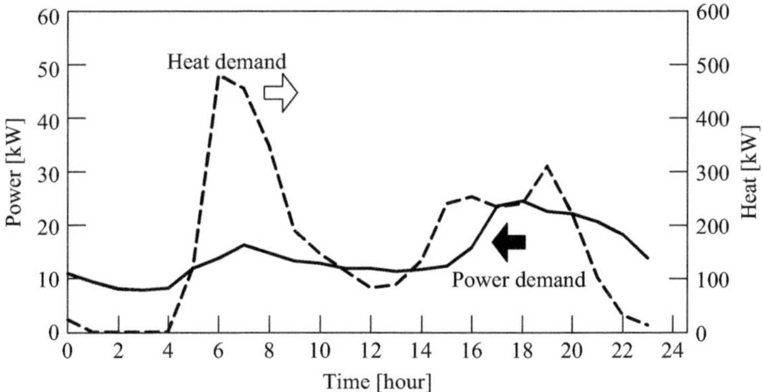

Figure 1.24 Energy demand of a 30-house microgrid in February in Sapporo, Japan [30]

power source is obtained by substituting the power-generation efficiency $\varphi_{fc,\eta_{fc,t}}$ (or $\varphi_{ge,\eta_{ge,t}}$) calculated from (1.17)–(1.19) into (1.20). $\dot{E}$ is the hourly electric power.

$$\eta_{fc,t} = \frac{E_{fc,t}}{C_{fc}} \quad \text{or} \quad \eta_{ge,t} = \frac{E_{ge,t}}{C_{ge}} \tag{1.16}$$

SOFC:

$$\varphi_{fc,\eta_{fc,t}} = -6.11 \times 10^{-5}\eta_{fc,t}^2 + 9.46 \times 10^{-3}\eta_{fc,t} + 0.167 \tag{1.17}$$

PEFC:

$$\varphi_{fc,\eta_{fc,t}} = -8.90 \times 10^{-6}\eta_{fc,t}^2 + 1.97 \times 10^{-4}\eta_{fc,t} + 0.2171 \tag{1.18}$$

G/E generator:

$$\varphi_{ge,\eta_{ge,t}} = -4.20 \times 10^{-6}\eta_{ge,t}^2 + 1.09 \times 10^{-3}\eta_{ge,t} + 0.20 \tag{1.19}$$

$$F_{fc,t} = \frac{\dot{E}_{fc,t}}{\varphi_{fc,\eta_{fc,t}}} \quad \text{or} \quad F_{ge,t} = \frac{\dot{E}_{ge,t}}{\varphi_{ge,\eta_{ge,t}}} \tag{1.20}$$

Equations (1.16)–(1.20) give the thermal power of each power source $H_{fc,t}$ (or $H_{ge,t}$), which is obtained from substitution of the value in (1.14)–(1.18).

SOFC:

$$H_{fc,t} = -2.45 \times 10^{-5}\eta_{fc,t}^2 + 3.78 \times 10^{-3}\eta_{fc,t} + 0.0667 \tag{1.21}$$

PEFC:

$$H_{fc,t} = -7.99 \times 10^{-6}\eta_{fc,t}^2 + 3.56 \times 10^{-3}\eta_{fc,t} + 0.177 \tag{1.22}$$

G/E generator:

$$H_{ge,t} = 1.22 \times 10^{-5}\eta_{ge,t}^2 - 2.23 \times 10^{-3}\eta_{ge,t} + 0.6013 \tag{1.23}$$

Figure 1.25 shows the example of the amount of global solar radiation in Sapporo in February. As the maximum in Figure 1.25 was 0.08 kW/m^2, the weather is a very bad example and the analysis example was investigated regarding the severest conditions of photovoltaics.

Operation of the heating equipment
The load factor of a heat pump is given by (1.24); (1.25) and (1.26) give the coefficient of performance (COP) and power consumption, respectively, for the heat pump shown in Figure 1.26. COP$_{hp,t}$ is obtained by substituting the load factor from (1.24) into (1.25). The power consumption is obtained by substituting COP$_{hp,t}$ and $H_{hp,t}$ into (1.26).

$$\eta_{hp,t} = \frac{H_{hp,t}}{C_{hp}} \tag{1.24}$$

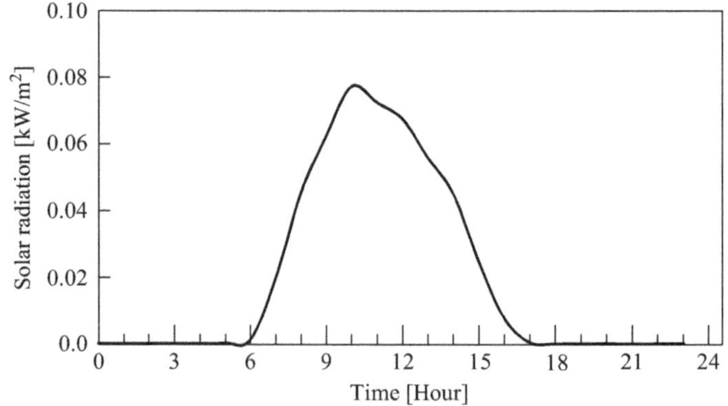

Figure 1.25   Solar radiation during February in Sapporo [9]

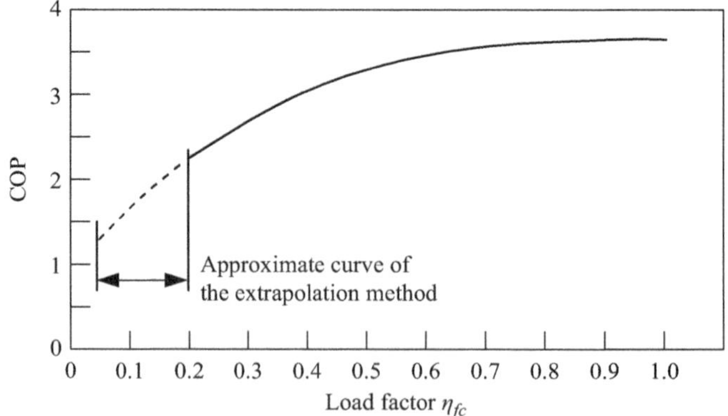

Figure 1.26   Performance of a heat pump system (Mitsubishi Electric, 2016)

$$\mathrm{COP}_{hp,t} = 2.70 \times 10^{-6} \eta_{hp,t}^3 - 8.11 \times 10^{-4} \eta_{hp,t}^2 + 0.0813 \eta_{hp,t} + 0.879 \tag{1.25}$$

$$E_{hp,t} = -\frac{H_{hp,t}}{\mathrm{COP}_{hp,t}} \tag{1.26}$$

When the thermal power of the power source exceeds heat demand ($H_{fc,t} > H_{\mathrm{need},t}$ or $H_{ge,t} > H_{\mathrm{need},t}$), the heat pump stops, and the excess heat ($H_{st,\mathrm{in},t} = H_{fc,t} - H_{\mathrm{need},t}$ or $H_{st,\mathrm{in},t} = H_{ge,t} - H_{\mathrm{need},t}$) is stored in the heat storage tank. However, when the total amount of excess heat exceeds the storage capacity C_{st}, it is transferred to the air from a radiator ($H_{\mathrm{rad},t}$). On the other hand, when the thermal power of a power source is less than heat demand ($H_{fc,t} < H_{\mathrm{need},t}$ or $H_{ge,t} < H_{\mathrm{need},t}$), the stored heat $H_{st,\mathrm{out},t} \cdot \phi_{st,\mathrm{out}}$ is supplied to the demand side. When the combined thermal output from the power source and the heat storage tank do not fulfil heat demands ($H_{fc,t} + H_{st,\mathrm{out},t}\varphi_{st,\mathrm{out}} < H_{\mathrm{need},t}$ or $H_{ge,t} + H_{st,\mathrm{out},t}\varphi_{st,\mathrm{out}} < H_{\mathrm{need},t}$), the heat pump compensates by producing heat $H_{hp,t}$. When $H_{hp,t}$ is fixed, the power consumption $E_{hp,t}$ of the heat pump is determined by (1.24)–(1.26).

1.3.3.5　Analysis conditions

The efficiency of each piece of equipment in the microgrid in Figure 1.21 is listed in Table 1.3. The differences between operation methods based on the type of power source are computed by assuming that the electrical and heat losses are the same; they are not taken into consideration during the analysis. Nor does the analysis consider changes in the conversion efficiency of the solar cell depending on the surface temperature.

　　The battery can respond to a time shift of electric-power supply and demand, and control of a short time of fluctuations by photovoltaics. However, in the analysis of this chapter, only the time shift of electric-power supply and demand is taken into consideration in the evaluation of battery capacity. As the electric-power quality (frequency, voltage, higher harmonic wave) of a microgrid is influenced by electric-power fluctuations, the evaluation of the battery capacity based on the stability of the electric power requires several samples taken at 10 ms intervals. In the analysis case of this chapter, to investigate the operation plan of a microgrid, sampling time is set into every hour. To take into consideration control of the electric-power fluctuation with a battery, it is necessary to shorten the sampling period further.

1.3.3.6　The level of the design parameters

The L_{18} array can accommodate eight design parameters and is used for the orthogonal array-GA hybrid analysis method in this example. The design parameters are (A)–(G) in Table 1.4. The solar cell [design parameter (A)] has two levels, and each of the other equipment [design parameters (B)–(G)] has three levels. The rated capacity of each piece of equipment, the amount of battery discharge $E_{btd,t}$ and the amount of charge $E_{btc,t}$ are all design parameters. The capacity C_{pv} of the solar cell has a maximum (the second level) of 100 kW based on an average usage of 3.3 kW per residence. In the factorial-effect chart (Figure 1.27), design parameters with only

Table 1.4   Parameters of design and level

Parameters of design	First level	Second level	Third level
(A) Capacity of photovoltaic C_{pv} (kW)	0	100	
(B) Capacity of battery C_{bt} (kWh)	5	15	25
(C) Amount of battery discharge E_{btd} (kW)	0	4	8
(D) Amount of battery charge E_{btc} (kW)	0	4	8
(E) Capacity of fuel cell C_{fc} (kW)	350	400	450
Capacity of G/E generator C_{ge} (kW)	700	750	800
(F) Capacity of heat storage tank C_{st} (kWh)	0	100	200
(G) Capacity of heat pump C_{hp} (kW)	500	600	700

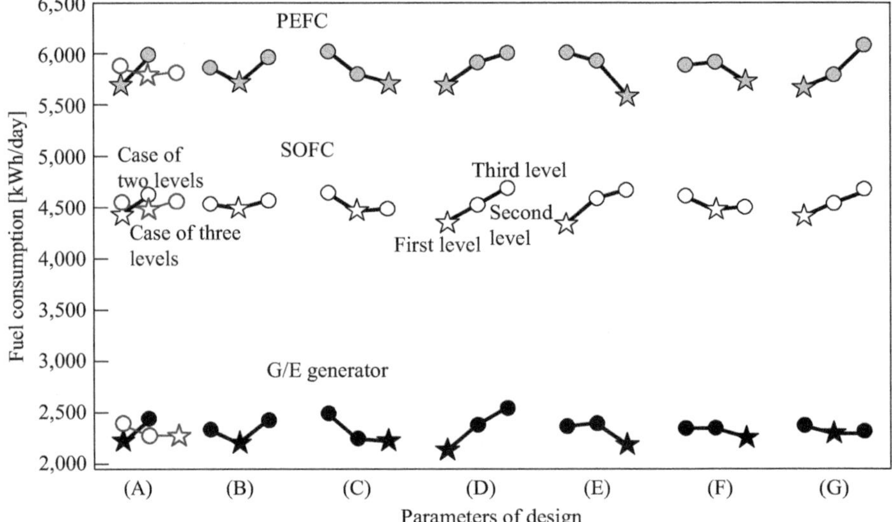

Figure 1.27   Factor effect chart

two levels cannot exhibit an inflection point, but design parameters with three levels can; therefore, we only apply the two levels of the orthogonal array to the design parameter without an inflection point, photovoltaics.

Design parameters (B), (E), (F) and (G) are the capacity C_{bt} of a battery, the capacity C_{fc} or C_{ge} of a power source, the capacity C_{st} of a heat storage tank and the capacity C_{hp} of a heat pump, respectively. The minimum (first level) of the heat storage tank is 0, indicating no heat storage. The first level of the amount of electric discharge of a battery $E_{btd,t}$ and the amount of charge $E_{btc,t}$ are also set to 0. The maximum of each design parameter was decided based on the maximum load of electric power and heat. The capacity of the G/E generator was determined based on the difference in power-generation efficiency between the G/E generator and the fuel cell shown in Figure 1.23.

1.3.3.7 Objective function (adaptive value)

Equation (1.27) presents an objective function for the system; in this analysis, it is defined as the minimisation of fuel consumption on a representative day. The fuel consumption of the system is due to the power supply as shown in Figure 1.21.

$$\sum_{t=0}^{24} f_{fc,t} \rightarrow \text{minimize} \quad \text{or} \quad \sum_{t=0}^{24} f_{ge,t} \rightarrow \text{minimize} \tag{1.27}$$

1.3.3.8 Initial values used for the orthogonal array-GA hybrid analysis

Each level value of each design parameter in Table 1.4 is described by the L_{18} orthogonal array in Figure 1.19. The factorial-effect chart for the design parameters (A)–(G) in an orthogonal array experiment is shown in Figure 1.27. The symbol '☆' denotes a level that most strongly contributes to the reduction of system-wide fuel consumption. The optimal solution is obtained by searching the GA within the neighbourhood of these level values. Based on the results of Figure 1.27, the chromosome model group of the initial generation is set as the values or ranges in Table 1.5. For example, the fuel consumption is reduced most for any power supply when the second level of the battery capacity C_{bt} is 15 kW [Figure 1.27(B)]. Accordingly, the range of battery capacity of the initial chromosome model is set as 7.5–17.5 kW.

1.3.3.9 Analysis parameters of the GA

There are 40 generations and 300 chromosomes, and the probability of mutations and cross-overs is 0.3. These solution parameters were decided by applying the trial-and-error method to the analysis.

Table 1.5 Analysis ranges of each design parameter

Parameters of design	SOFC	PEFC	G/E generator
(A) Capacity of photovoltaic C_{pv} (kW)	0	0	0
(B) Capacity of battery C_{bt} (kWh)	7.5–17.5	7.5–17.5	7.5–17.5
(C) Amount of battery discharge E_{btd} (kW)	2.0–6.0	Less than 8	Less than 8
(D) Amount of battery charge E_{btc} (kW)	0	0	0
(E) Capacity of fuel cell C_{fc} (kW)	Less than 350	Less than 450	
Capacity of G/E generator C_{ge} (kW)			Less than 800
(F) Capacity of heat storage tank C_{st} (kWh)	Less than 100	Less than 200	Less than 200
(G) Capacity of heat pump C_{hp} (kW)	Less than 500	Less than 500	550–650

1.3.4   Analysis results

1.3.4.1   Planning the equipment capacity

The analysis results are given in Table 1.6 when the analysis ranges in Table 1.5 are used as the initial values for the conventional GA. The capacity of the G/E generator is significantly larger than the SOFC and PEFC because operation of the G/E generator generates more heat. In contrast, fuel cells exhaust a small amount of heat, and the heat pump is required.

Neither photovoltaics nor a battery were used in the optimal operation configuration, regardless of the kind of power supply that was chosen. When the cycling losses of the battery are taken into account, the accumulation term becomes disadvantageous. The calculations involving photovoltaics were low because their capacity had only two levels in the orthogonal array. Therefore, the level of the design parameter (A) was changed to include three levels and the system was reanalysed. Setting the design parameter with two levels is very dangerous in the search for the optimal solution. Therefore, application of two levels should be avoided. The resulting generation number and fuel consumption are shown in Figure 1.28. Introducing photovoltaics decreases fuel consumption relative to the results shown in Table 1.6.

The optimal solution of each design parameter is shown in Table 1.7 for the scenario in which the second level of photovoltaics is 100 kW. Compared with the scenario in Figure 1.21, where photovoltaics are not used, the fuel consumption of the SOFC and PEFC decreased by 3.9% and 5%, respectively. Moreover, when the G/E generator was used, the total fuel consumption decreased by 22.6%. These findings indicate that equipment planning with only two levels decreases the analytical accuracy of the result.

1.3.4.2   Optimal operation

Power supply

The analysis of results for the optimal operation of power supplies and the heat pump are shown in Figure 1.29 based on the results of Table 1.7. The exhaust heat

Table 1.6   Optimum design parameters of proposal system

Parameters of design	SOFC	PEFC	G/E generator
(A) Capacity of photovoltaic C_{pv} (kW)	0	0	0
(B) Capacity of battery C_{bt} (kWh)	0	0	0
(C) Amount of battery discharge E_{btd} (kW)	0	0	0
(D) Amount of battery charge E_{btc} (kW)	0	0	0
(E) Capacity of fuel cell C_{fc} (kW)	146	142	
Capacity of G/E generator C_{ge} (kW)			769
(F) Capacity of heat storage tank C_{st} (kWh)	66	0	62
(G) Capacity of heat pump C_{hp} (kW)	469	454	0
Fuel consumption f_{fc}, f_{ge} (kWh)	3,210	4,692	1,910

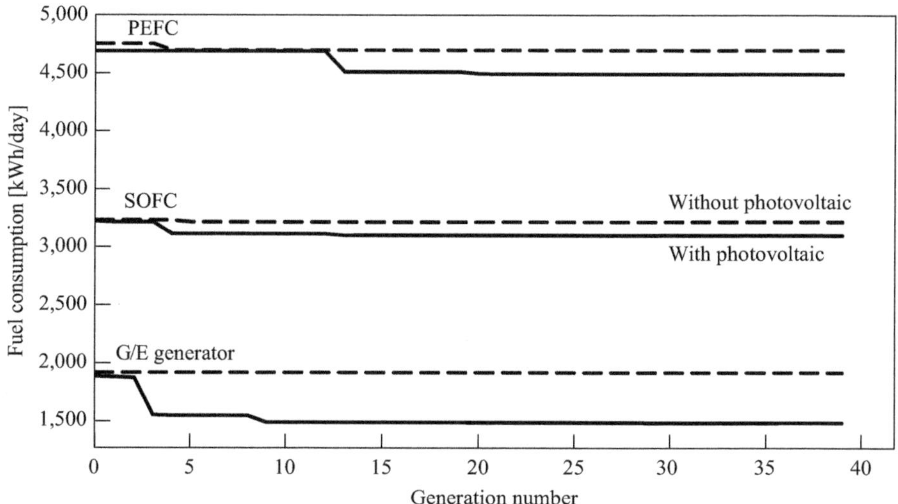

Figure 1.28   Fuel consumption for each generation

Table 1.7   Optimum design parameters of proposal system with photovoltaics

Parameters of design	SOFC	PEFC	G/E generator
(A) Capacity of photovoltaic C_{pv} (kW)	13	13	13
(B) Capacity of battery C_{bt} (kWh)	0	0	0
(C) Amount of battery discharge E_{btd} (kW)	0	0	0
(D) Amount of battery charge E_{btc} (kW)	0	0	0
(E) Capacity of fuel cell C_{fc} (kW)	146	143	
Capacity of G/E generator C_{ge} (kW)			776
(F) Capacity of heat storage tank C_{st} (kWh)	66	0	62
(G) Capacity of heat pump C_{hp} (kW)	470	454	0
Fuel consumption f_{fc}, f_{ge} (kWh)	3,087	4,457	1,478

output of each power supply and its load factor are shown in Figure 1.30. The pattern of the heat load strongly influences the operation method of the power supply because the heat demand pattern (Figure 1.24) is much larger than the electricity demand. Consequently, the load factor of the fuel cell in Figure 1.30(a) and (b) varies from 6% to 100% on a daily basis. In contrast, the operation of the G/E generator is planned so that a great amount of exhaust heat is produced with a relatively low load factor. The generator is planned by heat load following operation by a large equipment capacity. The output ratio of electric power and heat from the power supplies largely influences the composition and capacity of heating equipment.

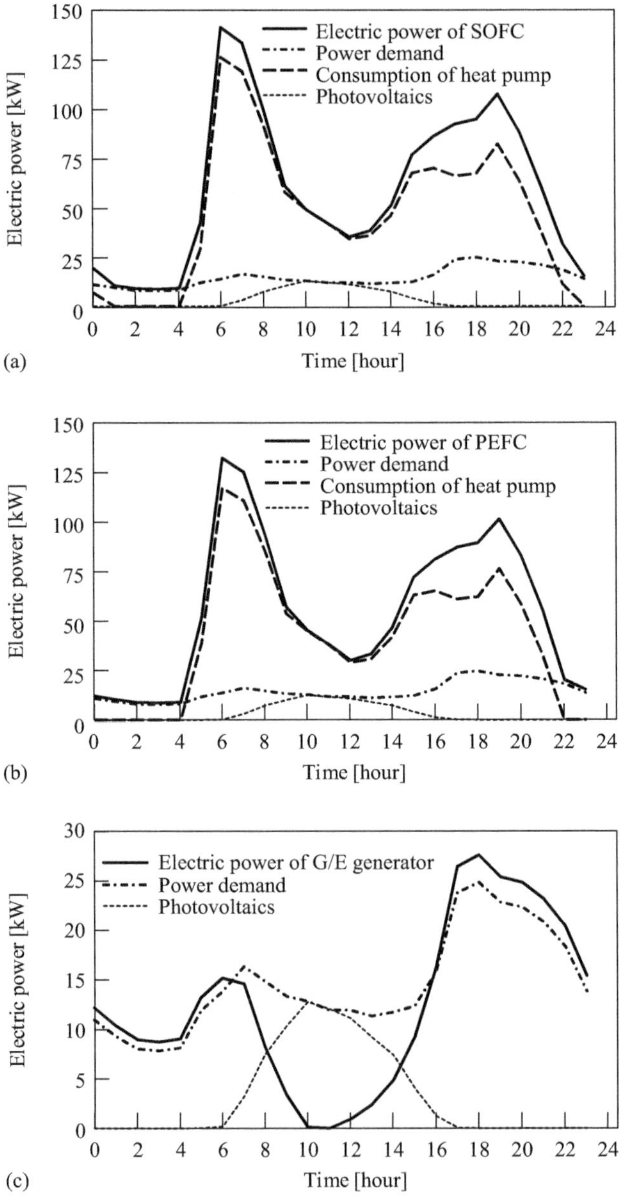

Figure 1.29   Optimised operational results of the power equipment: (a) SOFC, (b) PEFC and (c) G/E generator

Heating equipment

Figure 1.31 shows the results of the operation analysis of heat equipment. As there is less amount of exhaust heat with the SOFC and the PEFC to meet heat demand of the microgrid in this analysis case, the corresponding heat load is

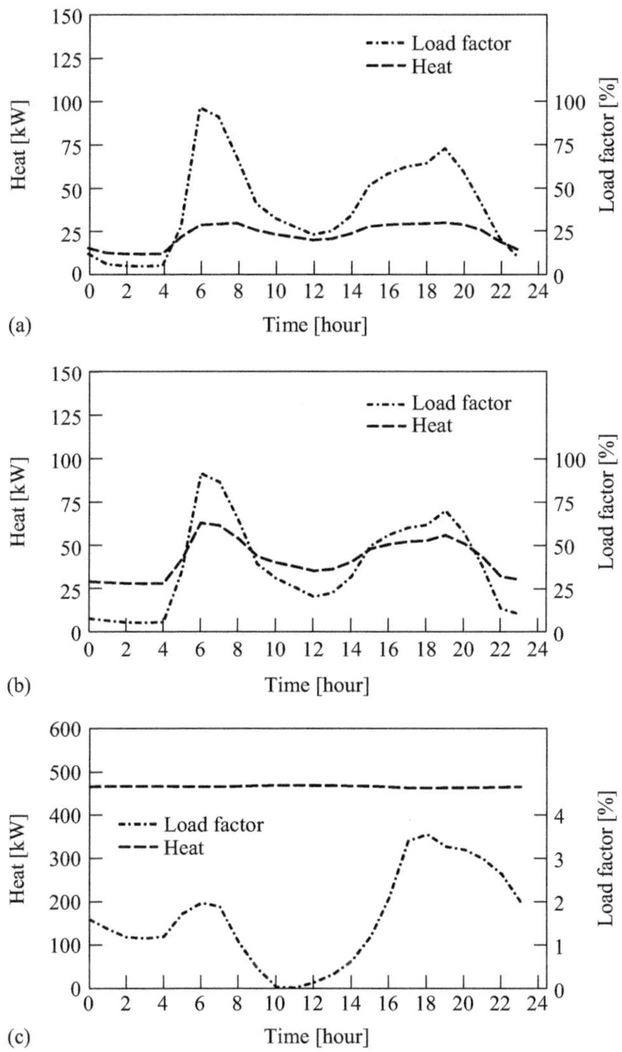

Figure 1.30   Operational planning and load factors of the power equipment: (a) SOFC, (b) PEFC and (c) G/E generator

largely based on the heat pump. Here, the heat exhaust means the thermal power of SOFC, PEFC or G/E generator. On the other hand, the exhaust heat of the G/E generator has exceeded the heat demand in almost all time periods. In the case of using G/E generator, the heat pump is not installed. The load factor of each fuel cell and COP of the heat pump are shown in Figure 1.32. The PEFC has more exhaust heat than the SOFC, so the load factor and COP of the PEFC are lower than those of the SOFC.

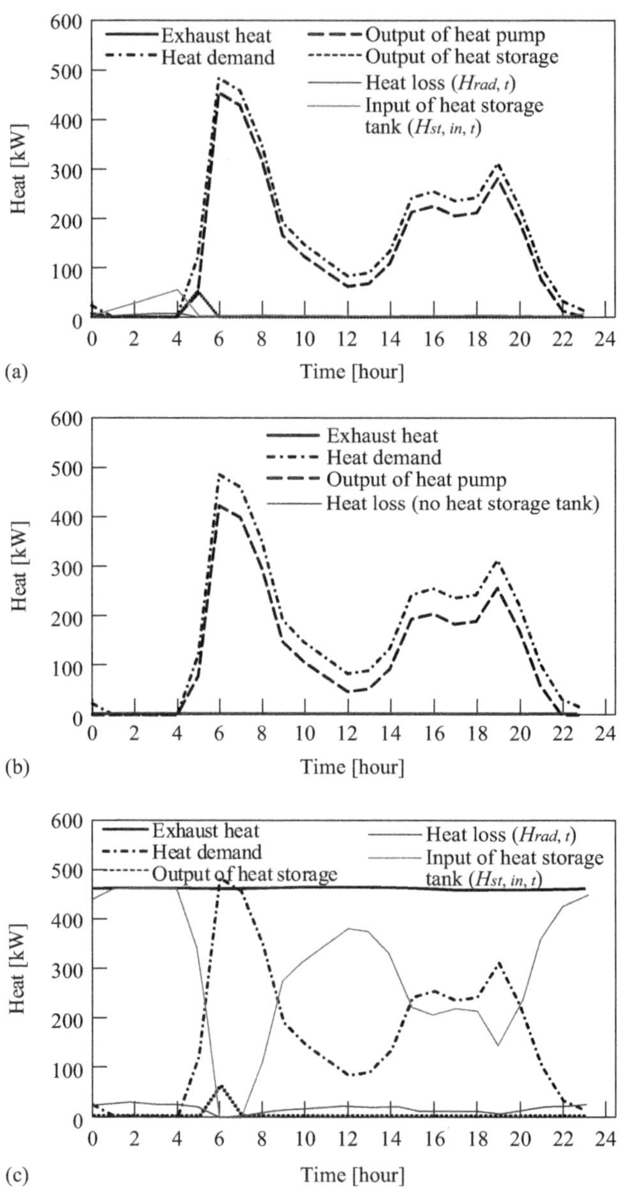

Figure 1.31   Results of the operation analysis of heating equipment: (a) SOFC, (b) PEFC and (c) G/E generator

1.3.4.3   Verification of the analysis algorithm

The frequency of convergence error between the conventional GA and the orthogonal array-GA hybrid analysis methods are shown in Figure 1.33. The analysis parameters are the generation number (40), the number of chromosomes (300), the

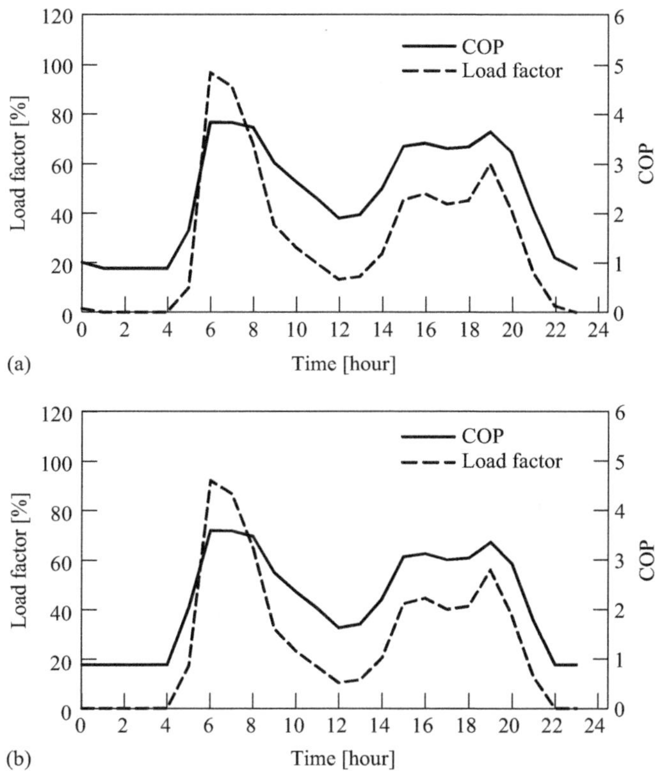

Figure 1.32   Analysis results of heat pump output and COP: (a) case of SOFC and (b) case of PEFC

probabilities of mutation and cross-over (0.3) and the number of trials (100). Conventional GA is applied to the analysis example and the solution fulfilled the objective function to be the optimal solution. As the conventional GA is an efficient algorithm for a random search by a computer, the solution very near the optimal solution is obtained through repeated calculation. The convergence error is a ratio of the difference of the optimal solution obtained by the repeated calculation of the conventional GA, and the analysis result obtained in 100 repetitions of each analysis method. Figure 1.33 shows the frequency of appearance for various levels of magnitude of the convergence error.

Analysis results for the SOFC and G/E generator always converge. However, the analysis results do not converge with respect to the PEFC; the convergence rate of the conventional GA is one in eight trials, and that of the proposed method was one in two trials. According to Figure 1.33, the convergence errors of the SOFC and PEFC are largely reduced by using the orthogonal array-GA hybrid analysis. Using the orthogonal array-GA hybrid analysis method with the SOFC, the convergence error is 2% or less for 73 of the 100 trials. On the other hand, the conventional GA provides a solution with the same convergence error for only 58 out of 100 trials.

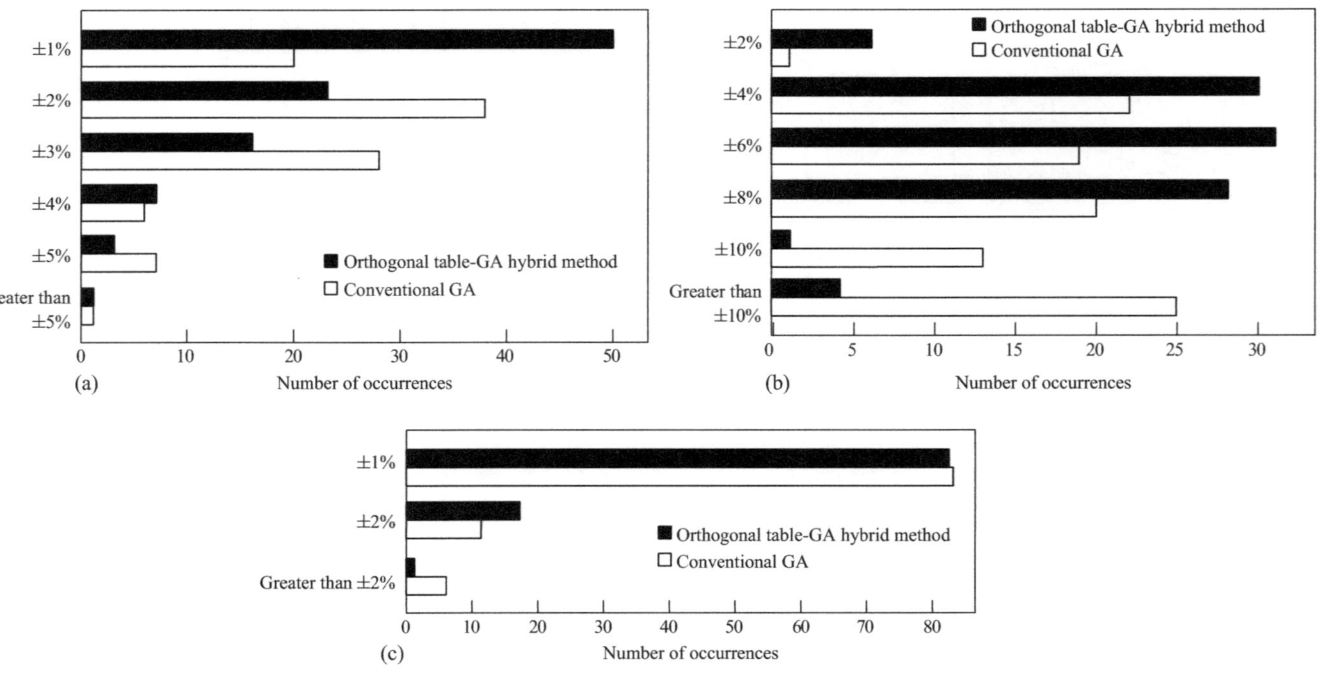

Figure 1.33   Convergence error of the optimal solution (for 100 trials): (a) SOFC, (b) PEFC and (c) gas engine generator

When the orthogonal array-GA hybrid analysis method is applied to the microgrid with the PEFC, the solution had less than 8% convergence error for 94 out of 100 trials. However, the same convergence error is obtained for conventional GA for only 65 out of 100 trials.

The convergence error of the G/E generator is sufficiently small using the conventional GA, so the proposed method is distinctly favourable in Figure 1.33(c). The analytic accuracy of the operation planning is influenced by the output characteristics of the power supply. As shown in Figure 1.28, the fuel consumption of the PEFC is the maximum. This is because the output ratio of heat and electric power of each power supply are largely different as shown in Figure 1.23. Even though the power generation using the G/E generator has most fuel consumption, the electric power consumption of the heat pump decreases with increasing exhaust heat. Therefore, the fuel consumption of the G/E generator is overall less than other power supplies. On the other hand, because the power-generation efficiency of the SOFC is high, fuel consumption is less than the PEFC. As a result, the fuel consumption of the PEFC increases more than other power supplies. As the objective function of the analysis example is minimisation of the fuel consumption (see (1.27)); the convergence error will increase relatively when the fuel consumption increases. As a result, it can be said that the convergence error of the PEFC falls rather than other power supplies as shown in Figure 1.33.

Optimisation of operation and equipment capacity for a microgrid is a non-linear problem that requires dynamic analysis with many variables. In this chapter, an orthogonal array-GA hybrid analysis method was proposed. According to the proposed method, the design parameters were chosen to be the installed capacity of each piece of equipment. An orthogonal array and factorial-effect chart were used for numerical simulations of the experimental design to select parameter values that were near the optimal solution. These design parameters were used as initial values for a conventional GA. Although the method proposed in this chapter is an improvement to the conventional GA, the strict optimal solution is not obtained. On the other hand, the real operation planning of an energy system contains the error of weather prediction (used for prediction of renewable energy sources power output), the disturbances regarding system control etc., for example. Therefore, when the errors included in the operation planning of a real energy system are taken into consideration, it is thought that the magnitude of analysis error by the proposal method is not acceptable. For this reason, we judged that the analytic accuracy of this proposal method can be applied industrially.

1.3.5 Conclusions

Generally, the output characteristics of the power sources are non-linear. Furthermore, because multiple power sources are used in compound energy system as microgrids, many variables must be considered to optimise the system. GAs provide a facile method for solving such problems and can be easily adapted to complicated energy systems. However, a large system requires lengthy analysis time, generates many suboptimal solution and unsatisfactory solutions of energy balance equations are obtained. As a solution, this chapter presented the orthogonal table – GA hybrid analysis method for planning the optimal operation of a compound energy system. The operation plan of each microgrid with an SOFC, PEFC and G/E generator, each with different output characteristics, was analysed to examine the analytical accuracy of the orthogonal array-GA hybrid method. The following conclusions were made.

The convergence error for the microgrid with an SOFC or PEFC was much less for the orthogonal array-GA hybrid analysis method than for the conventional GA. Using the orthogonal array-GA hybrid analysis method with an SOFC, the convergence error was 2% or less for 73 of the 100 trials. In contrast, the conventional GA provided a solution with the same convergence error for only 58 out of 100 trials. When the orthogonal array-GA hybrid analysis method was applied to a microgrid with a PEFC, the solution had less than 8% convergence error for 94 out of 100 trials. The same convergence error was obtained for conventional GA for only 65 out of 100 trials. The orthogonal array-GA hybrid analysis method exhibits a much higher analytic accuracy than the conventional GA. The analysis example of the microgrid indicates that the difference in the output characteristics of a power supply influences the analytic accuracy. These results can be directly applied to the operation of dynamic industrial scale energy grids with multiple heat and power sources.

Nomenclature

CT	cost [US dollar (USD)]
C	capacity (kW)
CT_k	annual maintenance cost of system (USD)
COP	coefficient of performance
E	electric power (kW)
$\dot{E}$	hourly electric power (kWh)
e	experiment number
FOB	objective function (USD)
f	fuel consumption (kWh)
f_{ek}	evaluation value of orthogonal array (kWh)
$\overline{f_{pr,\,ls}}$	average evaluation cost (kWh)
H	heat (kW)
Δh	heat consumption (kW)
I	the number of equipment with consumption of electric power
J	the number of equipment with consumption of heat
N	total number

m	number of generator
n	number of heat equipment
Δp	power consumption (kW)
t	sampling time (h)
x_p	level value (kW)

Greek characters

α	weight factor
ϕ	cost on greenhouse gas emission (USD)
φ	efficiency [–]
η	load factor [–]
λ	operating period (year)
θ	unit fuel price (USD/kWh)

Subscripts

bt	battery
btc	charge of battery
btd	discharge of battery
cd	power conditioner
conv	comparison system
cr	chromosome
fc	fuel cell (SOFC or PEFC)
ge	G/E power generator
h	heat equipment
hp	heat pump
i	number of equipment with consumption of electric power
j	number of equipment with consumption of heat
ls	level value of design parameter
m	number of electric generator
mh	month
n	number of heat equipment
needs	demand
p	generator
pb	payback
pl	number of design parameter
pv	photovoltaics
r	multiplier of 2
rad	heat radiation
st	heat storage tank

References

[1]   S. Abu-Sharkh, *et al.*, "Can microgrids make a major contribution to UK energy supply?", *Renewable and Sustainable Energy Reviews*, vol. 10, no. 2 (2006), pp. 78–127.

[2]   M. Muselli, G. Notton and A. Louche, "Design of hybrid-photovoltaic power generator, with optimization of energy management", *Solar Energy*, vol. 65, no. 3(1999), pp. 143–157.

[3]   Y. Ismail, Y. Kemmoku, H. Takikawa and T. Sakakibara, "An operating method for fuel savings in a stand-alone wind/diesel/battery system", *Journal of Japan Solar Energy Society*, vol. 28, no. 2(2002), pp. 31–38.

[4]   S. Obara, "Operating schedule of a combined energy network system with fuel cell" *International Journal of Energy Research*, vol. 30, no. 13(2006), pp. 1055–1073.

[5]   S. Obara and I. Tanno, "Fuel reduction effect of the solar cell and diesel engine hybrid system with a prediction algorithm of power generation", *Journal of Power and Energy Systems*, vol. 2, no. 4(2008), pp. 1166–1177.

[6]   S. Obara, "Dynamic operation plan of a combined fuel cell cogeneration, solar module, and geo-thermal heat pump system using genetic algorithm", *International Journal of Energy Research*, vol. 31, no. 13(2007), pp. 1275–1291.

[7]   K. Jorgensen, "Technologies for electric, hybrid and hydrogen vehicles: electricity from renewable energy sources in transport", *Utilities Policy*, vol. 16, no. 2(2008), pp. 72–79.

[8]   Homepage of Japan Meteorological Agency, 2007. Available from http://www.data.jma.go.jp/obd/stats/etrn/index.php.

[9]   NEDO Technical information data base, Standard meteorology and solar radiation data (METPV-3), 2009. Available from http://www.nedo.go.jp/database/index.html.

[10] S. Obara and I. Tanno, "Operation prediction of a bioethanol solar reforming system using a neural network", *Journal of Thermal Science and Technology*, vol. 2, no. 2(2007), pp. 256–267.

[11] K. Narita, "The research on unused energy of the cold region city and utilization for the district heat and cooling", Ph.D. Thesis, (1996), Hokkaido University.

[12] J. Pascual, J. Barricarte, P. Sanchis and L. Marroyo, "Energy management strategy for a renewable-based residential microgrid with generation and demand forecasting", *Applied Energy*, vol. 158, (2015), pp. 12–25.

[13] J. Sachs and O. Sawodny, "Multi-objective three stage design optimization for island microgrids", *Applied Energy*, vol. 165, (2016), pp. 789–800.

[14] C. N. K. Nair and L. Zhang, "Smart grid: future networks for New Zealand power systems incorporating distributed generation", *Energy Policy*, vol. 37, (2009), pp. 3418–3427.

[15] M. P. Anastasopoulos, A. C. Voulkidis, V. A. Vasilakos and G. P. Cottis, "A secure network management protocol for Smart Grid BPL networks: design, implementation and experimental results", *Computer Communications*, vol. 31, (2008), pp. 4333–4342.

[16] R. Hledik, "How green is the smart grid?, *The Electricity Journal*, vol. 22, (2009), pp. 29–41.

[17] S. M. Nazar and R. M. Haghifam, "Multiobjective electric distribution system expansion planning using hybrid energy hub concept", *Electric Power Systems Research*, vol. 79, (2009), pp. 899–911.

[18] Z. Yang, L. Wang and S. Li, "Investigation into the optimization control technique of hydrogen-fueled engines based on genetic algorithms", *International Journal of Hydrogen Energy*, vol. 33, (2008), pp. 6780–6791.

[19] L. Wang, M. He and Z. Yang, "Research on optimal calibration technology for hydrogen-fueled engine based on nonlinear programming theory", *International Journal of Hydrogen Energy*, vol. 35, (2010), pp. 2747–2753.

[20] S. Obara, S. Watanabe and B. Rengaraja, "Operation method study based on the energy balance of an independent microgrid using solar-powered water electrolyzer and an electric heat pump", *Energy*, vol. 36, (2011), pp. 5200–5213.

[21] G. Taguchi, "Experimental design, the third edition", Maruzen, ISBN-10: 4621082809, (2010) (in Japanese).

[22] Y. C. Lin and H. C. Lay, "Effects of carbonate and phosphate concentrations on hydrogen production using anaerobic sewage sludge microflora", *International Journal of Hydrogen Energy*, vol. 9, (2004), pp. 275–281.

[23] W. L. Yu, S. J. Wu and S. W. Shiah, "Parametric analysis of the proton exchange membrane fuel cell performance using design of experiments", *International Journal of Hydrogen Energy*, vol. 3, (2008), pp. 2311–2322.

[24] C. R. Dante., L. J. Escamilla, V. Madrigal, *et al.*, "Fractional factorial design of experiments for PEM fuel cell performances improvement", *International Journal of Hydrogen Energy*, vol. 28, (2003), pp. 343–348.

[25] S. Obara and S. Watanabe, "Study on the operation analysis of a compound energy system using orthogonal array-GA hybrid analyzing method", *Transactions of the Japan Society of Mechanical Engineers, Series B*, vol. 77, (2011), pp. 2004–2018 (in Japanese).

[26] S. Obara and S. Watanabe, "Optimization of equipment capacity and an operational method based on cost analysis of a fuel cell microgrid", *International Journal of Hydrogen Energy*, vol. 37, (2012), pp. 7814–7830.

[27] K. Maeda, K. Masumoto and A. Hayano, "A study on energy saving in residential PEFC cogeneration systems", *Journal of Power Sources*, vol. 95, (2010), pp. 3779–3784.

[28] T. Wakui and R. Yokoyama, "Optimal sizing of residential gas engine cogeneration system for power interchange operation from energy-saving viewpoint", *Energy*, vol. 36, (2011), pp. 3816–3824.

[29] NEDO Development of several 10 kW-class system of disk-type intermediate-temperature SOFC (FY2008), Final report 2009 (in Japanese).

[30] K. Narita, "The research on unused energy of the cold region city and utilization for the district heat and cooling", Ph.D. Thesis, (1996), Hokkaido University, Sapporo.

Chapter 2

Key concepts

Shin'ya Obara

2.1 Introduction: key concepts

Microgrids can operate connected or isolated from a larger or main power system. The microgrid interconnected to a large-scale commercial power system depends, in many aspects, on the interconnected system. The author believes that the study of an independent microgrid is a good method to develop various interconnected microgrids. The key concepts, in a microgrid, are better presented as examples, highlighting the main aspects and characteristics of this type of small grids.

Therefore, this chapter describes an example of a microgrid with a fuel cell combined cycle first. Next, the case of a microgrid with interconnected coal gasification fuel cell power generation and pumped hydropower generation is described.

- **First case**

The fuel cell triple combined cycle (SOFC-TCC) of rated power 1.4 MW consists of a solid oxide fuel cell (SOFC, 542 kW), a gas turbine (G/T, 550 kW) and a steam turbine (S/T, 308 kW). The relationship between the frequency deviation (based on the supply-and-demand difference) of an independent microgrid, which includes SOFC-TCC described above, and large-scale photovoltaics was investigated using numerical analysis (MATLAB$^®$/Simulink$^®$ R 2013a). The results show that following load fluctuations with SOFC and S/T required 1.8–2 h; this corresponds to governor-free control of G/T for load fluctuations of 2 h or less. Furthermore, frequency deviation of a microgrid with photovoltaics is strongly influenced by the magnitude of the inertia force of the G/T and S/T. The inertia changes the power characteristics (frequency) from cyclic fluctuation (change for several minutes or less) to sustained fluctuation (change exceeding 20 min). From the analysis, long-term supply-and-demand fluctuations, such as daily or seasonal fluctuations, are mainly controlled by output adjustment of the SOFC and S/T. The operation, controlled by setting of the governor-free control of the G/T and the inertia system of rotary machines, is appropriate for power fluctuations of cycles shorter than the long-term fluctuation cycle described above.

- **Second case**

Integrated coal gasification fuel cell (IGFC) combined cycle has significant advantages in power generation efficiency and reduced environmental impacts. Use of IGFCs has increased throughout the world, and its power generation efficiency is higher than that of a conventional combined cycle generator. This section presents a method of applying a system comprising renewable energy, pumped-storage power generation (PSPG), and IGFC as an independent microgrid. The IGFC is examined for its use as base load supply with a constant output. Investigation of potential technologies for interconnection with renewable energy has only recently begun. This study proposes an operational method for stabilising electric power supply and demand by adjusting the daily output of the IGFC and adjusting the input and output of the pumped-storage power station for each sampling period. Furthermore, when the power load pattern on a representative day in Hokkaido, Japan, was introduced into proposed microgrid, it was possible to establish the relation between the load factor of the IGFC and the power generation efficiency of each power source. Moreover, the study also investigated the economic efficiency of the proposed IGFC and integrated gasification combined cycle (IGCC), and confirmed the operational advantages of the IGFC.

2.2 Dynamic-characteristics analysis of an independent microgrid with an SOFC triple combined cycle

2.2.1 Introduction

Renewable energy sources cause large power fluctuation in a microgrid, which greatly restricts the amount of renewable energy sources that can be connected to a microgrid. However, a microgrid with high output renewable energy connected can control power fluctuations using the renewable energy sources and can minimise safety and environment issues, and provide fossil fuel saving. Many studies on control of power fluctuation of a microgrid have been reported [1–7]. The power fluctuation in an independent microgrid with a multi-axis type triple combined cycle solid oxide fuel cell (SOFC-TCC), a G/T, an S/T and a large-scale photovoltaic power plant is investigated using numerical analysis. Although some results from research on SOFC-TCC have been reported [8–10], investigation of control of the power fluctuations by interconnection of renewable energy and SOFC-TCC is not found.

Generally, electrical power changes over several minutes or less are described as cyclic fluctuation, and the changes from several minutes to about 20 min are described as short period fluctuation. Furthermore, the electricity change exceeding 20 min is described as sustained fluctuation. Cyclic fluctuations and a component of sustained fluctuation include fluctuations of renewable energy. Many system designs that consider fluctuations of photovoltaics have been reported [11–16]. Cyclic fluctuations of photovoltaics power supply can be caused by the weather: short period fluctuations and sustained fluctuations by the solar location and sustained fluctuations by the difference in seasons. Therefore, in this section, the effect of electric power stabilisation by adjustment of the output of the SOFC-TCC is investigated for the independent microgrid with SOFC-TCC and large-scale photovoltaics. Because the independent microgrid with renewable energy cannot absorb power fluctuations through interconnection with other power networks, stabilisation of the electricity of the microgrid is very difficult [17–21]. In order to control the power fluctuations by renewable energy, technology in which the difference of the supply-and-demand balance of the microgrid can be controlled at high speed needs to be introduced.

Although stabilisation of the power fluctuations with use of electricity storage equipment is most common [22–25], equipment cost is a major topic. The robustness of the electricity grid is increased, in this study, by the output adjustment through a governor-free control of the SOFC-TCC, and suitable setting of the inertia system of rotating machines (G/T and S/T) [26–29]. When the inertia system of rotary machines is set up appropriately, it will be expected at least that the cyclic fluctuation of electricity can be controlled. Moreover, the governor-free control of the rotary machines (G/T and S/T) is applied to the short-term fluctuation of electricity, and the governor-free control of the SOFC is applied to the sustained fluctuation. The objective of this study is to establish appropriate setting for the inertia system of the rotary machines and of SOFC-TCC and also to increase robustness of the electricity grid with large-scale photovoltaics by the governor free control of an SOFC, G/T and S/T, realising an independent microgrid without a battery system. Therefore, the difference of the electricity supply and demand of an independent microgrid (found from a frequency deviation) is investigated by modelling the SOFC-TCC and analysing the response characteristics of the system by MATLAB/Simulink 2013a. The investigation confirmed the relation between the magnitude of the load fluctuation added to the microgrid and the operation characteristics of the SOFC-TCC, the characteristics of load following and the magnitude of electricity supply-and-demand difference (frequency deviation). As a result, the magnitude of the large-scale photovoltaics, which can be connected to a microgrid with the SOFC-TCC, can be predicted.

2.2.2   *System configuration*

2.2.2.1   **SOFC triple combined cycle (SOFC-TCC)**

Standard operation

Figure 2.1 shows the triple combined cycle (SOFC-TCC) which consists of an SOFC, G/T and S/T, which are investigated in this paper. Although natural gas supplied to the anode of SOFC is reformed by an external reformer and part of the gas is discharged from the SOFC unused. On the other hand, high-temperature compressed air emitted from the G/T is supplied to the cathode of the SOFC. Although a part of oxygen in the compressed air supplied to cathode is consumed by the SOFC, a large quantity of excess air is discharged from the SOFC. The emission from the anode and cathode of the SOFC is supplied to a combustion chamber and mixed with natural gas from the reformer, and combustion gas of high

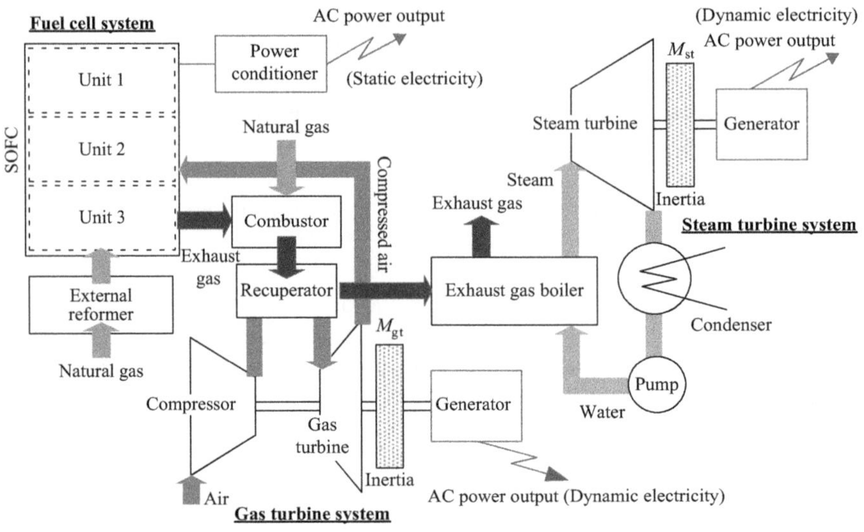

Figure 2.1   SOFC triple combined cycle

temperature and high pressure is generated. The combustion gas heats the air supplied to the G/T from the compressor, the high-temperature air discharged from the G/T is supplied to the cathode of the SOFC. The high-temperature emission discharged from the combustor is supplied to the exhaust heat boiler and serves as a heat source of steam used for the S/T. Moreover, the output power of the G/T and the S/T can be increased independently by supplying natural gas to the combustion chamber directly.

Dynamic electricity and static electricity

Because the G/T and S/T are rotary machines, they have an inertia force. Therefore, it is assumed that the electric power output of the G/T and S/T controls the short-time power fluctuations of the microgrid (dynamic control). Therefore, this paper shows relation between the inertia constants of the G/T and S/T shown in Figure 2.1 by M_{gt} and M_{st}, respectively, and the power fluctuations of the microgrid with a large-scale solar power system. M_{gt} and M_{st} are inertia constants that include the turbine, power generator and the flywheel of the G/T or the S/T. On the other hand, after the direct current power of the SOFC is converted into AC power by the power conditioner (DC–AC converter and inverter), the AC power is supplied to the microgrid. Because the electricity supplied to the microgrid from the SOFC does not have inertia force, there is almost no control effect on load changes (static electricity). Therefore, it is necessary to show clearly what stabilisation is possible through the set values of M_{gt} and M_{st}, and the governor-free control of SOFC-TCC on the power fluctuation caused by renewable energy, and load fluctuation from the demand side.

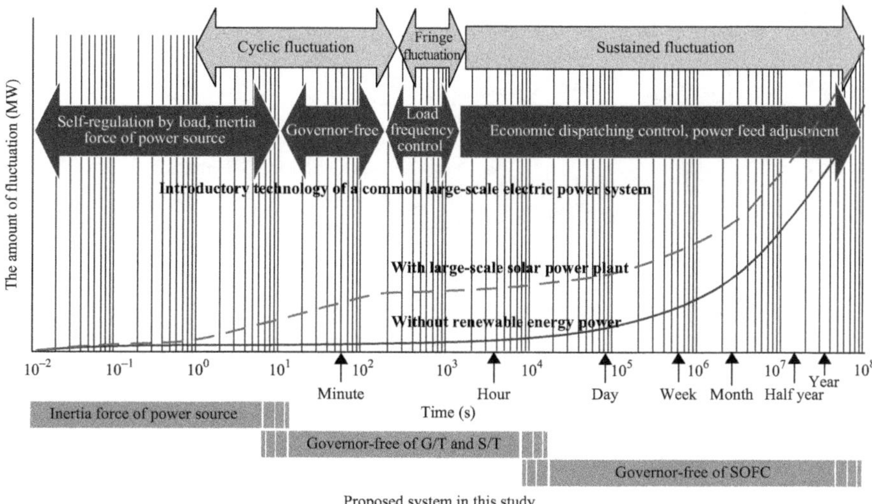

Figure 2.2   Electricity fluctuation and control technology

2.2.2.2   Dynamic characteristics of independent microgrid

Power fluctuations and control technique

Figure 2.2 shows the example of the power fluctuations of an independent microgrid. The continuous curve in the figure is the relation between the periodic gap of power fluctuation without renewable energy, and the amount of fluctuations. Fluctuations with photovoltaics contain sudden cyclic component due to the weather and also long-term fluctuations due to daily change or seasonal change (fringe fluctuation and sustained fluctuation). As a result, the amount of fluctuations of electricity increases as shown by the broken curve in Figure 2.2. In order to stabilise each fluctuation component described above, as in the common large-scale electric power system, two or more stabilisation technique shown in the figure are introduced.

Power fluctuations of proposed system

Figure 2.3 shows the example of power fluctuation of the proposed system. When there is no output power from the photovoltaics, the SOFC-TCC follows the load (continuous curve in the figure). On the other hand, the daytime load falls due to the output power of photovoltaics (middle under part in Figure 2.3) being added to the microgrid. When the load of the microgrid falls rapidly, the output power of the rotary machines (G/T and S/T) of the SOFC-TCC will be adjusted by governor-free control, but response time is dependent on the inertia forces of rotating machines. Because the output power of the SOFC is dependent on temperature control, response of the governor-free control of the SOFC requires long time compared with rotating machines. Therefore, the short-time power fluctuations of the microgrid are reduced by setting appropriate inertia constants of G/T and S/T of the SOFC-TCC. Longer-term power fluctuations are controlled by the governor-free

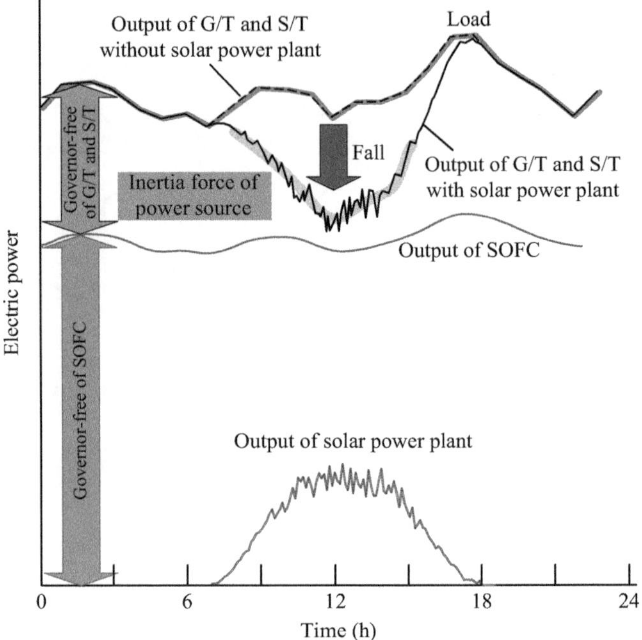

Figure 2.3   Operation model of SOFC triple combined cycle with large-scale solar power plant

control of the G/T. Furthermore, long-term fluctuations of the microgrid correspond to the governor-free control of the SOFC.

2.2.3   Modelling of equipment

2.2.3.1   Solid oxide fuel cell

Modelling of mass balance of the SOFC and energy balance is based on the past research [30–33].

Mass balance
Equation (2.1) is substance balance expression for the anode of the SOFC. i in (2.1) is an index of the substance shown in Table 2.1, and $n_{ad,rc}$ is the number of the chemical reactions at the anode and in the reformer shown in Table 2.2. On the other hand, (2.2) is substance balance expression for the cathode of the SOFC. i in (2.2) is an index of the substance shown in Table 2.1, and $n_{cd,rc}$ is a chemical reaction at the cathode shown in Table 2.3.

$$\frac{dm_{ad,i}}{dt} = \dot{m}_{ad,in,i} - \dot{m}_{ad,out,i} + \sum_{j=1}^{n_{ad,rc}} a_{ad,ij} r_{ad,j}, \quad i = 1,\ldots,7, \quad n_{ad,rc} = 4 \quad (2.1)$$

$$\frac{dm_{cd,i}}{dt} = \dot{m}_{cd,in,i} - \dot{m}_{cd,out,i} + \sum_{j=1}^{n_{ad,rc}} a_{cd,ij} r_{cd,j}, \quad i = 1,\ldots,7, \quad n_{cd,rc} = 1 \quad (2.2)$$

Table 2.1   Notation for components

i	1	2	3	4	5	6	7
Component	N_2	O_2	H_2	CH_4	H_2O	CO	CO_2

Table 2.2   Reaction at anode electrode and reformer

Reaction number	Anode reaction	Reaction rate
1	$H_2 + O^{2-} \rightarrow H_2O + 2e^-$	$r_{ad,1}$
2	$CH_4 + H_2O \leftrightarrow CO + 3H_2$	$r_{ad,2}$
3	$CO + H_2O \leftrightarrow CO_2 + H_2$	$r_{ad,3}$
4	$CH_4 + 2H_2O \leftrightarrow CO_2 + 4H_2$	$r_{ad,4}$

Table 2.3   Reaction at cathode electrode

Reaction number	Cathode reaction	Reaction rate
1	$0.5O_2 + 2e^- \rightarrow O^{2-}$	$r_{cd,1}$

Although an external reformer is introduced in Figure 2.1, mass balance of the external reformer is the same as (2.1). $r_{ad,2}$ to $r_{ad,4}$ in Table 2.2 gives the rate of each chemical reaction and the reaction velocity $r_{ad,1}$ of the anode obtained by (2.3). Moreover, the reaction velocity $r_{ad,2}$ to $r_{ad,4}$ of the reformer is calculated using (2.6) from (2.4) [31].

$$r_{ad,1} = r_{cd,1} = \frac{I}{2F} \tag{2.3}$$

$$r_{ad,2} = \frac{k_2/p_{ad,H_2}^{2.5}\left[p_{ad,CH_4}p_{ad,H_2O} - \left(p_{ad,H_2}^3 p_{ad,CO}/K_2\right)\right]}{DEN^2} \tag{2.4}$$

$$r_{ad,3} = \frac{k_3/p_{ad,H_2}\left[p_{ad,CO}p_{ad,H_2O} - \left(p_{ad,H_2}p_{ad,CO_2}/K_3\right)\right]}{DEN^2} \tag{2.5}$$

$$r_{ad,4} = \frac{k_4/p_{ad,H_2}^{3.5}\left[p_{ad,CH_4}p_{ad,H_2O}^2 - \left(p_{ad,H_2}^4 p_{ad,CO_2}/K_4\right)\right]}{DEN^2} \tag{2.6}$$

Here, DEN in (2.3)–(2.6) is given by (2.7), and $K_{ads,i}$ in (2.7) is given by (2.8). Moreover, rate coefficients for reforming reactions k_2, k_3 and k_4 in (2.3)–(2.6) are calculated by (2.9); the equilibrium constants K_2, K_3 and K_4 of the reaction numbers 2–4 in Table 2.2 are given by (2.10)–(2.12), respectively.

$$\mathrm{DEN} = 1 + K_{\mathrm{ads,CO}} p_{\mathrm{ad,CO}} + K_{\mathrm{ads,H_2}} p_{\mathrm{ad,H_2}} + K_{\mathrm{ads,CH_4}} p_{\mathrm{ad,CH_4}} \\ + K_{\mathrm{ads,H_2O}} p_{\mathrm{ad,H_2O}} / p_{\mathrm{H_2}} \tag{2.7}$$

$$K_{\mathrm{ads},i} = A_{K_{\mathrm{ads},i}} \exp\left(\frac{-\Delta \overline{h}_{\mathrm{ads},i}}{RT}\right), \quad i = \mathrm{H_2, CH_4, H_2O, CO} \tag{2.8}$$

$$k_j = A_{k_j} \exp\left(\frac{-E_j}{R \cdot T}\right), \quad j = 2, 3, 4 \tag{2.9}$$

$$K_2 = \exp\left(\frac{-26{,}830}{T} + 30.114\right) \tag{2.10}$$

$$K_3 = \exp\left(\frac{4{,}400}{T} - 4.036\right) \tag{2.11}$$

$$K_4 = \exp\left(\frac{-22{,}430}{T} + 26.078\right) \tag{2.12}$$

The molar flow rate discharged from the anode and cathode is obtained using choked exhaust flow equation, and when the pressure difference of the inlet port and the outlet port of each electrode is taken into consideration, the molar flow rate is given by (2.13) and (2.14) [34]. Moreover, (2.15) is a formula for the oxygen and fuel (hydrogen) utilisation factor for the cathode electrode and the anode electrode.

$$\dot{m}_{\mathrm{ad,out}} = \sqrt{k_{\mathrm{ad}}\left(p_{\mathrm{ad}} - p_{\mathrm{ad,out}}\right)} \tag{2.13}$$

$$\dot{m}_{\mathrm{cd,out}} = \sqrt{k_{\mathrm{cd}}\left(p_{\mathrm{cd}} - p_{\mathrm{cd,out}}\right)} \tag{2.14}$$

$$u_{\mathrm{O_2}} = 1 - \frac{\dot{m}_{\mathrm{out,O_2}}}{\dot{m}_{\mathrm{in,O_2}}}, \quad u_{\mathrm{H_2}} = 1 - \frac{\dot{m}_{\mathrm{out,H_2}}}{\dot{m}_{\mathrm{in,H_2}}} \tag{2.15}$$

Energy balance

Equation (2.16) is an energy balance equation of the whole SOFC [32,33]. The temperature differential (change over time) of the SOFC balances the sum total of enthalpy change at the anode and cathode, the energy change of chemical reactions, DC (direct-current) power output (P_{DC}), and thermal radiation and heat radiation of heat transmission ($P_{\mathrm{rad}}, P_{\mathrm{ht}}$). N in (2.16) is the number of components of the substances shown in Table 2.1, and M is the number of the chemical reactions at the anode and the reformer shown in Table 2.2.

$$C_S \frac{dT}{dt} = \sum_{i=1}^{N} \dot{m}_{\mathrm{ad,in},i} \left(\Delta \overline{h}_{\mathrm{ad,in},i} - \Delta \overline{h}_i\right) + \sum_{i=1}^{N} \dot{m}_{\mathrm{cd,in},i} \left(\Delta \overline{h}_{\mathrm{cd,in},i} - \Delta \overline{h}_i\right) \\ - \sum_{j=1}^{M} \Delta \overline{h}_{\mathrm{rc},j} r_{\mathrm{ad},j} - P_{\mathrm{DC}} - P_{\mathrm{rad}} - P_{\mathrm{ht}} \tag{2.16}$$

where $N = 7$, $M = 4$

Although P_{DC} in an (2.16) is obtained by (2.17), the stack voltage V of fuel cell is a difference of the open circuit voltage E_{ocv} and voltage loss V_{loss}, as shown in (2.18). Moreover, E_{ocv} can be calculated by (2.19), and the voltage loss V_{loss} is dependent on the current and temperature of the fuel cell, as shown in (2.20) [35].

$$P_{DC} = VI \tag{2.17}$$

$$V = E_{ocv} - V_{loss} \tag{2.18}$$

$$E_{ocv} = E_0 + \frac{RT}{2F} \ln\left(\frac{p_{ad,H_2} p_{ad,O_2}^{0.5}}{p_{ad,H_2O}}\right) \tag{2.19}$$

$$V_{loss} = c_1 I + c_2 T + c_3 \tag{2.20}$$

Equations (2.21) and (2.22) are the terms on heat radiation of SOFC in (2.16) and take the thermal radiation P_{rad} and the heat transmission P_{ht} in this section into consideration.

$$P_{rad} = A\varsigma\sigma\left(T^4 - T_s^4\right), \quad \sigma = 5.67 \times 10^{-8}\ \text{W}/\left(\text{m}^2\ \text{K}^4\right) \tag{2.21}$$

$$P_{ht} = Ah(T - T_s) \tag{2.22}$$

Stack unit
Supply of fuel and air to the SOFC assumes the parallel connection of a fuel cell stack, as shown in Figure 2.4. In this paper, the number of the fuel cell stacks connected in parallel is described as the number of units. As the heat capacity per unit decreases as the number of units increases and the rated power per unit becomes small, this simplifies the temperature control by air-flow rate control of SOFC described in Section 4. Therefore, if the number of units of SOFC increases, the short-term load following becomes possible. However, because the increase in the number of units leads to the increase in cost of the system, the maximum number of units in the analysis in this paper is set to 3.

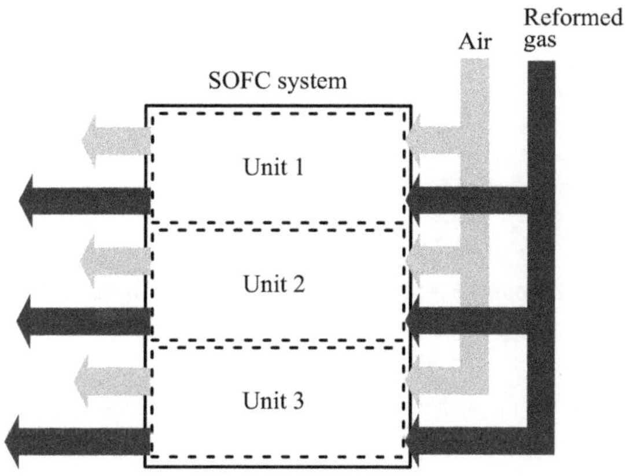

Figure 2.4   Parallel connection of SOFC

Transfer function
The transfer function on the output adjustment of SOFC is shown in (2.23). T_{fc} in (2.23) is a time constant of the output power of SOFC. Because the output power of SOFC is dependent on cell temperature, T_{fc} is decided from the rate of change of the cell temperature obtained from the energy balance (see (2.16)) of the SOFC. s is the Laplace operator in (2.23).

$$P_{\mathrm{fc,out}} = \frac{1}{1 + T_{\mathrm{fc}}s} m_{\mathrm{H_2,in}} \tag{2.23}$$

2.2.3.2   Gas turbine system

Relational expression of compressor
The outlet temperature $T_{\mathrm{ac,out}}$ of the air compressor of G/T is given by (2.24) using the outside air temperature T_{amb}. However, change of air volume in the compressor assumes adiabatic compression; η_{ac}, R_a, $W_{\mathrm{ac},a}$ and γ are compressor efficiency, compression ratio of the compressor, air-flow rate of the compressor and ratio of specific heat of air, respectively.

$$T_{\mathrm{ac,out}} = T_{\mathrm{amb}}\left\{1 + \frac{(R_a W_{\mathrm{ac},a})^{(\gamma-1)/\gamma} - 1}{\eta_{\mathrm{ac}}}\right\} \tag{2.24}$$

Gas turbine
Equation (2.25) is the inlet temperature $T_{\mathrm{gt,in}}$ of the G/T, and (2.26) is the outlet temperature $T_{\mathrm{gt,out}}$, where the subscript 'rat' shows rating, it is assumed that the flow of combustion gas is the same as the air-flow rate from the compressor. Furthermore, W_f in each equation is a fuel flow rate, $W_{\mathrm{ac},a}$ is an air-flow rate and η_t in (2.26) is the turbine efficiency. Change of the combustion gas volume in the G/T is assumed to be adiabatic expansion.

$$T_{\mathrm{gt,in}} = T_{\mathrm{ac,out}} + \left(T_{\mathrm{gt,in,rat}} - T_{\mathrm{ac,out,rat}}\right)\frac{\dot{m}_f}{\dot{m}_{\mathrm{ac},a}} \tag{2.25}$$

$$T_{\mathrm{gt,out}} = \left\{T_{\mathrm{ac,out}} + \left(T_{\mathrm{gt,in,rat}} - T_{\mathrm{ac,out,rat}}\right)\frac{\dot{m}_f}{\dot{m}_{\mathrm{ac},a}}\right\}\left\{1 - \left(1 - \frac{1}{(R_a \cdot \dot{m}_{\mathrm{ac},a})^{(\gamma-1)/\gamma}}\right)\eta_t\right\} \tag{2.26}$$

Transfer function
Equation (2.27) is the transfer function of the G/T, which uses the time constant T_{ac} of the compressor.

$$P_{\mathrm{gt}} = \frac{K_{\mathrm{gt}}\left\{\left(T_{\mathrm{gt,in}} - T_{\mathrm{gt,out}}\right) - \left(T_{\mathrm{ac,out}} - T_{\mathrm{outside}}\right)\right\}W_a}{1 + T_{\mathrm{ac}}s} \tag{2.27}$$

2.2.3.3 Steam turbine

Relational expression

The model of the S/T is Rankine cycle, and as shown in (2.28), the external work l_{st} of the S/T is obtained by excluding power consumption l_{pump} of the circulating pump from the quantity of heat P_{boiler} supplied to the exhaust gas boiler. Moreover, $\eta_{th,st}$ in (2.28) is theoretical thermal efficiency.

$$l_{st} = P_{boiler}\eta_{th,st} + l_{pump} \tag{2.28}$$

Transfer function

Because steam from the exhaust gas boiler is input into a turbine after going via steam pipes etc., the output power of the S/T has a time-lag (steam-receiver model). Therefore, in this paper, the time constant of the steam receiver is expressed by T_v, and the transfer function is given by (2.29). Moreover, the transfer function of the output power of the S/T is given by (2.30) using the delay time constant T_{boiler} of the exhaust-heat-recovery boiler, and the steam-turbine power coefficient K_{st}.

$$P_{out} = \frac{1}{1 + T_v s} P_{in} \tag{2.29}$$

$$P_{st} = \frac{K_{st} T_e W_v}{1 + T_{boiler} s} \tag{2.30}$$

2.2.3.4 Heat exchange

Equation (2.31) is a heat balance equation of the heat exchanger. Moreover, the temperature change of heat and cooling of the fluid by a heat exchanger is obtained from (2.32) and (2.33).

$$P_{he} = K_h A_{he} \Delta T \tag{2.31}$$

$$\frac{dT_{out,hot}}{dt} = \frac{1}{\tau}\left(T_{in,hot} - T_{out,hot} + \frac{-P_{he}}{\dot{m}_{hot} C_{p,hot}}\right) \tag{2.32}$$

$$\frac{dT_{out,cold}}{dt} = \frac{1}{\tau}\left(T_{in,cold} - T_{out,cold} + \frac{-P_{he}}{\dot{m}_{cold} C_{p,cold}}\right) \tag{2.33}$$

2.2.3.5 Photovoltaics

Figure 2.5 shows the example of the test results of the amount of global solar radiation distributed in the range of several kilometres and measured simultaneously [36]. The each curve in the figure are test results of each of the 20-set global-solar-radiation meter. The thick solid curve is the output power in the case of interconnecting the 20-set measuring instruments. Maximum-output fluctuation according to the one-set of the global-solar-radiation meter in the example of Figure 2.5 constitutes about 70% of rated power in several seconds. However, when many photovoltaics are interconnected, the cyclic fluctuation of the global solar radiation for less than 10 s will be reduced to 20% or less of rated power.

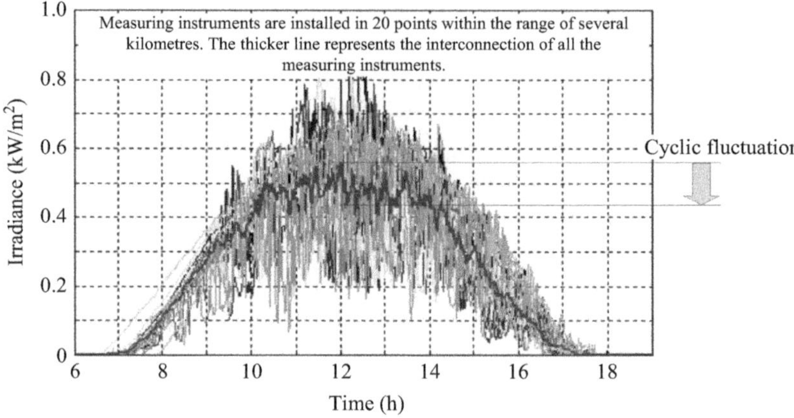

Figure 2.5   Experimental results of cyclic fluctuation of global solar radiation

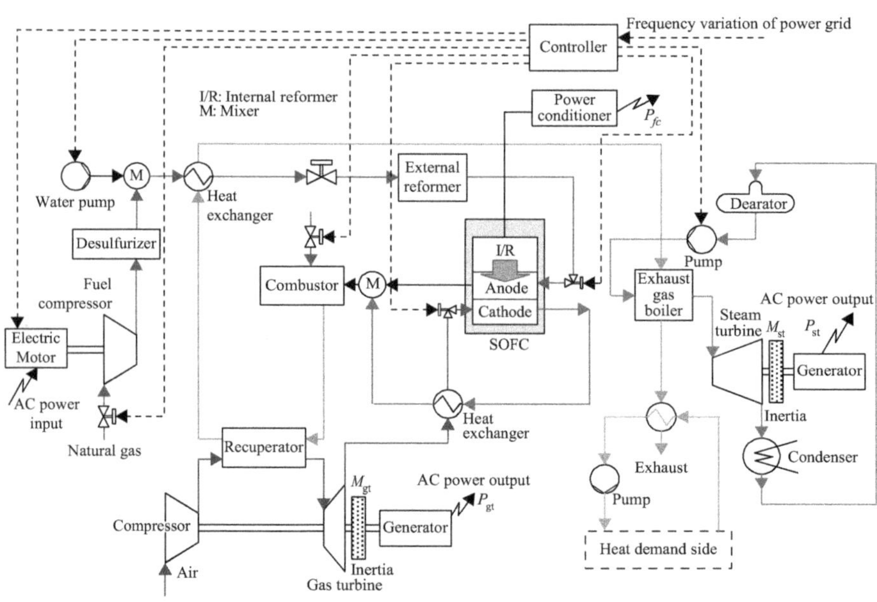

Figure 2.6   Block diagram of the SOFC triple combined cycle generation plant

2.2.4   System configuration of SOFC triple combined cycle

2.2.4.1   System configuration and control method

Figure 2.6 is detailed plan of the system configuration of the proposed SOFC-TCC. The models in Section 2.3 are used for material balance, energy balance and response characteristic (transfer function). Moreover, SOFC consists of three units,

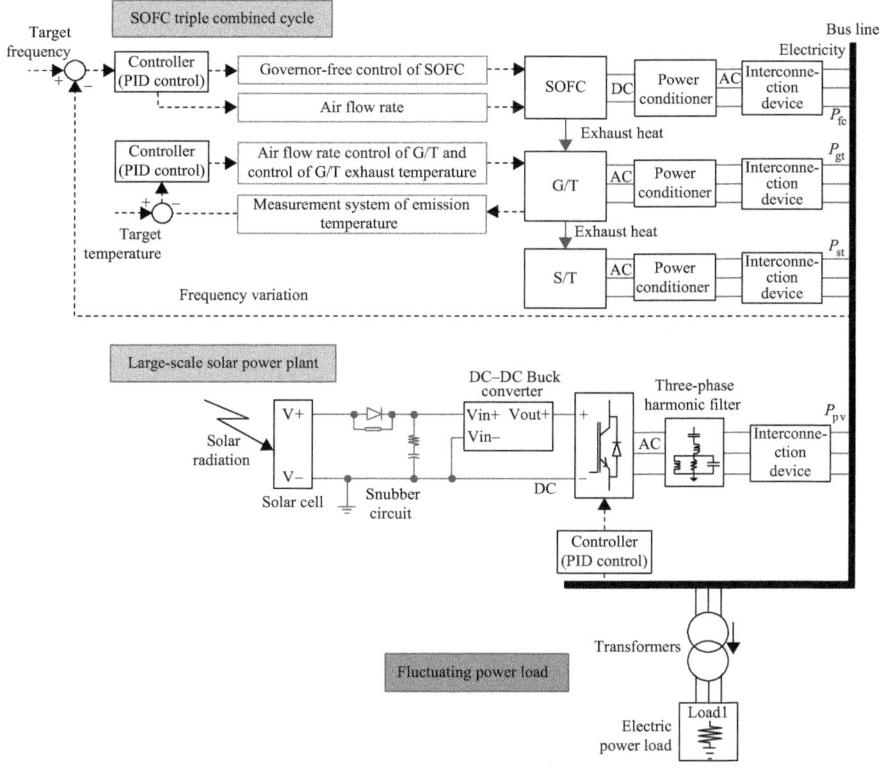

Figure 2.7   Whole block diagram of proposed microgrid

and the combined cycle of G/T and S/T is a multi-axis type. Furthermore, the inertia constants of the G/T and S/T are M_{gt} and M_{st}, respectively.

Figure 2.7 is a control block diagram of the whole independent microgrid containing SOFC-TCC, large-scale photovoltaics and power load. The output power of SOFC is adjusted with the governor-free control of reformed gas and air-flow rate control. The compressed air supplied to the G/T is heated by the exhaust heat of SOFC and combustion of the natural gas m_{cb}. Although the output power of SOFC and the exhaust heat of G/T are adjusted with PID control equipment, operation of S/T is not controlled.

2.2.4.2   Power controls and frequency changes

The supply-and-demand difference from the balance of production-of-electricity by the SOFC-TCC and the power load appear as a frequency deviation of the microgrid, as shown in (2.34). Therefore, the production of electricity of the SOFC-TCC can be adjusted by measuring the frequency deviation of the microgrid and using it as input to the PID control equipment in Figure 2.7. Figure 2.8 is a block diagram of the power controls of the proposed SOFC-TCC. Figure 2.8(a) shows the power controls of the SOFC-TCC; Figure 2.8(b) is a block diagram showing the

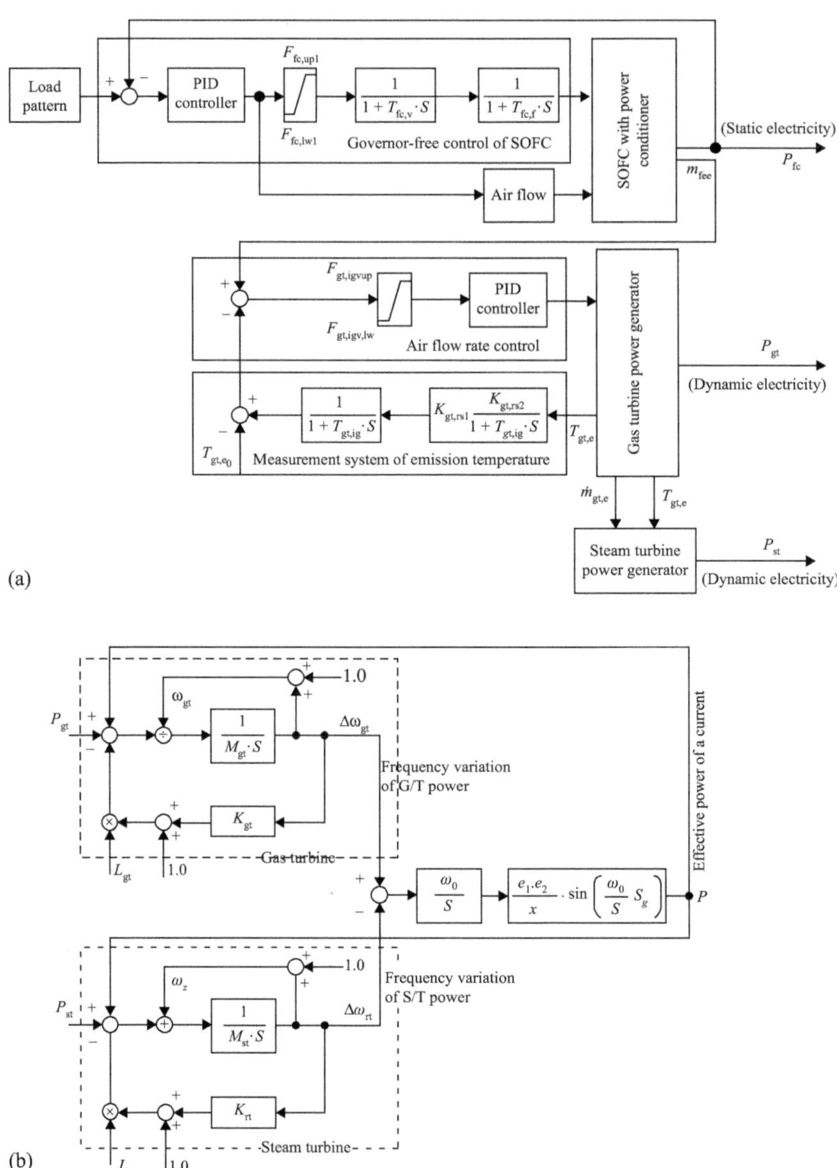

Figure 2.8　Control block diagram of SOFC triple combined cycle: (a) power output model of each power generator and (b) variability model of system frequency

connection between the frequency and effective power of the microgrid [37]. The frequency deviation $\Delta\omega$ of the microgrid is dependent on the difference $\left(P_g - P_l\right)$ of the electrical power output of a power generator and the load, and the inertia constant M, as shown in (2.34). Equation (2.35) is a differential equation of the inertia constant M. The inertia constants M_{gt} and M_{st} of the G/T and S/T are contained in Figure 2.8(b), and the short-time power fluctuation of the G/T and S/T is controlled by changing these values.

$$\Delta\omega = \frac{1}{Ms}\frac{1}{\omega}\left(P_g - P_l\right)38 \tag{2.34}$$

$$M = J\omega = \frac{1}{2}a_r^2 m_r \tag{2.35}$$

2.2.5 Analysis conditions

2.2.5.1 Equipment specifications and time constant

Table 2.4 is the specifications of the important equipment of the SOFC-TCC assumed in analysis in this paper [38,39]. The total production-of-electricity efficiency of the SOFC-TCC at the time of rated operation is 65%, and the productions of electricity efficiencies of the SOFC, G/T and S/T are based on the calorific value of the natural gas at the reformer inlet of 26%, 25% and 14%, respectively. Table 2.5 shows time constants etc. of the transfer function of important equipment, shown in Figure 2.8 [40–42]. p.u. in Table 2.5 is a notation for the per-unit system.

Table 2.4 Rated state of each equipment

System rated power	1.4 MW
Rated power of SOFC	542 kW
Rated power of G/T	550 kW
Rated power of S/T	308 MW
System frequency	50 Hz
SOFC solid heat capacity	600 J/(kg K)
Operation conditions	
Anode pressure	0.3 MPa
Cathode pressure	0.3 MPa
SOFC total current	870 A
SOFC cell voltage	0.657 V
SOFC temperature	1113 K
SOFC stack unit power	181 kW
The number of stack unit	3 set
Methane mass flow rate	7.4 g/s
Air mass flow rate	277 g/s
Fuel utilization rate	0.85
Oxygen utilization rate	0.23
Recycle ratio	0.38
Steam/methane ratio	2
Outside air temperature t_{amb}	288 K
Compressor outlet temperature $t_{ac,out}$	684 K
G/T entrance temperature $t_{gt,bin}$	1573 K
Pressure ratio of compressor R_{ac}	15
Ratio of specific heat γ	1.4
Efficiency of compressor η_{ac}	85%
Turbine efficiency of G/T η_t	85%

Table 2.5   Design parameters and time constants

Equipment		Governor-free control of G/T		Emission temperature control of G/T	
Power coefficient of gas turbine K_{gt}	0.0021	Fuel upper limit $F_{gt,up2}$	1.0 p.u.	Radiation shield coefficient $K_{gt,rs1}$	0.8
Steam-turbine power coefficient K_{st}	0.000391	Fuel lower limit $F_{gt,lw2}$	0.0 p.u.	Radiation shield coefficient $K_{gt,rs2}$	0.2
Delay time constant of compressor T_{ac}	0.2 s	Possible range of fuel regulation $K_{gt,f}$	0.77	Time constant of radiation shield $T_{gt,rs}$	15 s
Delay time constant of exhaust gas boiler T_b	300 s	Fuel flow rate at the time of no-load $K_{gt,f0}$	0.23	Time constant of thermostat $T_{gt,ig}$	2.5 s
Governor-free control of SOFC		Time constant of flow control valve $T_{gt,v}$	1.0 s	Integral control gain $T_{gt,tc}$	3.3
Fuel upper limit $F_{fc,up1}$	1.0 p.u.	Fuel system time constant $T_{gt,f}$	0.4 s	Time constant of integral control $T_{gt,t}$	250 s
Fuel lower limit $F_{fc,lw2}$	0.0 p.u.	*Air flow rate control of G/T*		Control-signal upper limit of temperature of exhaust gas $F_{gt,up3}$	1.05
Time constant of flow control valve $T_{fc,v}$	1.0 p.u.	Temperature setting bias $B_{gt,tp}$	5 K	Control-signal lower limit of temperature of exhaust gas $F_{gt,lw3}$	0
Fuel system time constant $T_{fc,f}$	0.4 p.u.	Integration time constant $T_{gt,igv}$	773 s	Reference temperature of exhaust gas $t_{gt,et0}$	858 K
Load and speed control of G/T		Open speed upper limit of air flow rate valve (AFV) $F_{gt,igv,up}$	0.01 p.u./s		
Speed-regulation gain $K_{gt,sp}$	25	Closed speed upper limit of AFV $F_{gt,igv,lw}$	−0.01 p.u./s		
Time constant of governor $K_{gt,f}$	0.05 s	Opening upper limit of AFV $F_{gt,up4}$	1.0 p.u.		
Load high limit setting $F_{gt,up1}$	1.05 p.u.	Opening lower limit of AFV $F_{gt,lw4}$	0.72 p.u.		
Load low limit setting $F_{gt,lw1}$	−0.1 p.u.				

Table 2.6   Parameters of PID controller

Number of SOFC	K_P	K_I	K_D	F_{cl}
1 set	0.0001157	1.010×10^{-10}	1.689	0.0001740
2 set	0.000673	4.565×10^{-9}	1.694	0.0008674
3 set	0.001442	1.575×10^{-8}	2.300	0.002186

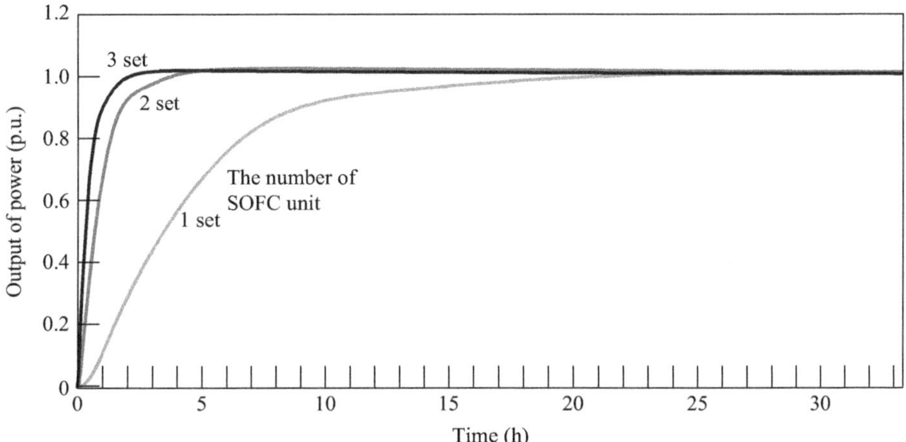

Figure 2.9   Dynamic characteristics of SOFC system

2.2.5.2   Parameters of PID control equipment

The output power of the SOFC model is adjusted by PID control of the governor and air flow in Figures 2.7 and 2.8. Equation (2.36) shows the relationship between each parameter of PID control, and the output power, and shows the values of these parameters used in this analysis in Table 2.6. Each parameter in Table 2.6 is chosen to converge a transient overshoot in a very short time. MATLAB/Simulink 2013a is used for the analysis of the system.

$$U(s) = K_P + K_I \frac{1}{s} + K_D \frac{F_{cl}}{1 + F_{cl}(1/s)} \tag{2.36}$$

2.2.6   *Dynamic-characteristics analysis of an SOFC*

Figure 2.9 shows the output characteristics from a cold start of the SOFC system, which combines the cell stack of SOFC and the reformer. The vertical axis of the figure is expressed as per-unit on the basis of the rated power (542 kW) of the SOFC. When the fuel cell stack is connected in parallel, the convergence time to rated power is shortened. Moreover, Figure 2.10 shows the analysis results of the number of units and response characteristics when the step load pattern of Figure 2.10(a) is connected to the SOFC system. When the number of SOFC units

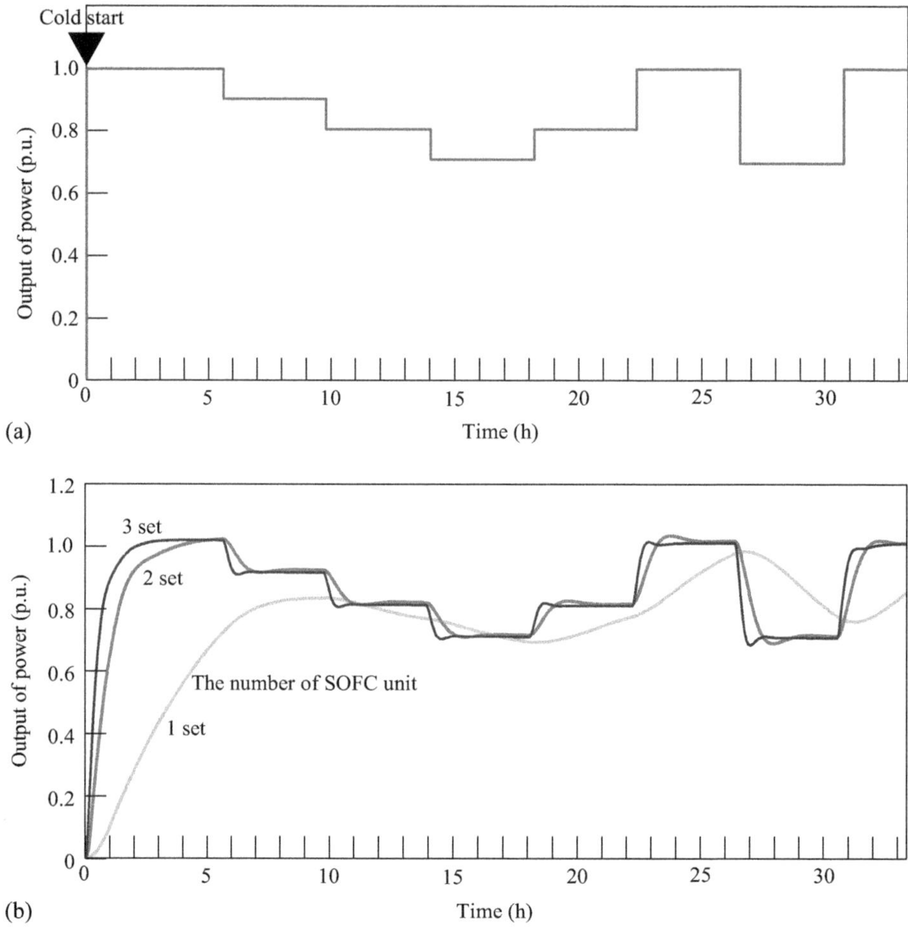

Figure 2.10   Dynamic characteristics of SOFC triple combined cycle: (a) load pattern and (b) response of SOFC system

is 1, the time-lag of the temperature control of the SOFC is large; therefore, it is difficult to follow the step load in Figure 2.10(a). As time-lag occurs between the step load and the response even when the number of SOFC units is 2, the number of units of SOFC is set as 3 in the following analysis.

2.2.7   *Dynamic characteristics of SOFC triple combined cycle (SOFC-TCC)*

2.2.7.1   SOFC, G/T, load distribution of S/T

Figure 2.11 shows the output characteristics from cold start of the SOFC-TCC in $M_{gt} = 15$ and $M_{st} = 10$. The per-unit system based on 1.4 MW of rated power of the SOFC-TCC is used for Figure 2.11. The power outputs of each of the generators as a proportion of the rated power of the SOFC-TCC are SOFC $= 0.4$, G/T $= 0.38$

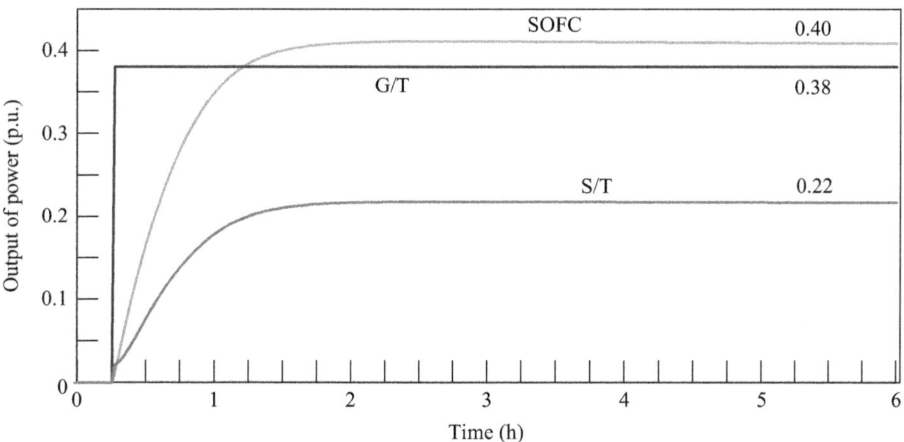

Figure 2.11   Output ratio of SOFC triple combined cycle

and $S/T = 0.22$. As the temperature rise of the cell stack takes long time, achieving the rated power of the SOFC takes a long time. Moreover, speed of response of the S/T is slow because of the steam-receiver model as described in Section 3.3.2. As shown in Figure 2.11, the settling time of SOFC and S/T (about 1.8–2 h) is long compared with the G/T operation. Therefore, for power fluctuations of the electricity demand and the photovoltaics of less than 2 h, the load following operation of the G/T becomes important, and the output adjustments of the SOFC and S/T are effective on the long-term power fluctuation between one day or seasonally.

2.2.7.2   Load response characteristics

Figure 2.12 shows the power supply pattern for a representative day of every month for the Hokkaido Electric Power [43]. Largest fluctuations of the power load for one day are on the August representative days; there is about 40% difference between day and night. Moreover, fluctuations of the power load for one year is the largest during 5:00 of June representative days, and 18:00 of February representative days, and there is about 44% of difference. Therefore, the step load shown in Figure 2.13(a) is input to the SOFC-TCC from the magnitude of the load fluctuation of Figure 2.12. The output characteristics [Figure 2.13(b)] and the frequency deviation [Figure 2.13(c) and (d)] of each power generator are obtained from the analysis, where Figure 2.13(a) and (b) shows with the per-unit system based on 1.4 MW, rated power of the SOFC-TCC. Because the speed of response of the SOFC and S/T is very slow, when step load is added to the system, reduction of the power variation of 2 h or less mainly corresponds to the response of the G/T operation as shown in Figure 2.13(b). On the other hand, as shown in Figure 2.13(c) and (d), the characteristics of the frequency deviation of about 600 s (10 min) of sampling time from 9,000 to 9,600 s change greatly with the inertia constants M_{gt} and M_{st} of the G/T and S/T. Therefore, inertia constants have large influence on the cyclic fluctuation and fringe fluctuation (Figure 2.2) of the microgrid. Although the ranges of

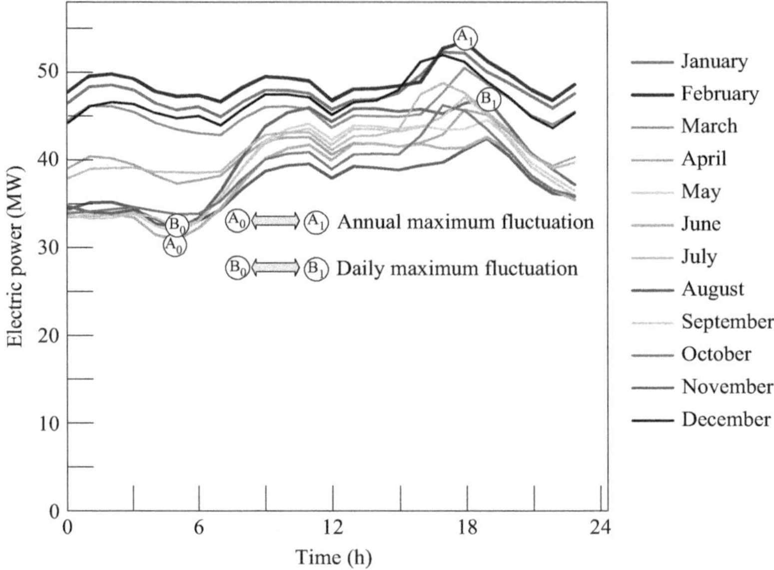

Figure 2.12   Electric power demand pattern

the inertia constant of common G/T and S/T are 10–15 s, the acceptable frequency deviation in Japan is ±0.2 Hz; therefore, the values of M_{gt} and M_{st} in Figure 2.13(c) are not suitable for the microgrid with the large step fluctuation shown in Figure 2.13(a). Moreover, with the values of M_{gt} and M_{st} shown in Figure 2.13(d), frequency deviations served are smaller than ±0.2 Hz. Therefore, when fluctuations of the large step load shown in Figure 2.13(a) are expected for the microgrid, it is necessary to set large values for M_{gt} and M_{st}. As described above, the characteristics of cyclic fluctuations (electricity change for several minutes or less) to fringe fluctuations (electricity change for 20 min or less) can be adjusted with setting of the inertia constant of the G/T and S/T. Therefore, it is necessary to take into consideration the governor-free control of the G/T, and the suitable setting of the inertia force of the G/T and S/T for the cyclic fluctuation and fringe fluctuation, and the sustained fluctuation of 2 h or less.

2.2.7.3   Load response characteristics with photovoltaics

Figure 2.14 shows the analysis results of output characteristics [Figure 2.14(b)] of each power generator and frequency deviation [Figure 2.14(c) and (d)] of the microgrid with large-scale photovoltaics and the SOFC-TCC. The maximum output of the assumed photovoltaics is the same value as the full power (1.4 MW) of the SOFC-TCC. Therefore, Figure 2.14 shows the extreme case where the greatest load change theoretically permitted by the SOFC-TCC is added. Moreover, Figure 2.14(a) and (b) shows the per-unit system based on 1.4 MW, rated power of the SOFC-TCC.

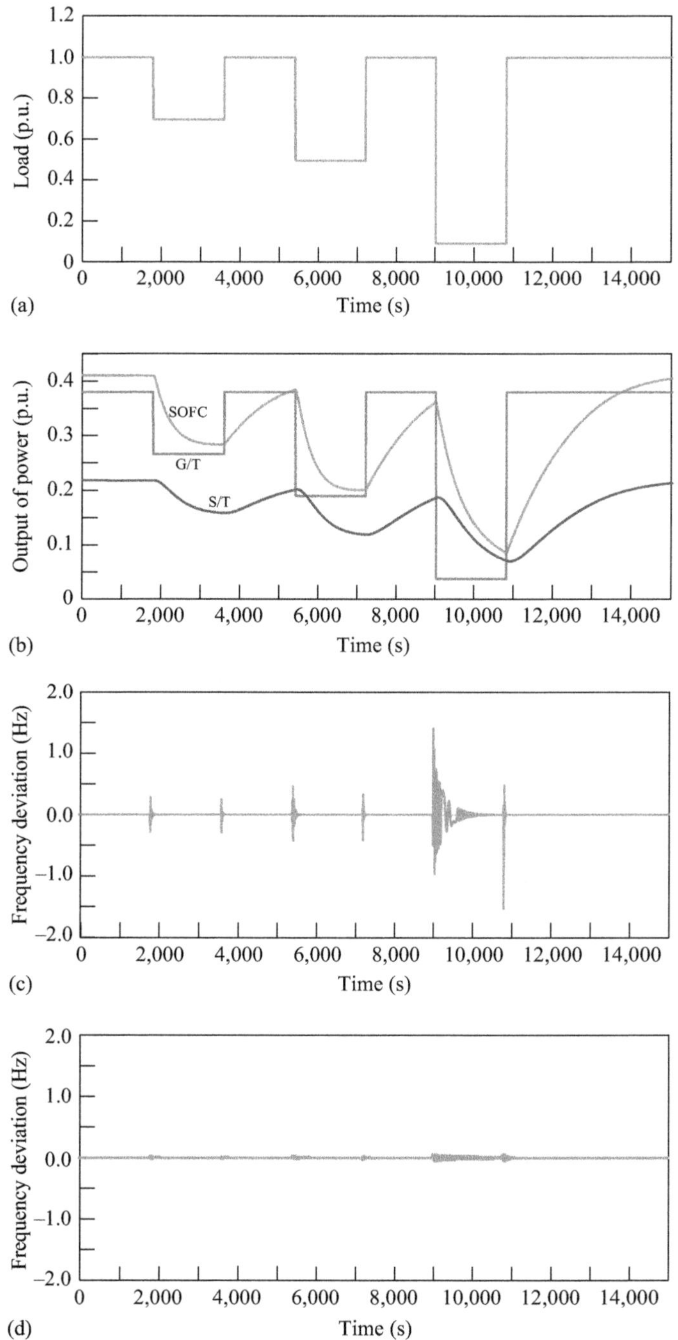

Figure 2.13   *Dynamic characteristics of SOFC triple combined cycle: (a) load pattern; (b) output from each generator ($M_{gt} = 15$, $M_{st} = 10$); (c) $M_{gt} = 15$, $M_{st} = 10$; and (d) $M_{gt} = 75$, $M_{st} = 50$*

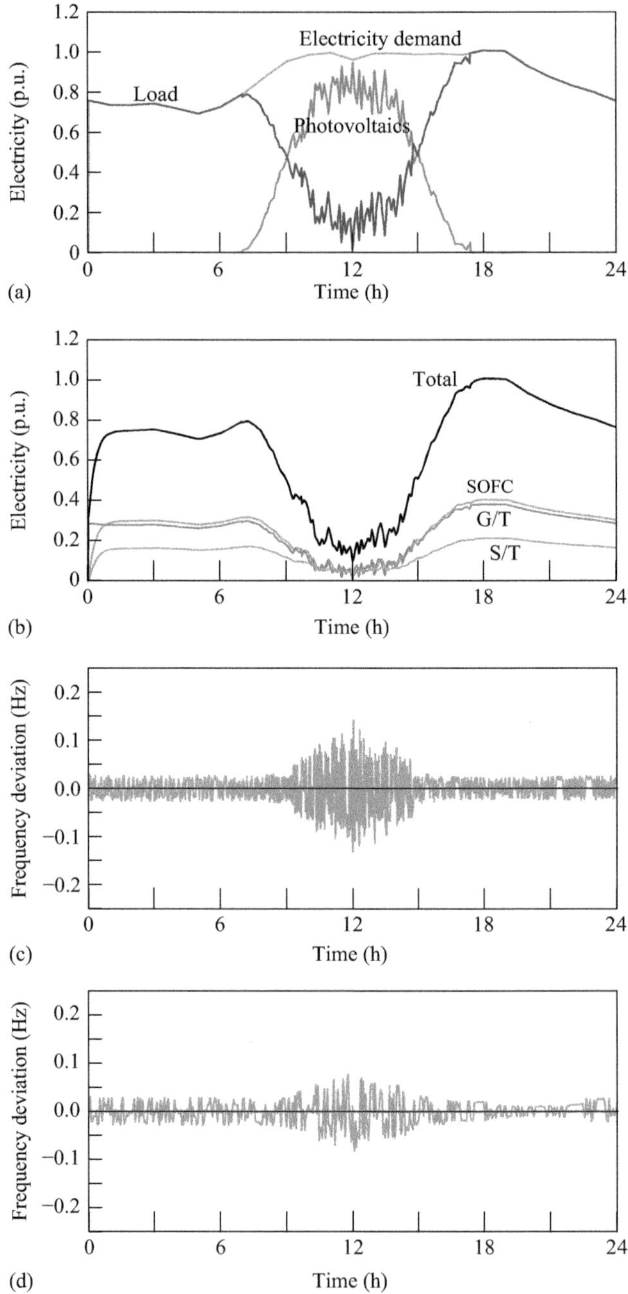

Figure 2.14   Dynamic characteristics of SOFC triple combined cycle with large-scale photovoltaics: (a) load pattern; (b) output from each generator ($M_{gt} = 15$, $M_{st} = 10$); (c) $M_{gt} = 15$, $M_{st} = 10$; and (d) $M_{gt} = 75$, $M_{st} = 50$

Electricity is supplied to the microgrid with the electricity demand pattern of August representative day with largest daily fluctuation shown in Figure 2.12 from the 1.4 MW distributed solar cell [accordingly, with average amount of insolation in Figure 2.5 (broken curve)] and enlarged capacity of Figure 2.5 and from the SOFC-TCC. The output of the SOFC-TCC is red curve excluding the output power of photovoltaics (orange curve) and the electricity demand pattern (continuos curve) in Figure 2.14(a). Figure 2.14(b) and (c) shows the results with $M_{gt} = 15$ and $M_{st} = 10$. Figure 2.14(b) shows the output characteristics of each power generator, and Figure 2.14(c) shows the frequency deviation of the microgrid. As shown in Figure 2.14(c), the frequency deviation is about ±0.14 Hz at the maximum. On the other hand, Figure 2.14(d) shows about ±0.07 Hz at the maximum frequency deviation with $M_{gt} = 75$ and $M_{st} = 50$. From the results of Figure 2.14(c) and (d), when a large-scale photovoltaic is interconnected to the microgrid, it turns out that long-term fluctuation exceeding 20 min depends on the values of the inertia force of the G/T and S/T.

As in Figure 2.14, the relation between the inertia constant of the rotating machines and frequency deviation is shown in Figure 2.15. The horizontal axis (output of large-scale photovoltaics) in the figure is a relative rate when the maximum output power of the photovoltaics shown in Figure 2.14(a) is set as 1.0. Although the range of the acceptable value of the frequency deviation is ±0.2 Hz, in the case of $M_{gt} = 3$ and $M_{st} = 2$, the frequency deviation is just within the limit in Figure 2.15. Furthermore, in order for the frequency deviation (accordingly, supply-and-demand difference of electricity) of the microgrid to come within the acceptable range, connection of distributed photovoltaics is limited in the case shown in Figure 2.5.

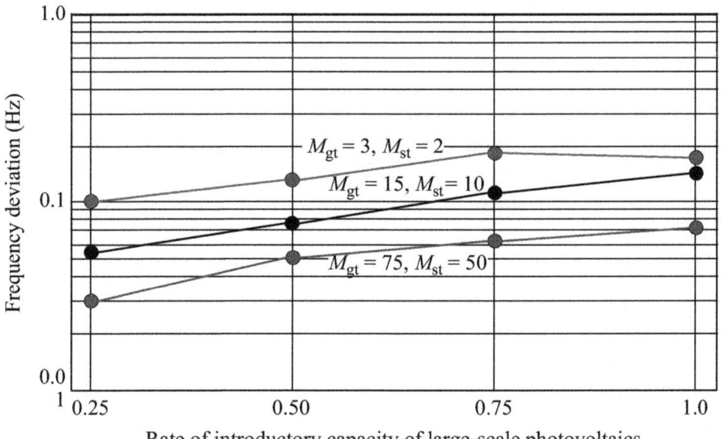

Figure 2.15 Relation between the amount of introduction of photovoltaics and frequency deviation

2.2.8   Conclusions

The electricity stabilisation at the time of introducing large-scale photovoltaics into an independent microgrid with triple combined cycle generator which consists of an SOFC (542 kW), a G/T (550 kW) and an S/T (308 kW) was considered. SOFC supplies electricity to the microgrid through an inverter and the G/T and S/T supply electricity to the microgrid with a synchronous generator. The following conclusions were obtained from this study.

1. The following load takes SOFC and S/T about 1.8–2 h. Therefore, the SOFC and S/T can correspond to long-term fluctuations of the electricity demand of the microgrid and photovoltaics output for one day or seasonally. On the other hand, the delayed response of the SOFC and S/T is too slow for power fluctuation for 2 h or less, it is necessary to follow the load with the governor free control of the G/T.
2. Setting of the inertia constant of the G/T and S/T showed that the power characteristics of cyclic fluctuation (change for several minutes or less) to fringe fluctuation (change for 20 min or less) of the microgrid could be damped. Therefore, when the conclusion described in (1) is taken into consideration, in addition to the governor free control of the G/T, it is necessary to set up the inertia constant of the G/T and S/T appropriately to damp the power fluctuations for 2 h or less.
3. Furthermore, when interconnecting larger-scale photovoltaics to the microgrid, the inertia force of the G/T and S/T showed clearly to have influence on long-term power fluctuations with a period exceeding 20 min.

2.3   Performance evaluation of an independent microgrid comprising an integrated coal gasification fuel cell combined cycle, large-scale photovoltaics and a pumped-storage power station

2.3.1   Introduction

The Japanese power industry became completely deregulated from April 2016, and the expansion of the distributed energy systems by local utilities is expected. For the next generation of electric power systems in Japan, a mix of conventional concentrated electric power and a new distributed energy system is anticipated. Although an interconnection between other electric power systems and an independent system to form a distributed energy system is possible, any distributed energy system can make effective use of the exhaust heat from an electric power plant. It is necessary to stabilise power fluctuations resulting from renewable energy use and the electric power load in a distributed energy system. High speed control of the supply–demand balance requires control technology for the power source to be developed. However, when electric power fluctuations are controlled by a thermoelectric power plant and the load factor of the power source is changed, a drop in the efficiency due to partial load operation occurs.

Control of the power changes due to renewable energy using PSPG has been investigated in Japan and the introduction of a combined cycle generation with coal gasification technology is expected [44]. Recently, several approaches were implemented for the development of the combined power system with coal gasification in Japan: IGCC demonstration plant [45]; clean coal technologies, such as the United States, Japan and Italy [46]; and energy flow of advanced IGCC [47]. In the IGFC combined cycle, hydrogen produced by coal gasification is supplied to an SOFC and a G/T is operated using the exhaust heat and unused hydrogen of the SOFC. Use of the IGFC to operate an S/T using the exhaust heat from the G/T is extremely promising with regard to power generation efficiency and reduction of environmental impacts: production of pure H_2 from gasified coal and capture most of the generated CO_2 [48]; system analysis of IGFC with energy recuperation [49]; steam co-gasification using biomass and coal [50]; greenhouse gas reduction and cost of hydrogen production [51]; and economic evaluations of gasification power plants [52]. With efficiency higher than the conventional combined cycle, the power generation system with coal gasification is expected to become the key technology of future thermal power generation. Although it is possible that the maximum gross thermal efficiency of the IGFC reaches 70%, the IGFC is considered to be the base supply for uniform output power. Many performance evaluation studies of SOFC, IGFC and IGCC and reduction of their environmental impact have been reported: prospect of low-temperature operation of SOFC [53]; evaluation of substitute natural gas production from coal gasification [54]; economic analysis of advanced IGCC and IGFC [55]; economic performance of novel IGFC with carbon capture [56]; and water–gas shift using synthesis gas from coal gasification [57]. An example of a relevant investigation into the output adjustment using the governor control of the IGFC has not yet been identified. In this study, the power generation efficiency of the proposed system with the daily output adjustment is confirmed by modelling of the IGFC. Moreover, a system using IGFC as a base power source for an independent microgrid with photovoltaics and PSPG, the operational method and the initial capacity of equipment based on cost minimisation are investigated. As the output adjustment of the IGFC takes several hours [58], the power output of photovoltaics and the power demand for a representative day are predicted on the day before the operation. Operation of the IGFC (uniform output power) on the representative day is planned on the basis of the prediction results.

Most initial examples of an independent IGFC microgrid have not been reported, and there is little documented research on the interconnection with renewable energy [59]. However, the economic efficiency of the independent microgrid with the daily output adjustment of the IGFC based on supply and demand and the time shift of electricity using PSPG has been reported. For the proposed microgrid with a power load pattern on a representative day in Hokkaido, Japan, numerical simulation was conducted to compare the cost when IGFC or an IGCC is installed. Moreover, optimal facility planning and the operation and for assessing power generation efficiency and cost are determined. Based on these results, the effects of the IGFC independent microgrid is investigated.

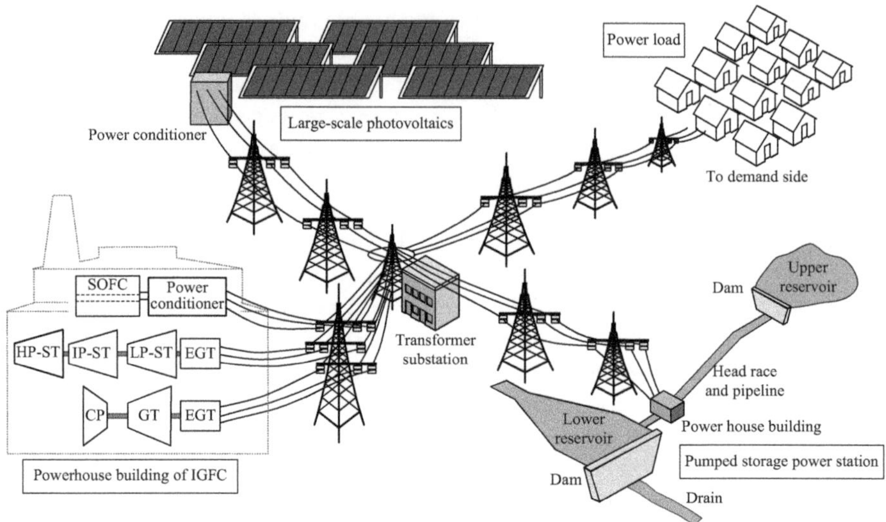

Figure 2.16   Proposed independent power system with an IGFC, PV and pumped storage power station

2.3.2   Materials and methods

2.3.2.1   Overview of the system

Components

Figure 2.16 shows the configuration of the proposed independent microgrid using an IGFC supplying a small-to-medium-size city. A large-scale solar power station, PSPG and IGFC are interconnected, and electric power is supplied through a substation. Although the IGFC is operated using constant daily load, the demand-side power load and the output from photovoltaics are predicted on the previous day. The output power of the IGFC converts the supply–demand difference for each sampling time using PSPG. Moreover, the capacity of each device is determined on the basis of the goals of minimising fuel and equipment costs. The IGFC is used as a source of uniform daily base power output, and the PSPG provides a source of peak power to be reflected in the supply–demand change. When the power output of the photovoltaics is high, or when the power load is low, the IGFC operates at a less efficient partial load. Therefore, in this study, the power generation efficiency of the whole IGFC was obtained by modelling the relation of efficiency and load factor of a coal gasifier, an SOFC, a G/T and an S/T. Together, these comprise the IGFC and provide a load factor for use in the model equations.

Operational method

Figure 2.17 shows the operational method for the proposed system on a representative day. Figure 2.17(a) shows the method using photovoltaics. Figure 2.17(b) shows the method with the output power of photovoltaics. The operational plan for the IGFC predicts the power demand and the output of the photovoltaics on the day

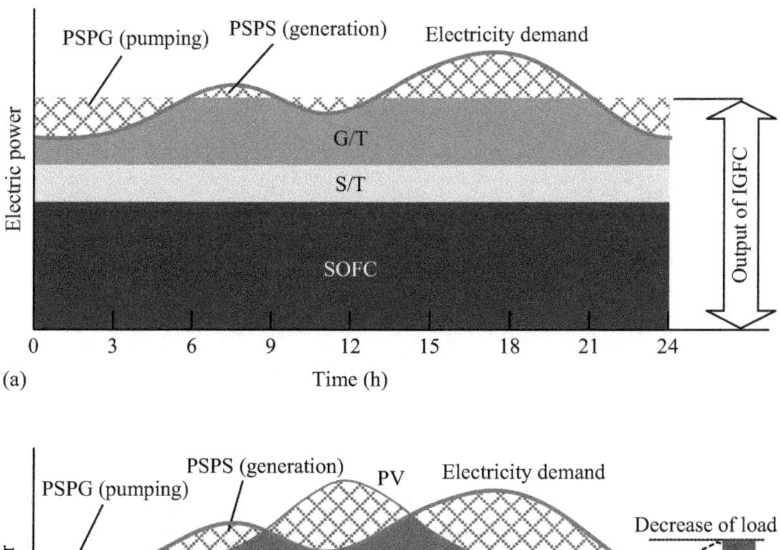

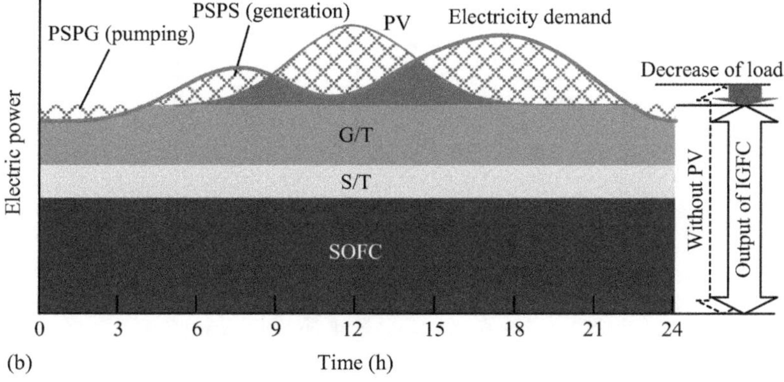

Figure 2.17   Output adjustment for the IGFC: (a) without PV output and (b) with PV output

before the representative day. By predicting the results using the analytic algorithm of the operational method, the output power of the IGFC on the representative day can be determined. As the IGFC operates with uniform output power on a representative day, the power outputs of the SOFC, G/T and S/T introduced into the IGFC are constant. The surplus power load is used for pumping up of the PSPG. Figure 2.17(a) and (b) shows that when the power load of the IGFC and photovoltaics is insufficient, the supply–demand balance is maintained by the power output of the PSPG. The optimal capacity of the large-scale solar power system, PSPG and IGFC in the independent microgrid is determined. In addition, the system's operational method is determined using an analytic algorithm. Moreover, the power generation efficiency, fuel cost and equipment cost of the proposed microgrid are clarified.

Operation of the system under partial load

Figure 2.18 shows the system configuration of the proposed IGFC. This system comprises an air supply unit (ASU), a gasifier unit, an SOFC, a G/T, a heat

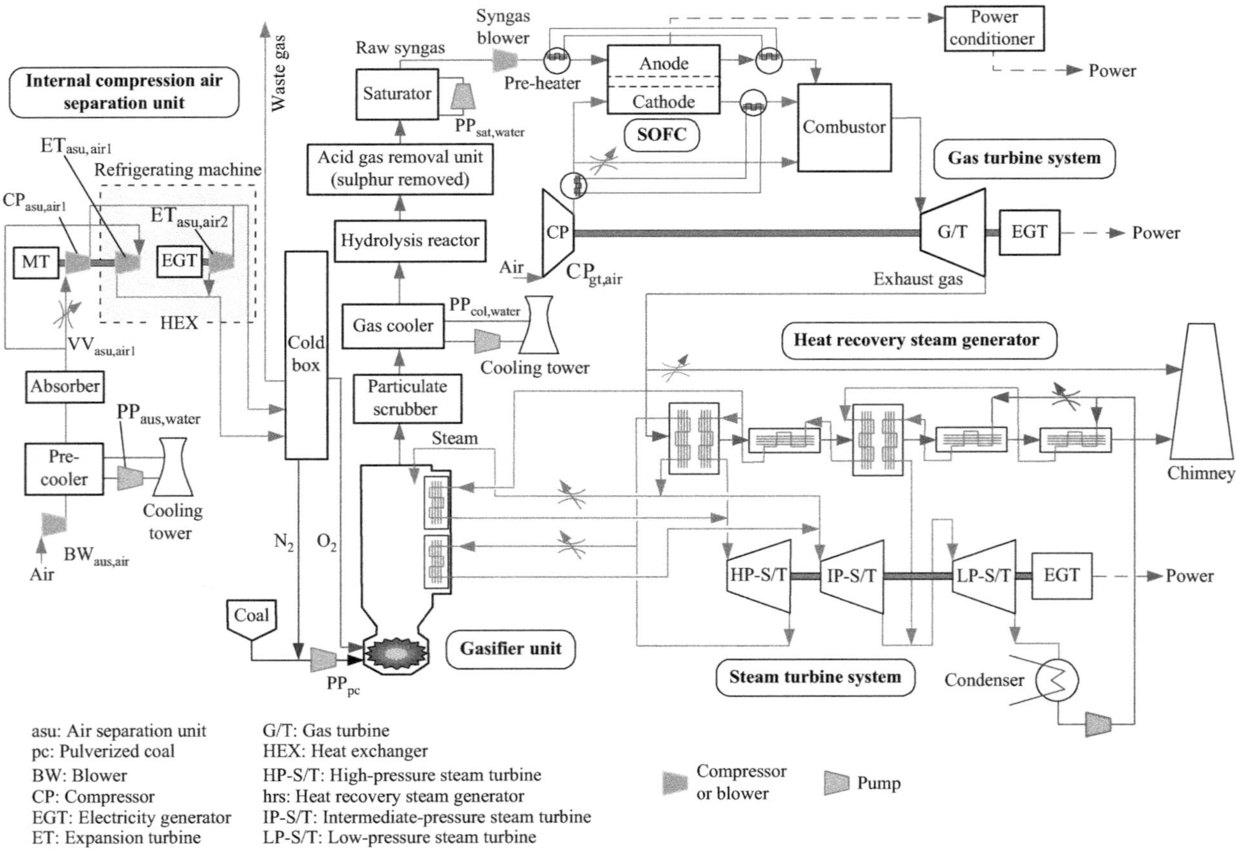

Figure 2.18 Proposed IGFC combined cycle

recovery steam generator and an S/T. The system configuration of the gasifier unit, the heat recovery steam generator and the S/T system is model of the authors' assumptions. To operate the system as shown in Figure 2.18 using a minimum load factor of 65%, the efficiency in each device is reduced. The ASU is a system configuration that assumes that the internal compression type air separation unit is integrated with liquefying process [60]. The quantity of air supplied from a blower $BW_{ASU,air}$ in the ASU changes with the load factor in the system. Furthermore, the power consumption $CP_{ASU,air1}$ of a compressor changes with the magnitude of load. Moreover, based on the load factor of the system, powdered coal is supplied to the gasifier with the nitrogen produced in the ASU. As raw syngas from the gasifier contains impurities and greenhouse gases, it is necessary to introduce processes that eliminate or reduce these undesired substances. As a result, hydrogen containing only trace amounts of impurities can be supplied to the SOFC.

Although high-temperature compressed air is supplied to the SOFC by the air compressor $CP_{gt,air}$, the operating temperature of the SOFC changes because the air temperature of the $CP_{gt,air}$ decreases when the load factor of the proposed system reduces. Therefore, when the load factor, rather than the rating, of the proposed system reduces, the power generation efficiency of the SOFC decreases. Moreover, as the calorie supply to the G/T reduces, the overall efficiency of the G/T reduces. As a result, the exhaust heat from the G/T is supplied to a heat recovery steam generator, and the output of the S/T changes.

2.3.2.2   Output model for the IGFC

Air supply unit

High-purity heat integrated air separation column is assumed in this study; two columns are installed in ASU, HPC is high-pressure column, and LPC is low-pressure column. The air is compressed to about 0.580 MPa by the compressor $CP_{ASU,air}$ and as a result its temperature increases to about 101 K. The compressor increases the pressure of LPC to 0.1157 MPa. ORLA and ORVA in Figure 2.18 are oxygen-rich liquid air and oxygen-rich vapour air, respectively. Moreover, OP, LLNP and NP are oxygen product, low-purity liquid nitrogen product and nitrogen product.

Figure 2.19 shows the equilibrium state between the air liquefaction and cold box. Equations (2.37) and (2.38) show the mass balance of the air liquefaction and

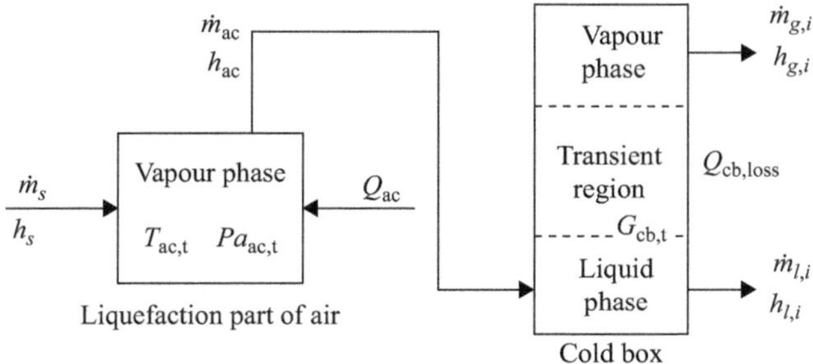

Figure 2.19   General schematic of the air separation unit equilibrium stage

cold box, respectively. The subscript i in (2.37) and (2.38) is an index of a substance, $i = 1$ shows oxygen, $i = 2$ shows nitrogen, and $i = 3$ shows other gases. Moreover, (2.39) and (2.40) represent the heat balance of the air liquefaction and cold box, respectively.

$$\dot{m}_s \cdot M_{s,i} = \dot{m}_{cb} \cdot M_{cb,i} \tag{2.37}$$

$$\dot{m}_{cb} \cdot M_{cb,i} = \dot{m}_l \cdot M_{l,i} + \dot{m}_g \cdot M_{g,i} \tag{2.38}$$

$$\dot{m}_s \cdot h_s + \dot{Q}_{cb} = \dot{m}_{cb} \cdot h_{cb} \tag{2.39}$$

$$M_{cb} \cdot C_{p,cb} \cdot \frac{dT}{dt} = \dot{m}_{l,i} \cdot h_{l,i} - \dot{m}_{g,i} \cdot h_{g,i} - \dot{Q}_{cb,loss} \tag{2.40}$$

Because liquid air stored in the cold box corresponds to the load change of the IGFC for several minutes or less, the dynamic characteristics of the air compressor is ignored. On the other hand, (2.40) represents the dynamic characteristics of the raw material fluid in the cold box.

As the control of number of units with two or more compressors is introduced into the proposed system, it is assumed that the reduction in efficiency of the ASU can be generally controlled. Equation (2.41) represents the power consumption P_{BW} of the blower $BW_{ASU,air}$ of the ASU. The terms Q_{BW}, Pa_{BW} and η_{BW} in (2.41) are the amount of supplied air flow, air pressure and the efficiency of the blower, respectively. Equation (2.42) represents the power consumption $P_{con,l,air}$ of the compressor $CP_{ASU,air1}$, wherein $CP_{ASU,air}$ supplies liquefied gas to the cold box. The terms $H_{sh,air}$ and $H_{lh,air}$ in (2.42) are the heating value of liquefaction of air – the sensible heat accompanying the liquefaction of air and the liquefaction latent heat of air, respectively. The term $\eta_{com,l,air}$ is the efficiency of the compressor. The power consumption of the ASU is P_{BW} and $P_{com,l,air}$.

$$P_{BW} = Q_{BW} \cdot Pa_{BW} \cdot \left(\frac{1}{\eta_{BW}}\right) \tag{2.41}$$

$$P_{con,l,air} = \frac{H_{l,air}}{\eta_{con,l,air}} = \frac{\left(H_{sh,air} + H_{lh,air}\right)}{\eta_{con,l,air}} \tag{2.42}$$

Coal gasifier

Solid-phase reactions in a coal gasification furnace are the coal thermal decomposition and char gasification reaction. Gases generated through the reactions described previously progresses to a gas-phase reaction [60]. Therefore, it is necessary to investigate the dynamic characteristics of the thermal decomposition reaction, char gasification reaction and gas-phase reaction in the reaction model of the coal gasification furnace. However, because the temperature inside the gasifying furnace is very high, the thermal decomposition reaction is completed instantaneously. Therefore, only the char gasification reaction and gas-phase reaction affect the dynamic characteristics of the coal gasification furnace.

Table 2.7 Fundamental reaction formulas for coal gasification

---

1. Thermal decomposition

$$\text{Coal} \rightarrow \text{CH}_4 + \text{C(s)}$$

2. Oxidation reaction

$$\text{C(s)} + \text{O}_2 \rightarrow \text{CO}_2, \text{C(s)} + 1/2\text{O}_2 \rightarrow \text{CO}$$

3. Reaction with carbon dioxide

$$\text{C(s)} + \text{CO}_2 \rightarrow 2\text{CO}$$

4. Reaction with steam

$$\text{C(s)} + \text{H}_2\text{O} \rightarrow \text{CO} + \text{H}_2$$
$$\text{C(s)} + 2\text{H}_2\text{O} \rightarrow \text{CO}_2 + 2\text{H}_2$$
$$\text{CO} + \text{H}_2\text{O} \rightarrow \text{CO}_2 + \text{H}_2$$

5. Reaction with hydrogen

$$\text{C(s)} + 2\text{H}_2 \rightarrow \text{CH}_4, \text{CO} + 3\text{H}_2 \rightarrow \text{CH}_4 + \text{HO}_2$$

Mass balance

When coal is heated (Table 2.7), its thermal decomposition occurs first. Next, solid coal char (C(s)) and pyrolysis gas are generated, the latter containing methane, low-grade hydrocarbon gas and so on. When oxygen, vapour, carbon dioxide, hydrogen and other gases react with the coal char, hydrogen and carbon monoxide are produced, which are the main components of fuel gas. Equation (2.43) expresses the reaction of gas A with B, generating substance C. When the reaction velocity (see (2.43)) departs from the equilibrium value, the reaction rate can be approximated as an exponential of C_A and C_B in the Arrhenius equation shown in (2.44). When the reaction is close to the equilibrium state, its rate can be found from the partial gas pressures (Pa_A, Pa_B), as shown in (2.45). The index j in r_j (see (2.44) and (2.45)) is the number of the target reaction formula, and $A_{i,j}$ is a pre-exponential factor.

$$A + B \rightarrow C \tag{2.43}$$

$$\frac{dC_C}{dt} = A_{f,j} \cdot \exp\left(-\frac{E_j}{R \cdot T}\right) \cdot C_A{}^m \cdot C_B{}^n = r_j \cdot C_A{}^m \cdot C_B{}^n \tag{2.44}$$

$$\frac{dPa_C}{dt} = A_{f,j} \cdot \exp\left(-\frac{E_j}{R \cdot T}\right) \cdot Pa_A \cdot Pa_B = r_j \cdot Pa_A \cdot Pa_B \tag{2.45}$$

Reactions 1–3 ($j = 1, 2, 3$) in Table 2.8 are the char gasification reactions with oxygen, carbon dioxide and vapour, and reactions 4–8 ($j = 4, 5, 6, 7, 8$) are the gas-phase reactions. r_1 to r_8 in Table 2.8 refers to the reaction rate for the Arrhenius equation described in (2.43)–(2.45) [61–67].

Table 2.8   Important chemical reactions [57,62,63,65–68]

Reaction no.	Reaction	Rate (mol/s)	Heat
Thermal decomposition reaction			
	Coal $(C_mH_nO_l) \rightarrow CH_4 + C(s)$		
Char gasification reactions			
$j = 1$	$C(s) + 1/2O_2 \rightarrow CO$	$r_1 = 5.67 \times 10^9 \times e^{(-1.60 \times 10^5/(RT))}$	-111 kJ/mol
2	$C(s) + CO_2 \leftrightarrow 2CO$	$r_2 = 1.6 \times 10^{12} \times e^{(-2.24 \times 10^4/(RT))}$	172 kJ/mol
3	$C(s) + H_2O \rightarrow H_2 + CO$	$r_3 = 1.33 \times 10^3 \times e^{(-1.75 \times 10^4/(RT))}$	131 kJ/mol
Gas-phase reactions			
4	$H_2 + 1/2O_2 \rightarrow H_2O$	$r_4 = 1.00 \times 10^{14} \times e^{(-4.20 \times 10^4/(RT))}$	-242 kJ/mol
5	$CO + 1/2O_2 \rightarrow CO_2$	$r_5 = 2.20 \times 10^{12} \times e^{(-1.67 \times 10^5/(RT))}$	-283 kJ/mol
6	$CH_4 + 1/2O_2 \rightarrow CO + 2H_2$	$r_6 = 3.00 \times 10^8 \times e^{(-1.26 \times 10^5/(RT))}$	-35.7 kJ/mol
7	$CO + H_2O \leftrightarrow CO_2 + H_2$	$r_7 = 2.78 \times 10^3 \times e^{(-1.26 \times 10^4/(RT))}$	-41.1 kJ/mol
8	$CH_4 + H_2O \leftrightarrow CO + 3H_2$	$r_8 = 4.40 \times 10^{11} \times e^{(-1.68 \times 10^5/(RT))}$	206 kJ/mol

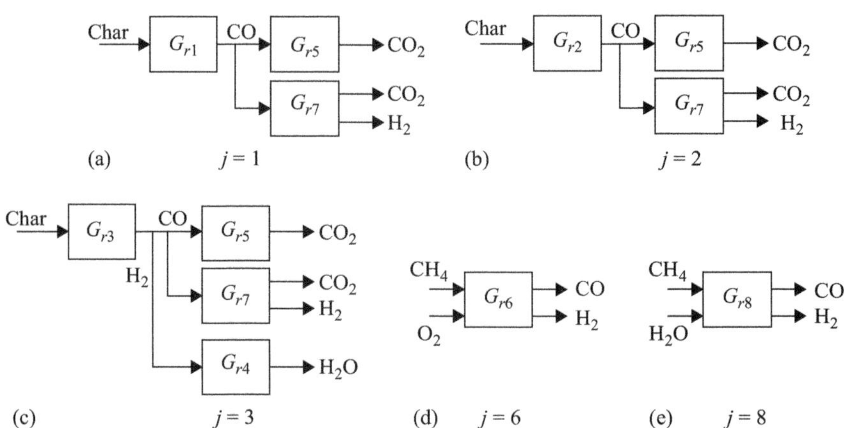

Figure 2.20   Transfer-function blocks of the chemical reaction in the coal gasification

The shift in equilibrium constant for the coal gasification reaction in the coal gasifier can be expressed by the equation as follows [68,69].

$$K_e = \frac{\dot{m}_{CO_2} \cdot \dot{m}_{H_2}}{\dot{m}_{CO} \cdot \dot{m}_{H_2O}} = \exp\left(-3.689 + \frac{4019}{T_g}\right) \tag{2.46}$$

where m_j is the molar quantity of the gas constituent j, and T_g is the gas temperature.

The chemical reactions in the coal gasification furnace (Table 2.8) are the char gasification reaction and gas-phase reaction, which occur consecutively. The transfer functions block diagram (Figure 2.20) present the chemical reactions and reaction path referring to Table 2.8. G_{r_1} to G_{r_8} in Figure 2.20 correspond to the transfer functions for the reaction rates r_1 to r_8, as shown in Table 2.8.

Heat balance

Equation (2.47) reflects the heat balance in the gasification reaction and gas-phase reaction of a char, considering the enthalpy transportation for substance *i*, the heat quantity during the chemical reaction, the amount of heat transferred between the coal particles and fluid, the furnace wall and the heat loss. In addition, *i* is a substance output from and input into the system (Table 2.9; $i = 1, 2, 3, \ldots, 6$). Moreover, $\dot{m}_{cr,j}$ is the reaction rate of the chemical reaction *j* (Table 2.8). The convective heat transfer q_c (see (2.48)) between gas and coal particles and the radiative heat transfer q_c (see (2.49)) between the furnace wall and fluid are considered for heat exchange [64]. The Nusselt number in (2.48) is calculated from the Ranz–Marshall correlation (see (2.50)) [65].

$$\sum_{i=1}^{N_{cg}} \left(m_i \cdot C_{cg}\right) \frac{dT_{pc}}{dt} = \sum_{i=1}^{N_{cg}} \dot{m}_{cg,in-out,i} \cdot h_{cg,in-out,i} - \sum_{j=1}^{N_{cg,rc}} r_{cg,cr,j} \cdot \dot{m}_{cr,j}$$

$$+ (q_C + q_r) - q_{loss}, \quad N_{cg} = 6, \quad N_{cg,cr} = 8 \qquad (2.47)$$

$$q_c = \pi \cdot d \cdot \lambda_g \cdot Nu \cdot (T - T_{pc}) \qquad (2.48)$$

$$q_r = \sigma \cdot \varepsilon_{wall} \cdot S_{wall} \cdot \left(T_{wall}^4 - T_g^4\right) \qquad (2.49)$$

$$Nu = 2 + 0.6 \cdot Re^{1/2} \cdot Pr^{1/3} \qquad (2.50)$$

The Central Research Institute of the Electric Power Industry of Japan (Figure 2.21), the efficiency of coal gasifiers is increasing to the point that the load factors

Table 2.9   Notes on the required components

i	1	2	3	4	5	6
Component	C	O_2	H_2	CO	N_2	H_2O

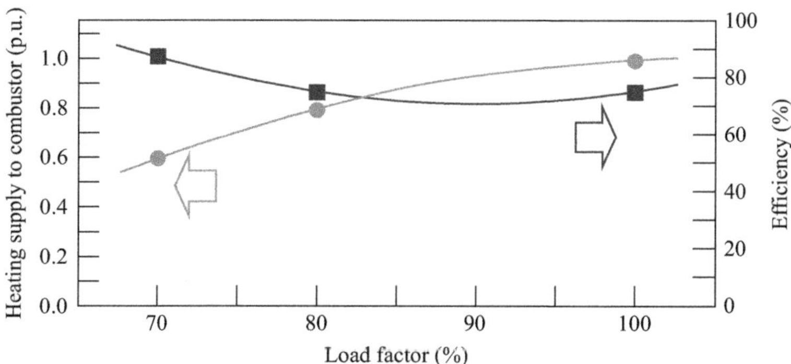

Figure 2.21   Gasifier unit efficiency

Table 2.10   Conditions of the coal gasifier operation

Load factor (%)	100	80	70
Air ratio	0.472	0.529	0.608
Combustion temperature		1,847 K	
Pressure of combustor		2.1 MPa	

reduced [60]. Basic specifications of the gasifier are shown in Table 2.10. The legend for the vertical axis in Figure 2.21 uses the unit method (p.u.) based on rated power. The efficiency of a common coal gasifier ranges from 75% to 77% [70,71].

Dynamic characteristics

The dynamic characteristics of the gasifying furnace can be approximated either by the chemical reaction model or heat transfer model; long response times are required to control the reaction conditions. Therefore, in the analysis of the gasifying furnace response characteristics, a model with a long response time is used to estimate the rate of the consecutive reactions shown in Figure 2.20 and the heat balance of (2.47).

Solid oxide fuel cell

As shown in (2.51), the cell voltage V_{cell} of the SOFC is the value excluding the overvoltage V_{ohm} of the ohmic potential drop, the activation overpotential V_{act} and the concentration overpotential V_{conc} from the open voltage V_{op}. Here, the concentration overpotential V_{conc} is the sum of the overvoltage $V_{\text{conc,ad}}$ of the anode and the overvoltage $V_{\text{conc,cd}}$ of the cathode.

$$V_{\text{cell}} = V_{\text{op}} - \left(V_{\text{ohm}} + V_{\text{act}} + V_{\text{conc}}\right) = V_{\text{op}} - \left(V_{\text{ohm}} + V_{\text{act}} + V_{\text{conc,ad}} + V_{\text{conc,cd}}\right)$$

$$(2.51)$$

Gas turbine

Figure 2.22(a) shows the exhaust heat system of the SOFC and G/T. Figure 2.22(b) shows a temperature–entropy chart, and Figure 2.22(c) is a pressure-specific volume diagram. As shown in Figure 2.22(a), the atmospheric air (operating point 2) compressed by the compressor (CP) is supplied to the cathode in the SOFC. Furthermore, the cathode exhaust gas (operating point $2'$) in the SOFC burns existing hydrogen in the anode exhaust gas from the SOFC. The combustion exhaust gas (operating point 3) with this high temperature and high pressure is supplied to the G/T. As shown in Figure 2.22(b), the heat supplied to the G/T system contains the exhaust heat ($q_{\text{SOFC,cd,ex}}$) from the SOFC and the exhaust gas ($q_{\text{gt,comb}}$) from the combustor. Although the continuous line in Figure 2.22(b) is an ideal cycle (1-$2'$-3-$4'$), when the adiabatic efficiency ($\eta_{\text{gt,cp}}, \eta_{\text{gt,tb}}$) of the CP and the turbine is considered, the cycle of the broken line (1-2-3-4) is obtained. In contrast, Figure 2.22(c) shows that the output of the G/T excludes area ($Pa_{\text{hm,3}}$-2-1-$Pa_{\text{hm,4}}$) from area ($Pa_{\text{hm,3}}$-3-4-$Pa_{\text{hm,4}}$).

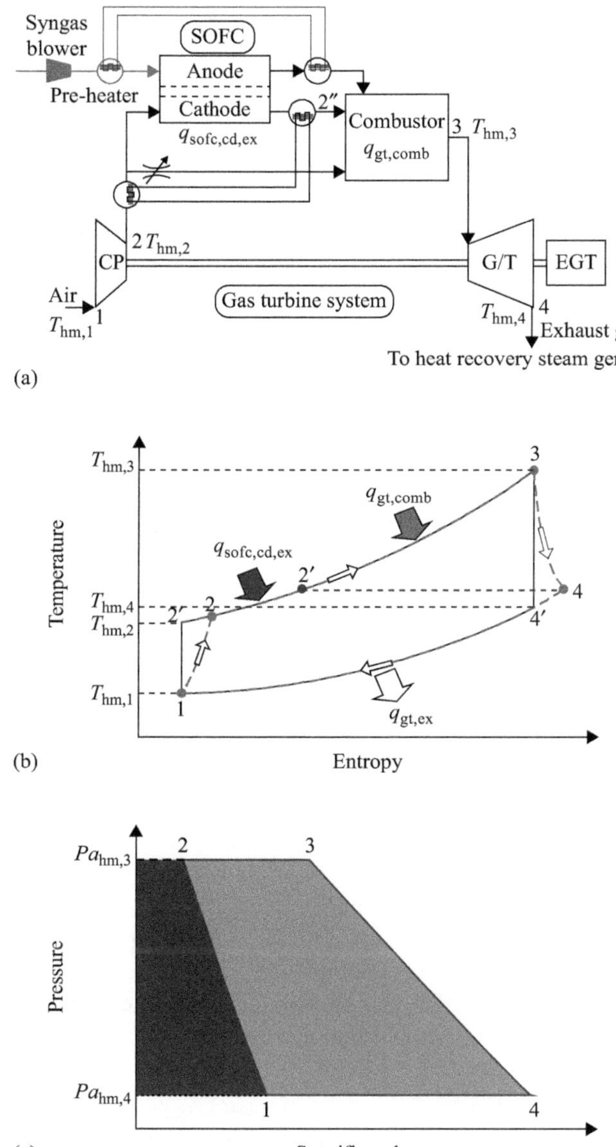

Figure 2.22   Components of the SOFC exhaust heat and G/T system: (a) system configuration, (b) temperature entropy diagram and (c) temperature entropy diagram

The thermal efficiency of the G/T is derived in the following equation by the amount of exhaust heat from the fuel cell $q_{SOFC,cd,ex}$, the heating value of the exhaust gas in the combustor $q_{gt,comb}$ and the amount of heat radiation $q_{gt,ex}$, respectively.

$$\eta_{gt,a} = 1 - \frac{q_{gt,ex}}{q_{SOFC,cd,ex} + q_{gt,comb}} = 1 - \frac{T_2 - T_1}{T_3 - T_4} = 1 - \frac{T_1}{T_3} \cdot \gamma^{(\kappa-1)/\kappa} \qquad (2.52)$$

Equations (2.53) and (2.54) represent the heat insulation thermal ratio θ and the thermal ratio τ, respectively. Equation (2.55) is the efficiency η_{gt} of the G/T.

$$\theta = r_p^{(\kappa-1)/\kappa} \qquad (2.53)$$

$$\tau = \frac{T_3}{T_1} \qquad (2.54)$$

$$\eta_{gt} = \frac{\left[\left(\tau \cdot \eta_{gt,cp} \cdot \eta_{gt,tb}\right)/\theta\right] - 1}{[(\tau - 1)/(\theta - 1)] \cdot \eta_{gt,cp} - 1} \qquad (2.55)$$

Heat recovery steam generator
As the heat transfer area of the heat recovery steam generator is constant, when the exhaust heat temperature is the same as the temperature of the heat recovery area, radiation loss increases according to the amount of exhaust heat. Therefore, when the amount of exhaust heat from the G/T changes with change in the load factor of the system, the radiation loss increases depending on the amount of exhaust heat. Therefore, in this analysis, radiation loss is calculated using a loss coefficient for the amount of exhaust heat from the G/T.

Steam turbine
Figure 2.23(a) shows the steam power generation system comprising the gasifier unit, the heat recovery steam generator and the S/T. Figure 2.23(b) shows the temperature–entropy chart. Steam is generated and supplied to the S/T from the G/T exhaust heat and gasifier unit. The S/T contains three types of high pressure (HP-S/T), interior pressure (IP-S/T) and low voltage (LP-S/T).

2.3.2.3   Method of analysis

Equation (2.56) represents the electric power output P_{SOFC} of the SOFC. The P_{SOFC} is obtained by multiplying the difference between the power consumption P_{ASU} in the air and the calorific power of fuel P_{fuel} by the efficiency of the gasifier unit, the SOFC cell and a power conditioner (η_{gu}, η_{SOFC} and η_{eq}). The power consumption P_{ASU} in the air separator is the sum of the power consumption P_{BW} in the blower and the power consumption P_{con} of the compressor.

$$P_{SOFC} = (P_{fuel} - P_{ASU}) \cdot \eta_{gu} \cdot \eta_{SOFC} \cdot \eta_{eq}$$

$$= [P_{fuel} - (P_{BW} + P_{con})] \cdot \eta_{gu} \cdot \eta_{SOFC} \cdot \eta_{eq} \qquad (2.56)$$

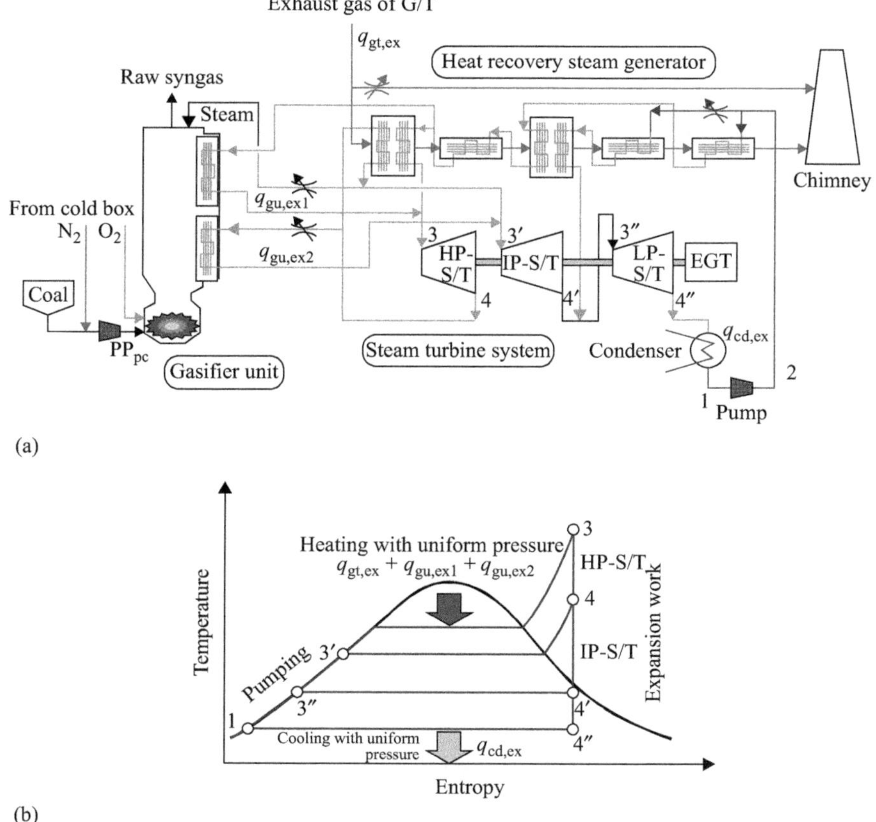

Figure 2.23 Proposed S/T system: (a) system configuration and (b) temperature entropy diagram

Equation (2.57) is the electric power output P_{gt} of the G/T. P_{gt} is derived by multiplying the SOFC exhaust heat $q_{SOFC,ex}$ by the efficiency η_{gt} of G/T and the efficiency η_{egt} of a generator. Equation (2.58) is the electric power output P_{st} of the S/T. P_{st} is obtained by multiplying the sum total $(q_{gt,ex} + q_{gu})$ of the G/T exhaust heat and gasifier unit by the efficiency η_{hrsg} of the heat recovery steam generator, the efficiency η_{st} of the S/T and the efficiency η_{egt} of the generator.

$$P_{gt} = q_{SOFC,ex} \cdot \eta_{gt} \cdot \eta_{egt} \tag{2.57}$$

$$P_{st} = (q_{gt,ex} + q_{gu}) \cdot \eta_{hrsg} \cdot \eta_{st} \cdot \eta_{egt} \tag{2.58}$$

Equation (2.59) is the overall efficiency η_{system} of the proposed IGFC based on the results of (2.56) and (2.58).

$$\eta_{system} = \frac{(P_{SOFC} + P_{gt} + P_{st})}{P_{fuel}} \tag{2.59}$$

2.3.2.4   Expression of relations

Equipment cost

The process for estimating power generation efficiency and fuel cost during operational planning of the proposed system is described in Figure 2.17. Estimated costs are in US dollars (USD).

As shown in (2.60), the equipment cost UC_{system} of the proposed system reflects equipment unit price of photovoltaics UC_{pv}, proposed IGFC UC_{IGFC}, PSPG UC_{PSPG} and electric equipment (power transmission and service wires, system interconnection equipment, protection instrument, etc.), UC_{pf} and each rated installed capacity s_{pv}, s_{IGFC}, s_{PSPG} and s_{pf}.

$$C_{system} = s_{pv} \cdot UC_{pv} + s_{IGFC} \cdot UC_{IGFC} + s_{PSPG} \cdot UC_{PSPG} + s_{pf} \cdot UC_{pf} \qquad (2.60)$$

Fuel cost

As shown in (2.61), the fuel cost FC_{day} for the representative day of operation is based on the amount of consumption of coal $D_{coal,t}$ for each sampling time t and the unit price of the coal UC_{coal}.

$$FC_{day} = uc_{coal} \cdot \sum_{t}^{Day} D_{coal,t} \qquad (2.61)$$

Objective function

The objective function is the minimisation of the electric power unit price UF_{system}. This is derived by dividing the sum of the equipment unit cost UC_{system} and the fuel cost $UC_{fuel,day}$ during the reference period RP by the total production of electricity (PT_{system}). As shown in (2.62), the equipment plane and the operational method of the proposed system are reflected as the minimum.

$$UF_{system} = \frac{s \cdot UC_{system} + RP \cdot FC_{day}}{PT_{system}} \rightarrow \text{Minimize} \qquad (2.62)$$

Electric power balances

Equation (2.63) represents the power balances of the system. The terms P_{pv}, P_{IGFC} and P_{PSPG} are the power output of photovoltaics, the proposed IGFC and PSPG, respectively. The terms P_d and P_{loss} are the power load and electric power losses during transmission. The term t is sampling time.

$$P_{pv,t} + P_{IGFC} + P_{PSPG,t} = P_{d,t} + P_{loss,t} \qquad (2.63)$$

2.3.2.5   Flow of the analysis algorithm

The flow of the algorithm is based on the output adjustment of the IGFC shown in Figure 2.17. The representative day for operational planning is described as follows. First, the power load $P_{d,t}$ of the representative day, the electric power loss $P_{loss,t}$ and the production of electricity $P_{pv,t}$ of photovoltaics are predicted on the day preceding the representative day. When the operating power P_{IGFC} of the IGFC on the representative day is planned arbitrarily, as reflected in (2.53), the planning $P_{\text{PSPG},t}$ for the output of the PSPG on the representative day is obtained. Furthermore, based on the results of the operation planning of $P_{\text{pv},t}$, P_{IGFC}, $P_{\text{PSPG},t}$ on the representative day, equipment and fuel costs are obtained as presented in Sections 2.4.1 and 2.4.2 The maximum output of the large-scale solar power system, IGFC, PSPG and electric equipment on the representative day is determined by the installed capacity (s_{pv}, s_{IGFC}, s_{PSPG}, and s_{eq}) in each device.

2.3.3   *Example of the proposed microgrid analysis*

2.3.3.1   **IGFC and conventional IGCC**

For the purpose of evaluating the cost of the proposed IGFC, the difference with the conventional IGCC is determined. Therefore, for the proposed microgrid using the IGFC shown in Figure 2.16, the introductory case of the IGCC, rather than the IGFC, is investigated. The analytic model of the IGCC excludes the SOFC from the model of the proposed IGFC described in Section 2.2. The model for the gasifying furnace, G/T, heat recovery steam generator and S/T of the IGFC and IGCC is the same as that presented in Section 2.2. In the model, the specifications of the gas product required for the gasifier differs in the IGFC and IGCC. The throughputs of air in the ASU also differ.

2.3.3.2   **Conditions for analysis**

Specifications for the IGFC

Table 2.11 shows the rated power and power generation efficiency of each power source. Total power of 100 MW was obtained using the model formula for each type of equipment described in Section 2.2. Calculations for efficiency of the IGFC are described in Section 2.3.1. Table 2.12 summarises the rated power and power generation efficiency of each power source in the IGCC at the point when the IGFC had the same rated power. As the temperature of the IGCC combustion gas supplied to the G/T is high compared with the IGFC, the power generation efficiency of the G/T and S/T differs substantially from the IGFC.

Table 2.11   Configuration of the proposed IGFC at a rated power of 100 MW

System rated power	100 MW (1.0 p.u.)
• Rated power of SOFC	73.2 MW (0.732 p.u.)
• Rated power of G/T	16.8 MW (0.168 p.u.)
• Rated power of S/T	10.0 MW (0.10 p.u.)
Power generation efficiency (rated power, HHV)	
• Total	53.7%
• SOFC (not include fuel utilisation rate)	60.2%
• G/T	38.0%
• S/T	39.6%
System frequency	50 Hz

*Table 2.12   Configuration of the conventional IGCC at a rated
power of 100 MW*

System rated power	100 MW
• Rated power of G/T	73.2 MW
• Rated power of S/T	26.8 MW
Power generation efficiency (rated power, HHV)	
• Total	42.4%
• G/T	54.3%
• S/T	45.9%
System frequency	50 Hz

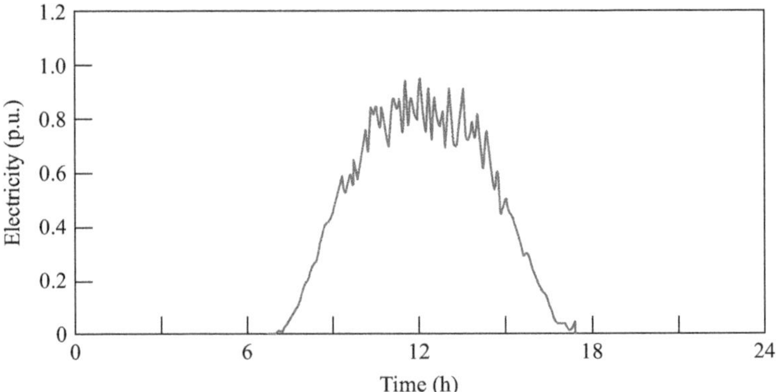

*Figure 2.24   Example of photovoltaic output pattern during August in Hokkaido,
Japan*

Output characteristics of the photovoltaics

Figure 2.24 shows the output pattern of the photovoltaics based on the amount of insolation on the representative day in August in Hokkaido, Japan. The value of the output is shown in units of rated power. The output pattern of the photovoltaics introduced into the proposed microgrid is shown in Figure 2.24.

Specifications for the IGFC

Air supply unit

The power consumption of the ASU blower and compressor is obtained using (2.37) and (2.38). Table 2.13 summarises the setting conditions for each type of equipment.

Gasifier

The relation of the gasifier load factor and efficiency uses the model shown in Figure 2.21 [62]. As shown in the table, the minimum load factor for the gasifier is 60%.

Table 2.13   Specifications for the ASU

Pressure of air blower	100 kPa
Efficiency of air blower	50.0%
Efficiency of liquefaction compressor of air	65.0%

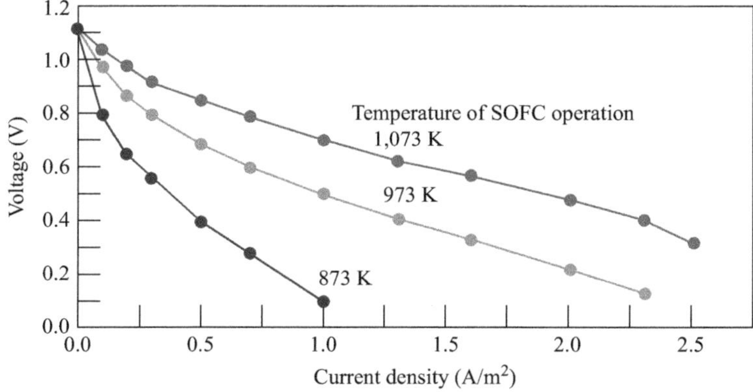

Figure 2.25   SOFC cell performance

Table 2.14   Specifications for the SOFC

SOFC cell voltage	0.705 V
The number of cells	100 cells/stack
• The number of stack unit	120 set
Area of cell	1 m^2
• Specific resistance	0.5 Ω cm
Cell thickness	2 mm
Fuel utilisation rate	0.85
Oxygen utilisation rate	0.23
• Efficiency of inverter	0.95

Solid oxide fuel cell

Figure 2.25 shows the cell output characteristics of the SOFC used for the analysis presented in this paper [72]. As the output characteristics of the SOFC are strongly dependent on cell temperature, it is necessary to determine the cell voltage with an arbitrary temperature. The cell voltage with an arbitrary temperature gives a value excluding the overvoltage of the ohmic potential drop, activation overpotential and concentration overpotential from the open voltage, as shown in (2.47). Table 2.14 summarises the specifications of the SOFC used for the analysis.

Gas turbine

Although the temperature of combustion gas at the combustor outlet changes with the load factors in the system, the inlet port temperature of the G/T in both the IGFC

Table 2.15   Specifications for the G/T

	IGFC	IGCC
Outlet temperature of combustor (K)	$\leq 1,473$	$\leq 1,873$
Pressure ratio	15	15

Table 2.16   Specifications for the S/T

Outlet temperature of low pressure turbine	423 K
Outlet temperature of condenser	313 K
• Inlet temperature of turbine	773 K
Inlet pressure of turbine	10 MPa
• Efficiency of generator	0.95

Table 2.17   Estimated costs of equipment and fuel

IGFC	1,738 USD/kW [73]
IGCC	1,491 USD/kW [74]
Photovoltaics	2,000 USD/kW
PSPG	5,000 USD/kW
Coal	62 USD/ton, 0.00858 USD/kWh

and IGCC is shown in Table 2.15. As shown in the table, the compression ratio of the IGFC and IGCC are the same. The efficiency of the power generator is 0.95.

Steam turbine
As shown in Table 2.16, the operating conditions of the S/T for both the IGFC and IGCC are the same. The table also shows the values for other specifications.

2.3.3.3   Cost

Table 2.17 summarises the estimated equipment unit costs for the conventional IGCC, the proposal IGFC, the photovoltaics and the PSPG used in the analysis. Based on the analysis, the estimated cost of power transmission and service wires, a system interconnection device, a protection instrument and other equipment are similar under the same power load. Moreover, because estimated costs are compared on a relative basis, the cost of electric facilities is not included in the analysis.

2.3.3.4   Electric power supply pattern

Figure 2.26 shows the electric power supply pattern for a representative day in August provided by the Hokkaido Electric Power Co. Inc. [75]. Fluctuations in the power load are largest on that day, with a 40% difference in day and night. In the analysis, the most severe conditions are given for one year, based on the load pattern for the representative August day (Figure 2.26).

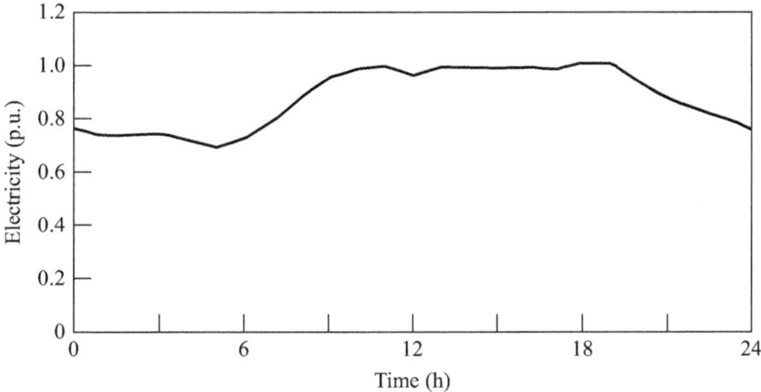

Figure 2.26 Electric power supply pattern during August in Hokkaido, Japan

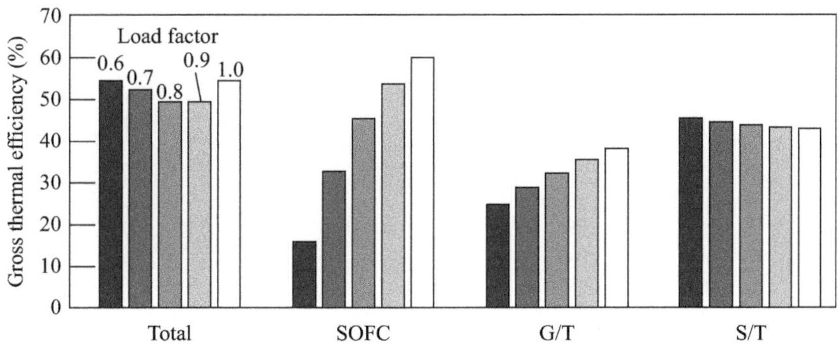

Figure 2.27 Results of generator efficiency analysis

2.3.4 Results of analysis and discussions

2.3.4.1 Power generation characteristics

Overall efficiency

Figure 2.27 shows the results of the analysis for overall system efficiency and the gross thermal efficiency of the SOFC, G/T and S/T. The load factor is a ratio of the value of the load to the maximum power of the IGFC. High overall efficiency is obtained during reduced loads (load factor 0.6) and high loads (load factor 1.0). In the case of a load factor of 0.8, the overall efficiency of the system is at a minimum. Figure 2.27 shows that, when the load factor of the IGFC increases, the power generation efficiency of the SOFC and G/T increases, and the power generation efficiency of the S/T reduces. As the load factor of the IGFC increases, and as the air temperature of the compressor and the reaction temperature of the SOFC rise, the production of electricity significantly increases. Moreover, when the load factor of the IGFC is low, the exhaust heat temperature of the SOFC and the efficiency of the G/T reduce. In contrast, as the amount of exhaust heat supplied to the S/T from the G/T increases, the production of electricity from the S/T increases. Figure 2.27 shows that on the basis of the results described earlier, overall efficiency, including the partial load operation of the IGFC, shows unique characteristics. However, the IGFC with a partial load needs careful coordination of the capacity of the G/T and S/T with the IGFC constant loads because the exhaust heat from the SOFC changes dramatically.

Output characteristics of the SOFC, G/T and S/T
As shown in (2.48), (2.52), (2.54), the cell voltage of the SOFC is dependent on cell temperature. Therefore, the relation between the IGFC load factor and SOFC operating temperature is investigated using numerical analysis. Figure 2.28 shows the results of the load factor analysis for the IGFC, the SOFC operating temperature and gross thermal efficiency. As shown in Figure 2.28, the operating temperature and the power generation efficiency of the SOFC are strongly dependent on the IGFC load factor. Furthermore, Figure 2.29 shows the relation between the IGFC load factor and SOFC cell output. Because the reaction temperature of the cell decreases when the load factor decreases from 1.0 to 0.6, the cell output drops by

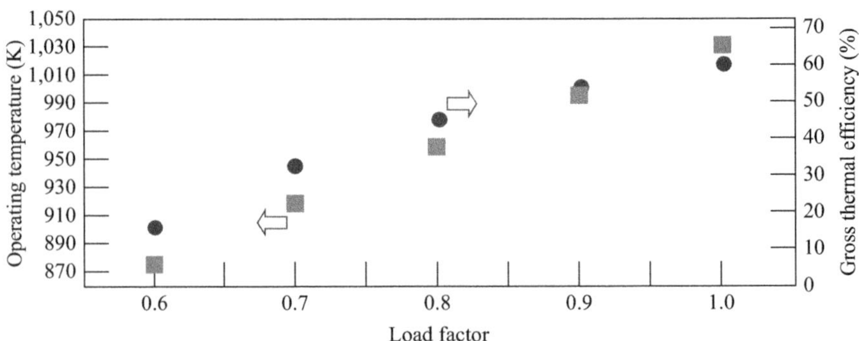

Figure 2.28   *Relation of the SOFC operating temperature to SOFC gross thermal efficiency*

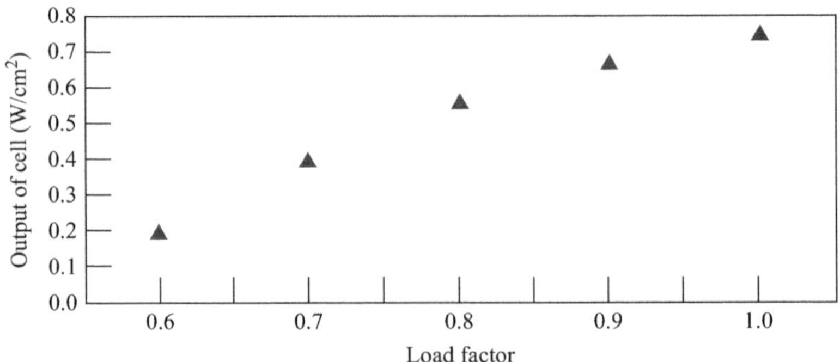

Figure 2.29   *Results of cell output analysis*

approximately 73%. As a result, as the IGFC load factor decreases, the amount of exhaust heat from the SOFC to the G/T increases. Because the exhaust heat temperature is low, the gross thermal efficiency of the G/T falls. In contrast, as the load factor of the IGFC decreases, the enthalpy of steam at the S/T condenser inlet port and the waste heat from the condenser increases. However, the power generation efficiency of the S/T increases due to the increase in the enthalpy of steam at the condenser inlet port.

2.3.4.2 Operation of the microgrid with the IGFC, large-scale photovoltaics and PSPG

Installation capacity of the IGFC, PV and PSPG
Figure 2.30 shows the analysis results of the operational method for the power load $P_{\text{load},t}$ during August in Hokkaido, the output of a large-scale solar power system $P_{\text{pv},t}$, the output of the IGFC $P_{\text{IGFC},t}$ and the input and output of the electric power from the PSPG $P_{\text{PSPG},t}$. In the analysis, the maximum $P_{\text{load},t}$ is set to 1.0 p.u. Figure 2.30 shows the results of analyses when the rated power of the large-scale solar power system is set to 1.0 p.u. [Figure 2.30(a)] or 0.6 p.u. [Figure 2.30(b)]. Based on the analysis described in Section 2.5, the rated power of the IGFC was 0.997 p.u. [Figure 2.30(a)] and 0.751 p.u. [Figure 2.30(b)]. Furthermore, the installed capacity of the PSPG is 1.09 p.u. [Figure 2.29(a)] and 0.395 p.u. [Figure 2.30(b)]. Although there is a 1.67× difference in the rated power of the photovoltaics of Figure 2.30(a) and (b), when the installed capacity of each PSPG is investigated, a 2.76× difference is measured. In the analysis, equipment capacity was used to satisfy energy balance (see (2.57) and (2.75)). However, when the equipment cost of the photovoltaics is less than that of the PSPG, it is advantageous to further increase the amount of power generated by photovoltaics. In this case, a portion of the surplus electric power from photovoltaics does not store electricity using the PSPG (parallel off).

Fuel costs
Figure 2.31 shows the estimated fuel costs of the IGCC and proposed IGFC. Figure 2.31(a) shows the relation between load factor and fuel cost; it shows that the fuel cost for the IGFC changes with the load factor slowly. In contrast, although the fuel cost of the IGCC exceeds the IGFC when the load factor is 0.79 or more, the fuel cost in the range of other load factors is lower than the IGFC. As shown in Figure 2.27, the gross thermal efficiency of the SOFC significantly falls in relation to the G/T during the partial load operation of the IGFC. Therefore, the IGCC comprising the G/T and S/T is determined to have a power generation efficiency higher than that of the IGFC in the partial load operational zone. Figure 2.31(b) shows the relation of the installed capacity of the large-scale solar power station to fuel costs of the IGCC and IGFC. Fuel cost is reduced when the use of photovoltaics increases with any power source for the IGFC and IGCC. Based on the difference in overall efficiency, fuel costs for the IGFC are lower than those for the IGCC.

Equipment planning and costs
The power balances of the system are derived using (2.63). If the load pattern and installation capacity of the large-scale solar power station are calculated, the

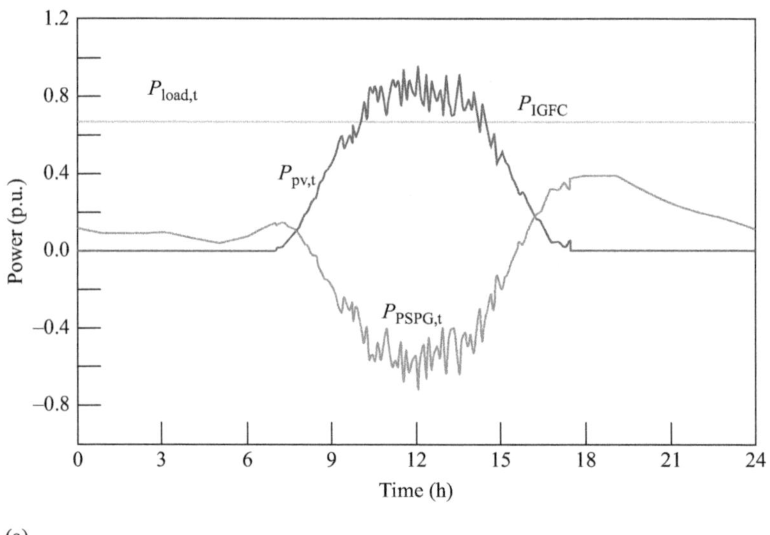

(a)

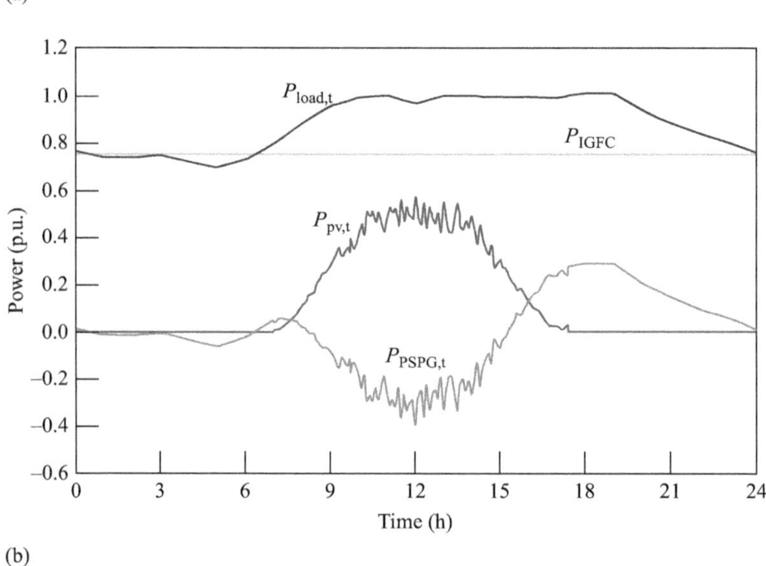

(b)

*Figure 2.30   Results of operational analysis of the proposal independent
microgrid: (a) $P_{pv,t} = 10$ p.u. and (b) $P_{pv,t} = 0.6$ p.u*

installation capacity of the IGFC and PSPG can be determined. Figure 2.32 shows
the analytical results for the installed capacity of the large-scale solar power station,
IGFC and PSPG. Although the capacity of IGFC decreases when the use of photovoltaics increases, the capacity of PSPG for adjusting the supply–demand balance of
electric power increases significantly. As shown in Figure 2.32, the cost of equipment for all the power sources is based on the unit price for each type of equipment.

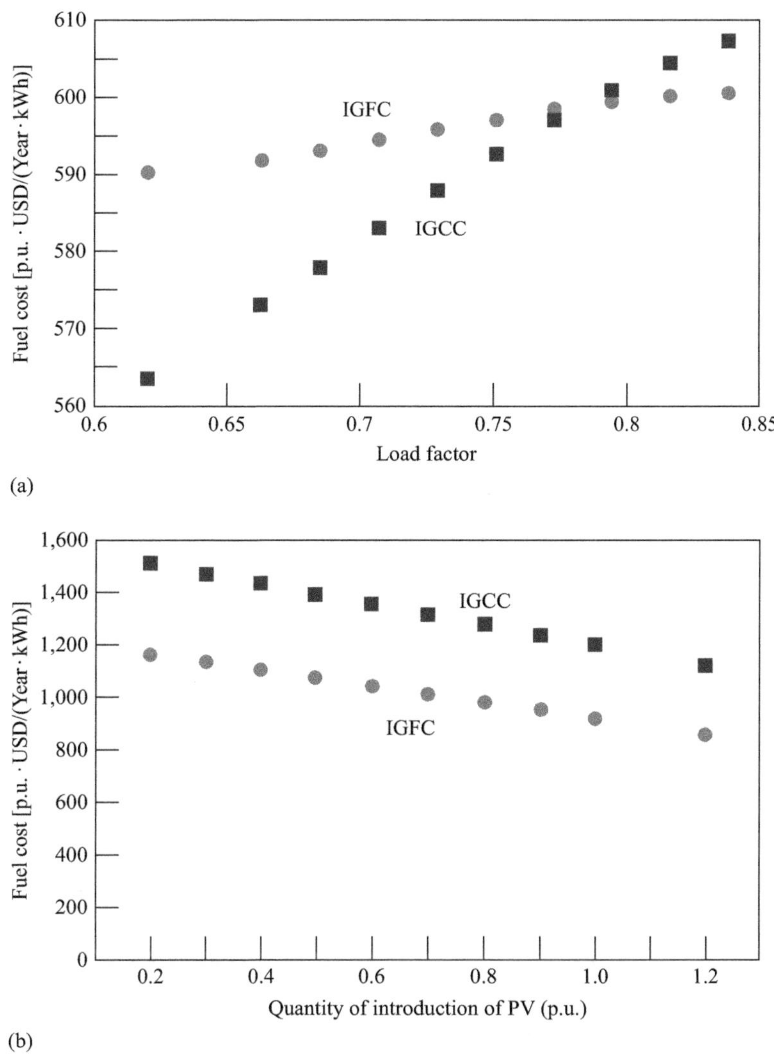

(a)

(b)

Figure 2.31   Fuel cost of the proposed independent microgrid: (a) load factor characteristics and (b) characteristics of photovoltaic system use

Figure 2.33 shows the analytical results for the cost of the independent microgrid based on an electric power mix of the IGFC, large-scale photovoltaics and the PSPG. The cost of a transmission network and substation and the administrative expense for each type of equipment are not included. Figure 2.33(a) shows the cost of the facilities of an independent microgrid that has an IGFC or IGCC. When the unit costs of facilities shown in Table 2.17 are included, the cost of the IGFC is high. In contrast, Figure 2.32(b) shows the sum total of facilities costs from Figure 2.33(a) and the fuel cost assuming use of coal for 10 years. Although the

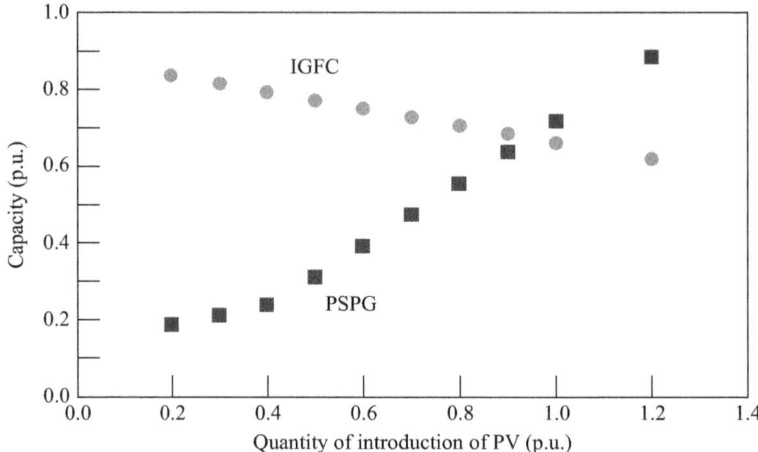

Figure 2.32   Results of the analysis of the installation plan for the PV, pumped hydropower generation and IGFC

cost of the facilities of the IGFC exceeds the IGCC, the cost of the IGFC is nearly the same as the IGCC based on the difference in gross thermal efficiency. In order for use of the IGFC to be more feasible than the IGCC, the power generation efficiency of the SOFC (60.2%) shown in Table 2.11 requires further improvement. The efficiency of the SOFC has a high potential to increase by approximately 5% in the future. Furthermore, when the SOFC operates for 10 years or more, it is expected that the total cost of the IGFC will be lower than that of the IGCC.

2.3.5   Conclusion

This section investigated the optimal equipment planning and operation method for a microgrid comprising an IGFC, large-scale photovoltaics and PSGP as independent distributed energy system. The following conclusions summarise the analytical results for power-generation efficiency, fuel cost and equipment cost of the proposed system.

1. Because the reaction temperature of SOFC decreases when the load factor of the IGFC decreases from 1.0 to 0.6, the cell output reduces by 73%. Furthermore, when the exhaust heat temperature of the SOFC decreases, the power generation efficiency of the G/T reduces.
2. In contrast, when the load factor of the IGFC reduces, the power generation efficiency of the S/T increases. The reason for this is that, because the enthalpy of steam in the condenser inlet port of the S/T increases with a drop in the load factor, the waste heat of the condenser increases. As a result, under conditions of full load, the overall efficiency of the IGFC is maximised. Furthermore, overall efficiency is achieved with a minimum load factor of 0.8 and improves as the load factor increases.

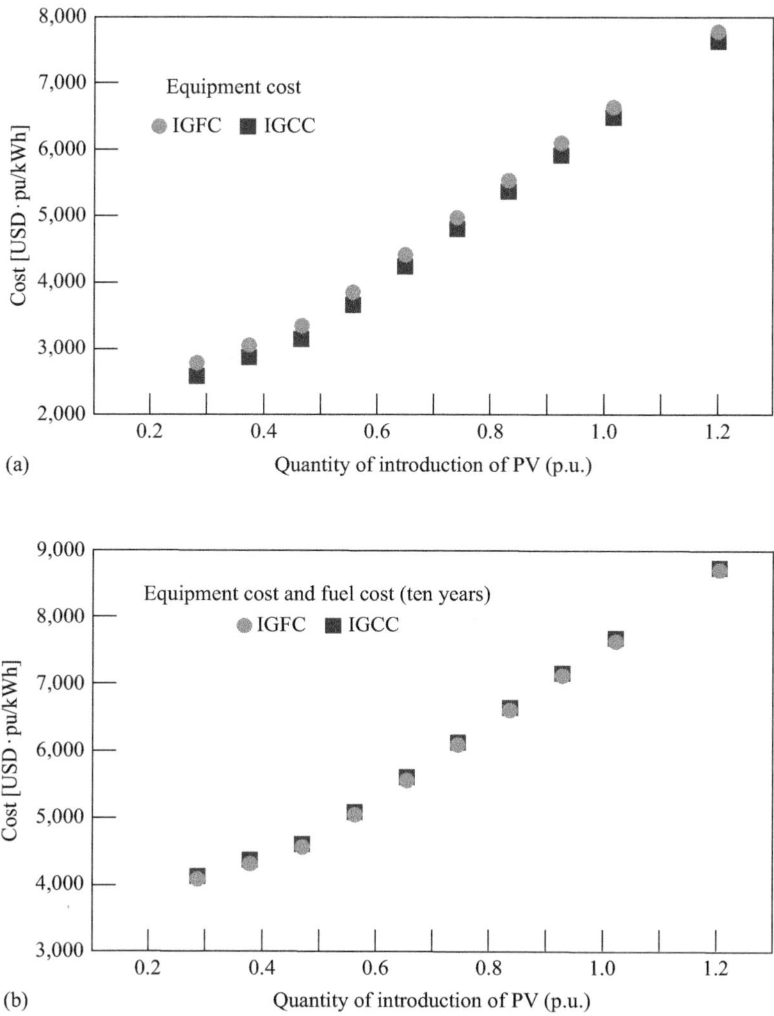

Figure 2.33   *Results of cost analysis: (a) equipment cost and (b) equipment and fuel cost over a 10-year period*

For the purpose of numerical analysis, the power-load pattern in Hokkaido, Japan on a representative day in August was used as input to the proposed microgrid system. The findings of that analysis are summarised below.

1.   When the installed capacity of photovoltaics was increased by a factor of 1.67, the installed capacity of the PSPG increased by a factor of 2.76. Therefore, when the ratio of equipment cost to cost of power generation using a photovoltaic power station is lower than that of the PSPG, the use of photovoltaics is increased.

2. As the SOFC with a high power generation efficiency is introduced, the fuel cost of the IGFC is less than that of the IGCC.
3. Although the equipment cost of the IGFC exceeds the IGCC, the total cost of equipment and fuel for the IGFC over a 10-year period is almost the same as that of the IGCC based on the difference in power generation efficiency. If the improvement of the power generation efficiency of the SOFC and extension of the period of operation are possible, the total cost of the IGFC will be less than that of the IGCC.

Nomenclature

A	SOFC surface area (m^2)
A_i	pre-exponential factor for i
a	stoichiometric matrix
a_r	radius of rotor plate (m)
C_p	specific heat at constant pressure (J/K)
C_S	heat capacity (J/K)
c	constant number
DEN	denominator
E	activity energy (J/mol)
E_{ocv}	open circuit voltage (V)
e_1, e_2	both-end voltage of equivalent interconnection line (p.u.)
F	faraday constant (96,485 sA/mol)
F_{cl}	filter factor of PID control
h	heat transfer coefficient [W/(m^2 K)]
$\Delta\bar{h}$	molar specific enthalpy (J/mol)
I	current (A)
J	moment of inertia (kg m^2)
K	equilibrium constant
K_{ads}	adsorption constant
K_{gt}	power coefficient of gas turbine
K_h	coefficient of overall heat transfer [W/(m^2 K)]
K_D, K_I, K_P	parameter of PID control
K_{st}	power coefficient of steam turbine
K_{gt}	power coefficient of gas turbine
k	rate coefficients for reforming reactions
L	load (p.u.)
l	external work rate (N m/s)
M	factor of inertia

m	number of moles (mol)
$\dot{m}$	moles rate (mol/s)
m_r	mass of rotor plate (kg)
$n_{ad,rc}$	number of chemical reactions in anode
$n_{cd,rc}$	number of chemical reactions in cathode
P	power (W), (p.u.)
p	pressure (Pa)
R	universal gas constant [J/(mol K)]
R_a	compression ratio
r	reaction rate (mol/s)
S_g	slip (p.u.)
s	Laplace operator
T	temperature (K)
T_e	exhaust gas temperature of SOFC (K)
T_i	delay-time constant for equipment i
T_s	surrounding temperature
ΔT	difference in temperature (K)
t	time (s)
u	utilisation
V	voltage (V)
W	quantity of flow (p.u.)
ASU	air separation unit
A_i	pre-exponential factor for i
BW	blower
C_p	specific heat at constant pressure [J/(g K)]
CP	compressor
C	heat capacity [J/(g K)]
D	amount of consumption (g/s)
d	characteristic length (m)
E	activity energy (J/mol)
FC	fuel cost [USD(US dollar)/day]
G	transfer function
G/T	gas turbine
H	heating value (W)
h	enthalpy (J/g)
M	mass (g) or (mol)
$\dot{m}$	moles rate (mol/s), mass rate (g/s)
Nu	Nusselt number

P	power (W), (p.u.)
Pa	pressure (Pa)
Pr	Prandtl number
PT	total production of electricity (Wh)
$\dot{Q}$	heat (W)
q_c	convective heat transfer (W)
q_r	radiative heat transfer (W)
R	universal gas constant [J/(mol K)]
Re	Reynolds number
RP	reference period (days)
r	reaction rate (mol/s)
r_p	pressure ratio
S	area (m^2)
s	rated power output (W)
T	temperature (K)
t	time (s)
UC	unit cost (USD/W)
uc	unit cost (USD/g)
V	voltage (V)

Greek characters

γ	ratio of specific heat
η	efficiency
ω	turbine revolving speed
ω_0	rated frequency
$\Delta\omega$	deviation of turbine revolving speed (=frequency deviation)
σ	Stefan–Boltzmann constant [5.67×10^{-8} J/(sm^2 K^4)]
ς	shaping factor
τ	time constant
ε	emissivity
η	efficiency
κ	ratio of specific heat
λ	thermal conductivity [W/(m K)]
θ	heat insulation thermal ratio
σ	Stefan–Boltzmann constant 5.67×10^{-8} [W/(m^2 K^4)]
τ	thermal ratio

Subscripts

ac	air compressor
amb	ambiance

DC	direct current
e	exhaust
fc	fuel cell
g	generation
gt	gas turbine
he	heat exchanger
ht	heat transfer
l	load
lw	lower
rad	heat radiation
st	steam turbine
up	upper
v	vapour
act	activation
ad	anode
ASU	air separation units
BW	blower
cb	cold box
cd	cathode
cg	coal gasifier
con	compressor
comb	combustor
conc	concentration
cp	compressor
cr	chemical reaction
egt	electric power generator
eq	power conditioner equipment
ex	exhaust
g	gas
gt	gas turbine
gu	gasifier unit
hm	heat medium
hrsg	heat recovery steam generator
l	liquid
lh	latent heat
ohm	ohmic
pc	pulverised coal
pf	electric power facility

PSPG   pumped-storage power generation
pv   photovoltaic
s   supply air
sh   sensible heat

References

[1]   C. S. A. Nandar, "Robust PI control of smart controllable load for frequency stabilization of microgrid power system", *Renewable Energy*, 2013;56:16–23.

[2]   I. Ngamroo, "Application of electrolyzer to alleviate power fluctuation in a stand alone microgrid based on an optimal fuzzy PID control", *International Journal of Electrical Power & Energy Systems*, 2012;43:969–976.

[3]   C. S. A. Nandar, "Robust PI control of smart controllable load for frequency stabilization of microgrid power system". *Renewable Energy*, 2013;56:16–23.

[4]   S. Vachirasricirikul and I. Ngamroo, "Robust controller design of heat pump and plug-in hybrid electric vehicle for frequency control in a smart microgrid based on specified-structure mixed H_2/H_∞ control technique", *Applied Energy*, 2011;88:3860–3868.

[5]   H. Jia, Y. Mu and Y. Qi, "A statistical model to determine the capacity of battery–supercapacitor hybrid energy storage system in autonomous microgrid", *International Journal of Electrical Power & Energy Systems*, 2014;54:516–524.

[6]   T. C. Ou and C. M. Hong, "Dynamic operation and control of microgrid hybrid power systems", *Energy*, 2014;66:314–323.

[7]   M. Saeji and I. Ngamroo, "Stabilization of microgrid with intermittent renewable energy sources by SMES with optimal coil size", *Physica C: Superconductivity*, 2011;471:1385–1389.

[8]   J. H. Choi, J. H. Ahn and T. S. Kim, "Performance of a triple power generation cycle combining gas/steam turbine combined cycle and solid oxide fuel cell and the influence of carbon capture", *Applied Thermal Engineering*, 2014;71:301–309.

[9]   A. Arsalis, "Thermoeconomic modeling and parametric study of hybrid SOFC–gas turbine–steam turbine power plants ranging from 1.5 to 10 MWe", *Journal of Power Sources*, 2008;181:313–326.

[10] Mitsubishi Heavy Industries, "Mitsubishi to develop SOFC-turbine triple combined cycle system", *Fuel Cells Bulletin*, 2012;7:5–6.

[11] M. Z. Daud, A. Mohamed and M. A. Hannan, "An improved control method of battery energy storage system for hourly dispatch of photovoltaic power sources", *Energy Conversion and Management*, 2013;73:256–270.

[12] S. Koohi-Kamali, N. A. Rahim and H. Mokhlis, "Smart power management algorithm in microgrid consisting of photovoltaic, diesel, and battery storage plants considering variations in sunlight, temperature, and load", *Energy Conversion and Management*, 2014;84:562–582.

[13] M. Datta, T. Senjyu, A. Yona and T. Funabashi, "A fuzzy based method for leveling output power fluctuations of photovoltaic-diesel hybrid power system", *Renewable Energy*, 2011;36:1693–1703.

[14] J. Marcos, O. Storkël, M. Arroyo, M. Garcia and E. Lorenzo, "Storage requirements for PV power ramp-rate control", *Solar Energy*, 2014;99:28–35.

[15] J. Wang, X. Li, H. Yang and S. Kong, "Design and realization of microgrid composing of photovoltaic and energy storage system", *Energy Procedia*, 2011;12:1008–1014.

[16] T. C. Ou and C. M. Hong, "Dynamic operation and control of microgrid hybrid power systems", *Energy*, 2014;66:314–323.

[17] M. Tao, Y. Hongxing and L. Lin, "Performance evaluation of a stand-alone photovoltaic system on an isolated island in Hong Kong", *Applied Energy*, 2013;112:663–672.

[18] M. Xiandong, W. Yifei and Q. Jianrong, "Generic model of a community-based microgrid integrating wind turbines, photovoltaics and CHP generations", *Applied Energy*, 2013;112:1475–1482.

[19] D. Manoj, S. Tomonobu, Y. Atsushi, F. Toshihisa and K. Chul-Hwan, "Photovoltaic output power fluctuations smoothing methods for single and multiple PV generators", *Current Applied Physics*, 2010;10:265–270.

[20] Z. Qi, C. M. Benjamin, T. Tezuka and T Ishihara, "An integrated model for long-term power generation planning toward future smart electricity systems", *Applied Energy*. 2013;112:1424–1437.

[21] S. Samir, D. Denkenberger and R. Williams, "Optimization of specific rating for wind turbine arrays coupled to compressed air energy storage", *Applied Energy*, 2012;96:222–234.

[22] D. Rodolfo, M. Juan, R. Lujano and J. Bernal-Agustín, "Comparison of different lead–acid battery lifetime prediction models for use in simulation of stand-alone photovoltaic systems", *Applied Energy*, 2014;115:242–253.

[23] V. Manuel Jesús, B. José Manuel and A. José Manuel, "Optimal sizing for UPS systems based on batteries and/or fuel cell", *Applied Energy*, 2013;105:170–181.

[24] K. Darcovich, E. R. Henquin, B. Kenney, I. J. Davidson, N. Saldanha and I. Beausoleil-Morrison, "Higher-capacity lithium ion battery chemistries for improved residential energy storage with micro-cogeneration", *Applied Energy*, 2013;111:853–861.

[25] X. Yinjiao, H. Wei, P. Michael and T. Kwok Leung, "State of charge estimation of lithium-ion batteries using the open-circuit voltage at various ambient temperatures", *Applied Energy*, 2014;113:106–115.

[26] De F Giovanni, M. Vincenzo and S. Ramteen, "Simulation of an electric transportation system at The Ohio State University", *Applied Energy*, 2014;113:1686–1691.

[27] M. Amita and R. Saroj, "Short term generation scheduling of cascaded hydro electric system using novel self-adaptive inertia weight PSO", *International Journal of Electrical Power & Energy Systems*, 2012;34:1–9.

[28]   W. Bignell, H. Saffron, T. Nguyen and W. D. Humpage, "Effects of machine inertia constants on system transient stability", *Electric Power Systems Research*, 1999;51:153–165.

[29]   M. R. Raghavan and R. Jayachandran, "Analysis of the performance characteristics of a two-inertia power transmission system with a plate clutch", *Mechanism and Machine Theory*, 1989;24:499–503.

[30]   R. Kandepu, L. Imsland, B. A. Foss, C. Stiller, B. Thorud, and O. Bolland, "Modeling and control of a SOFC-GT-based autonomous power system", *Energy*, 2007;32:406–417.

[31]   J. Xu and G. F. Froment, "Methane steam reforming, methanation and water–gas shift; I. Intrinsic kinetics", *AIChE Journal*, 1989;35:88–96.

[32]   P. Thomas, "Simulation of industrial processes for control engineers", Woburn, MA: Butterworth-Heinemann; 1999.

[33]   M. D. Lukas, K. Y. Lee and H. Ghezel-Ayagh, "An explicit dynamic model for direct reforming carbonate fuel cell stack", *IEEE Transaction on Energy Conversion*, 2001;16(3):289–295.

[34]   J. Padullés, G. W. Ault and J. R. McDonald, "An integrated SOFC plant dynamic model for power systems simulation", *Journal of Power Sources*, 2000;86;495–500.

[35]   B. Thorud, C. Stiller, T. Weydahl, O. Bolland and H. Karoliussen, "Part-load and load change simulation of tubular SOFC systems", *Proceedings of Fuel Cell Forum*, Lucerne, 28 June–2 July 2004.

[36]   Output fluctuation and alleviation method of PV, The National Institute of Advanced Industrial Science and Technology (AIST) of Japan, 2014, Retrieved from https://unit.aist.go.jp/rcpvt/ci/about_pv/output/fluctuation.html.

[37]   H. Taniguchi, "Electric-Power-System Analysis – Modeling and Simulation", *Ohmsha*, 2009 (in Japanese).

[38]   C. Stiller, B. Thorud and O. Bolland, "Safe dynamic operation of a simple SOFC/GT hybrid cycle", *Proceedings of ASME Turbo Expo*, 2005.

[39]   A. Hafsia, Z. Bariza, H. Djamel, B. M. Hocine, G. M. Andreadis and A. Soumia, "SOFC fuel cell heat production: analysis", *Energy Procedia*, 2011;6:643–650.

[40]   IEEE Working Group on Prime Mover and Energy Supply Models for System Dynamic Performance Studies, IEEE PES 94 WM 185-9 PWRS, 1994.

[41]   Task Force C 4-02-25: Modeling of gas turbines and steam turbines in combined cycle power plant, CIGRE Technical Brochure, 2003.

[42]   K. Baba and N. Kakimoto, "Dynamic behavior of combined cycle power plant for frequency drop", *IEEJ Transactions on Power and Energy*, 2002;122:392–400 (in Japanese).

[43]   Electric Power Supply Data in 2010. Hokkaido Electric Power Co. Inc.; 2014. http://www.hepco.co.jp/english/index.html.

[44]   O. Shinada, A. Yamada and Y. Koyama, "The development of advanced energy technologies in Japan: IGCC: a key technology for the 21st century", *Energy Conversion and Management*, 2002;43:1221–1233.

[45] H. Jaeger, "Japan 250 MW coal based IGCC demo plant set for 2007 start-up", Southport, USA: Gas Turbine World, 2006:1–4.

[46] A. Franco and A. R. Diaz, "The future challenges for clean coal technologies: joining efficiency increase and pollutant emission control", *Energy*, 2009;34:348–354.

[47] M. Kawabata, N. Iki, O. Kurata, *et al.*, "Energy flow of advanced IGCC With CO2 capture option", *ASME 2010 International Mechanical Engineering Congress and Exposition*, 2010;40456:551–558.

[48] K. Jordal, R. Anantharaman, T. A. Peters, *et al.*, "High-purity H_2 production with CO_2 capture based on coal gasification", *Energy*, 2015;88:9–17.

[49] R. Nomura, N. Iki, O. Kurata, *et al.*, "System analysis of IGFC with exergy recuperation utilizing low-grade coal", *ASME 2011 Turbo Expo: Turbine Technical Conference and Exposition*, 2011;4:243–251.

[50] S. Mohammad, M. Asnadi, J. R. Grace, *et al.*, "Biomass/coal steam co-gasification integrated with in-situ CO_2 capture", *Energy*, 2015;83: 326–336.

[51] A. Verma, B. Olateju and A. Kumar, "Greenhouse gas abatement costs of hydrogen production from underground coal gasification", *Energy*, 2015;85:556–568.

[52] C. C. Cormos, "Economic evaluations of coal-based combustion and gasification power plants with post-combustion CO_2 capture using calcium looping cycle", *Energy*, 2014;78:665–673.

[53] E. D. Wachsman, C. A. Marlowe and K. T. Lee, "Role of solid oxide fuel cells in a balanced energy strategy", *Energy & Environmental Science*, 2012;2:5498–5509.

[54] S. Karellas, K. D. Panopoulos, G. Panousis, A. Rigas, J. Karl and E. Kakaras, "An evaluation of substitute natural gas production from different coal gasification processes based on modelling", *Energy*, 2012;45:183–194.

[55] N. S. Siefert and S. Litster, "Exergy and economic analyses of advanced IGCC–CCS and IGFC–CCS power plants", *Applied Energy*, 2013;107: 315–328.

[56] A. Lanzini, T. G. Kreutz, E. Martelli and M. Santarelli, "Energy and economic performance of novel integrated gasifier fuel cell (IGFC) cycles with carbon capture", *International Journal of Greenhouse Gas Control*, 2014;26:169–184.

[57] S. W. Lee, J. S. Park, C. B. Lee, *et al.*, "H_2 recovery and CO_2 capture after water–gas shift reactor using synthesis gas from coal gasification", *Energy*, 2014;66:635–642.

[58] H. Apfel, M. Rzepka, H. Tu and U. Stimming, "Thermal start-up behaviour and thermal management of SOFC's", *Journal of Power Sources*, 2006;154:370–378.

[59] A. Choudhury, H. Chandra and A. Arora, "Application of solid oxide fuel cell technology for power generation – a review", *Renewable and Sustainable Energy Reviews*, 2013;20:430–442.

[60] H. Hashimoto, "Latest internal compression type air separation unit integrated with liquefying process", Technical Report of Taiyo Nippon Sanso Co., Ltd. 2012;31:48–49.

[61] J. C. Lee, H. H. Lee, Y. J. Joo, C. H. Lee and M. Oh, "Process simulation and thermodynamic analysis of an IGCC (integrated gasification combined cycle) plant with an entrained coal gasifier", *Energy*, 2014;64:58–68.

[62] C. D. Blasi, "Dynamic behaviour of stratified downdraft gasifiers", *Chemical Engineering Science*, 2000;55(15):2931–2944.

[63] C. K. Westbrook and F. L. Dryer, "Simplified reaction mechanisms for the oxidation of hydrocarbon fuels in flames", *Combustion Science and Technology*, 1981;27(1–2):31–43.

[64] H. Watanabe and M. Otaka, "Numerical simulation of coal gasification in entrained flow coal gasifier", *Fuel*, 2006;85(12–13):1935–1943.

[65] W. C. Macklin, "Comments on the general heat and mass exchange of spherical hailstones", *Journal of the Atmospheric Science*, 1963;21:227–228.

[66] H. Freund, "Gasification of carbon by CO_2: a transient kinetics experiment", *Fuel*, 1986;65(1):63–66.

[67] N. Wang and J. G. Brennan, "A mathematical model of simultaneous heat and moisture transfer during drying of potato", *Journal of Food Engineering*, 1995;24(1):47–60.

[68] C. Y. Wen and T. Z. Chaung, "Entrainment coal gasification modelling", *Industrial & Engineering Chemistry Process Design and Development*, 1979;18;684–695.

[69] B. Sun, Y. Liu, X. Chen, Q. Zhou and M. Su, "Dynamic modeling and simulation of shell gasifier in IGCC", *Fuel Process Technology*, 2011; 92:1418–1425.

[70] A. Valero and S. Usón, "Oxy-co-gasification of coal and biomass in an integrated gasification combined cycle (IGCC) power plant", *Energy*, 2006;31:1643–1655.

[71] S. Kajitani, S. Hara and H. Matsuda, "Gasification rate analysis of coal char with a pressurized drop tube furnace", *Fuel*, 2002;81:539–546.

[72] D. P. Bakalis and A. G. Stamatis, "Full and part load exergetic analysis of a hybrid micro gas turbine fuel cell system based on existing components", *Energy Conversion and Management*, 2012;64:213–221.

[73] Integrated Gasification Fuel Cell Performance and Cost Assessment, Report #:DOE/NETL-1361(2009), US Department of Energy.

[74] Electricity Market Module, Report #:DOE/EIA-0554(2010), US Department of Energy.

[75] Electric power supply data in 2010. Hokkaido Electric Power Co., Inc.; 2014. http://www.hepco.co.jp/english/index.html.

Chapter 3

Control and energy management system in microgrids

Wencong Su

3.1 Introduction

The U.S. Department of Energy defines a microgrid [1] as "a group of inter-connected loads and distributed energy resources (DERs) within clearly defined electrical boundaries that act as a single controllable entity with respect to the grid. A microgrid can connect and disconnect from the grid to enable it to operate in both grid connected and island mode." It is interesting to mention that the concept of a microgrid has been around for more than a hundred years. Around 1880, Thomas Edison founded and established the first investor-owned electric utility on Pearl Street in lower Manhattan of New York City [2]. On September 4, 1882, the Pearl Street power station went into operation. This small electric utility was allowed to operate its 27-t "Jumbo" constant-voltage dynamo (steam generator) and serve 82 local customers without being connected to a main grid, which did not exist yet. This investor-owned electric utility can be considered as the very first version of the microgrid, as shown in Figure 3.1.

In the early 1900s, the first state-wide regulation of electric utilities emerged [3]. Due to the evolution of interconnected power grids through long transmission lines, the electric utilities were moving from a microgrid-like independent system to a highly centralized and regulated one. Across the world, the development of the microgrid had been fairly silent until the early 2000s. In the past two decades, however, the original microgrid concept has drawn increased attentions to address the limited electricity access issues for in remote and less developed communities.

Microgrids are often the only practically possible solution or the most cost-effective way for these areas that are not connected with to the utility grid. In addition, the enhanced microgrid concept offers new socioeconomic benefits that have not even been imagined previously. For instance, non-traditional power generators (e.g., wind turbines, solar panels, and small-scale diesel generators) in microgrids are allowed to sell electricity to local consumers, ultimately boosting electricity market restricting activities. In addition, the microgrid no longer relies on a single power source; the on-site generation can be used as an emergency backup in the event of a blackout or brownout to mitigate the disturbance and improve power reliability.

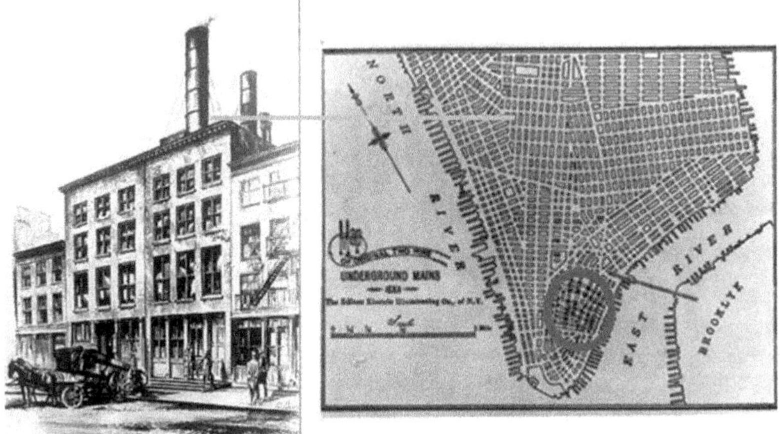

*Figure 3.1   Edison strategically located the station in the densely populated area
of lower Manhattan (Photo credit: ConEd)*

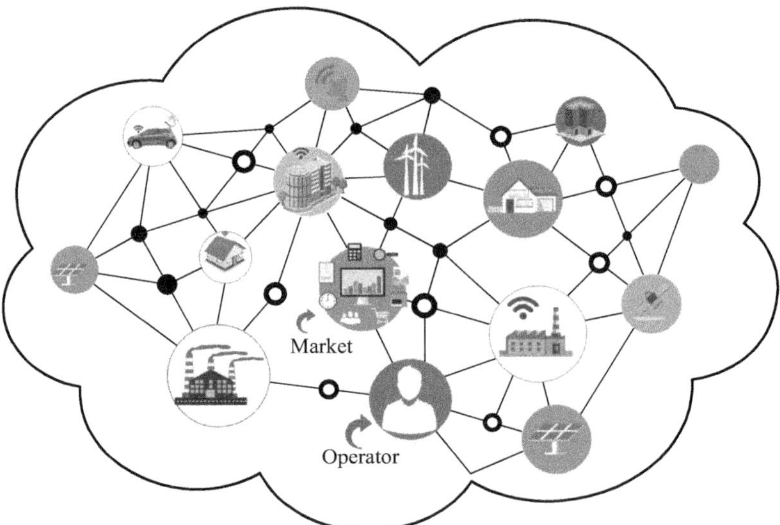

Figure 3.2   An illustrated microgrid system architecture

Figure 3.2 demonstrates the concept of modern microgrids. Technically
speaking, a modern microgrid is a small portion of a low-voltage distribution network that is located downstream from a distribution substation through a point of
common coupling [4]. Due to the nature of microgrid operations (e.g., ownership,
reliability requirement, locations), a major microgrid deployment is expected to
be carried out on university campuses and research institutions, military bases, and

industrial and commercial facilities. According to Navigant Research (formerly called Pike Research), the global microgrid capacity is expected to grow from 1.4 GW in 2015 to 7.6 GW in 2024 under a base scenario [5]. Modern microgrids not only offer great promise due to their significant benefits, but also result in tremendous technical challenges. There is an urgent need to investigate the sophisticated and state-of-the-art control and energy management systems (EMSs) in microgrids.

The remainder of this chapter is organized as follows: Section 3.2 discusses the protection and control aspects of a microgrid. Section 3.3 discusses the energy management aspect of a microgrid. Section 3.4 introduces the demand response (DR) and demand side management (DSM). Section 3.5 briefly discusses the home EMS (HEMSs). Section 3.6 presents the EMS with supervisory control and data acquisition (SCADA) system.

Section 3.7 covers the supporting infrastructure of a microgrid, including smart meters, advanced metering infrastructure (AMI), and communication infrastructure. Section 3.8 summarizes the major contributions of this chapter and briefly discusses the future research trends.

3.2   Protection and control of microgrids

The electrical energy generated by wind farms, solar energy, and even small local generators inside of the microgrids is reaching a considerable portion of the total produced energy in comparison to that of the previous decade. The presence of new energy sources, distributed storage, power electronic devices, and communication links make a power system's control and protection more complicated than before [6] because they impose considerable and fundamental changes in the configuration of the power system topology and power flow direction [7]. Thus, to enhance the power system visibility and controllability, more data, and communication links should be provided throughout the entire power system; however, this huge amount of data can also cause heavy computational burden, as well as possibly negatively impact the performance of protection schemes.

As mentioned several times, in recent decades, more attention has been given to the microgrids framework from the aspects of the market, control, management, reliability, etc. due to the active role of both the energy producers and consumers. A microgrid that could be a kind of smart grid provides us with more flexibility and reliability for control and protection of a power system. Live interaction between private commercial generators and controllable consumers is an inseparable part of smart grids that makes the power system more and more complex to handle. Thus, it is admittedly evident that conventional protection and control systems will not effectively work in a microgrid because they cannot satisfy all the control and protection requirements of such a dynamic and variable grid. The importance of the inescapable integration of communication and the physical energy network (i.e., the power system) needs to be taken into account as a way to reach an advanced and developed management system for future grid-connected microgrids [8–10].

3.2.1   Microgrids protection

Microgrids protection and reliability are the most serious challenges in the area of power system protection due to the complex nature of microgrids, two-way flow of both power and information in the power system, and presence of local distributed generations. One of the protection issues microgrids suffer from is islanding. Islanding will happen if a microgrid that includes distributed generation/storage and local loads becomes separated from the rest of the power grid. The island area brings its own problems, such as low power quality, safety matters, and overload issues [11].

Islanding detection methods are an immediate and effective solution for failure, disconnection and outage inside microgrids. Researchers have done great efforts to both introduce new detection methods and advance previous ones [12–15] to lessen the adverse effects of the islanding phenomenon. However, the smart technology applied to future microgrids provides effective tools and fundamentals to use islanding detection as a last resort to avoid any consequent trouble. As Fang *et al.* [16] classified two protection mechanisms that can be named as fault prevention and fault detection/recovery. The former mechanism is normally suitable as a preventive measure for avoiding any failure as much as possible, whereas the latter is used after failure to detect and remove the fault and recover the system in the shortest possible amount of time. Several approaches are proposed to improve the fault detection and recovery process. Tate *et al.* [17] introduced an algorithm that uses measured data of phasor measurement units to successfully detect line outages based on information regarding the system topology. It is mentioned in [18] that phasor measurement for detection of system, parameters' error is necessary because conventional measurement cannot provide a powerful tool to identify such certain error.

One of the approaches for enhancing the system reliability is self-healing ability. This self-healing ability helps the system to reconfigure itself based on the intended pattern and partitions the system into some small intentional island. This reconfiguration will be done against the disturbances that cannot be removed. Li *et al.* [19] proposed a controlled area partitioning algorithm to break down the whole system under danger into some island to minimum active and reactive power imbalances. The most important contribution of the proposed method is its improvement of the voltage profile of the partitioned subsystems compared with those of similar algorithms. In fact, cascading failure will be prevented and the impact of all disturbances will be restricted [16].

As previously mentioned, the flow of information through the communication network is important as same as flow of power through transmission line. Thus, any missing data, failure on communication channel or problem within smart meter could cause serious issues with regard to microgrids operation, visibility and controllability.

3.2.2   Control approach of microgrid control

In this chapter, different microgrid control methods ranging from conventional to recently introduced ones are studied and categorized into three major groups: centralized, decentralized, and distributed control methods.

In a power system, the control of generators and their economic dispatch (ED) can be carried out by a centralized method that much research has been dedicated to this approach [20]. In this approach, data and information are gathered from all over the system and will be processed in a central controller; then, a control signal will be transmitted directly to each agent through SCADA. SCADA is an advanced automation control system that centrally manages the control, gathering, and monitoring of an electrical power system's operation [21]. As can be seen from Figure 3.3, a two-way communication channel for each agent is required for transferring data to the center of the system, based on the nature of the centralized approach. Therefore, the system will face a barrier due to the numerous communication links that become necessary as the number of agents increases [22]. Moreover, the centralized method is not an effective or practical approach in the future grids because both the communication and electrical network topologies of a microgrid are always subject to change [23].

Using the centralized method for a large number of agents may not be a cost-effective solution due to the need for a high level of connectivity. Implementation of the centralized method, although theoretically easy, is still not a straightforward way to expand power systems in the future. However, it is obviously proven that centralized methods have currently become a mature and developed approach for power system and cannot be replaced completely by other new methods. Thus, any new method must be compatible with the current infrastructure of a power system

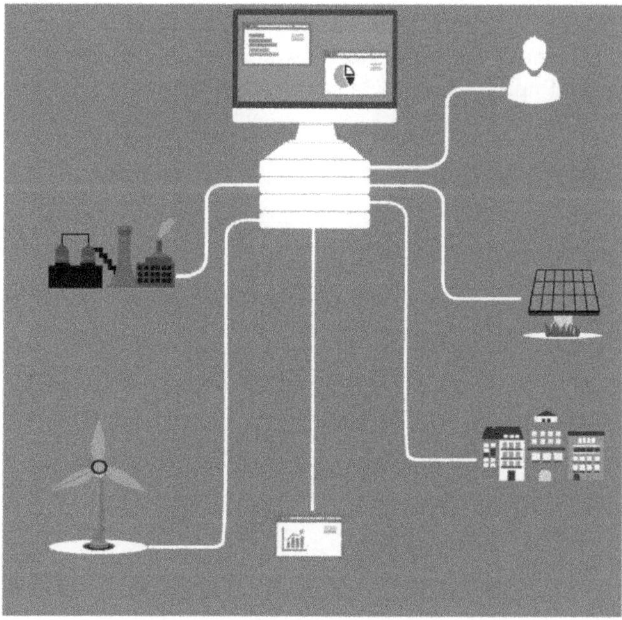

Figure 3.3   Centralized method

to work successfully with it. Solving an ED problem for a large-scale power system is a good example of widespread use of the centralized method.

As mentioned before, one type of control method is the decentralized method, in which each agent or subsystem has its own controller, meaning each agent makes decisions based on local measurements, such as voltage and frequency value, to get more profit from the market or more stability. In contrast to the centralized method, which needs global information to make a decision for the whole system, the decentralized method will not require agents to exchange all information with other agents or a central control, neither globally nor locally; however, some leader agents can send and receive information through the center.

Due to the absence of communication links, decentralized approach cannot guarantee that the system will reach the global optimization or stability; thus, they are used for local improvement. However, the decentralized method needs to be considered from different angles because each method has its own pros and cons. One of the important advantages of decentralized systems is their ability to protect the agents' privacy by secluding their private information [24–26]. A centralized and decentralized approach have been provided [27] to improve power system transient stability in a small-scale power system, and the authors also discuss and compare the performance of both methods.

Furthermore, a decentralized system is more stable than an identically connected centralized system; for instance, if some leaders lose their connection with other agents, we still have some decentralized systems that can remain stable. A simple representation of the decentralized method is shown in Figure 3.4.

Recently, painstaking research has been done to find an alternative method, or at least a short-term solution, to reduce some of the problems of the centralized method. Now, the control of a sophisticated system, such as a power system that uses a distributed control approach for achieving an optimal point, has drawn more and

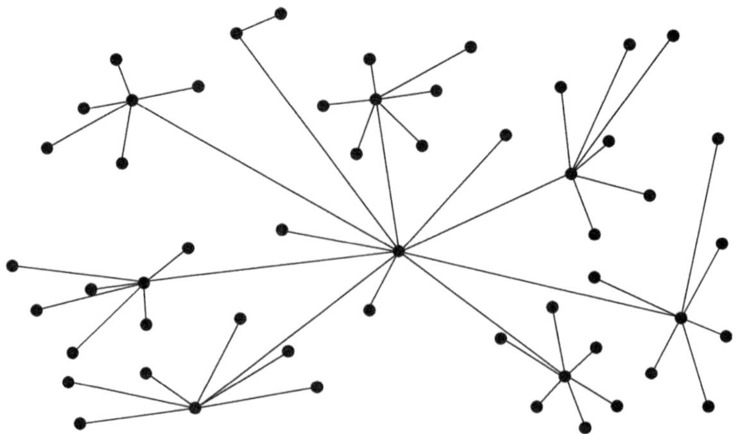

Figure 3.4   Decentralized methods schematic

more attention due to its considerable ability to extend a complex system more easily than a conventional central control approach can. In the distributed consensus-based control approach (commonly known as the distributed control approach), each agent (i.e., generators and users of a microgrid) uses local information provided by its neighbors and locally measured parameters, such as voltage and frequency. Figure 3.5 shows the concept of a distributed control system that uses local communication links. In the distributed approach, local agents will share their information among one another through two-way communication links.

In other words, in contrast to decentralized systems, in which agents use local measurements without sharing information with their neighbors, distributed systems provide a suitable environment that allows users to share information; thus, the distributed method can achieve global optimization, just like the centralized method.

According to new investigations conducted by different researchers, the distributed approach could be practical for handling the variable nature of the microgrid, which is due to its plug-and-play characteristic. In fact, this method will not be affected by changes to the smart grid topology due to the sharing of local information, and thus would be easy to extend as new agents arbitrarily connect to the network. From Figure 3.6, it can be shown how a simple microgrid can be equipped with the distributed method.

Solving ED problems is a famous and frequent use of the distributed algorithm in microgrids and power systems. In the electrical market, different participants, including consumers and prosumers, have their own private cost function affecting their actions in the market. It is evident that they must consider privacy principles if they want to get more satisfaction and benefits in the electrical competitive market. To a great extent, the distributed algorithm can preserve the agents' privacy because most of the important information, at least, will not be released and shared globally.

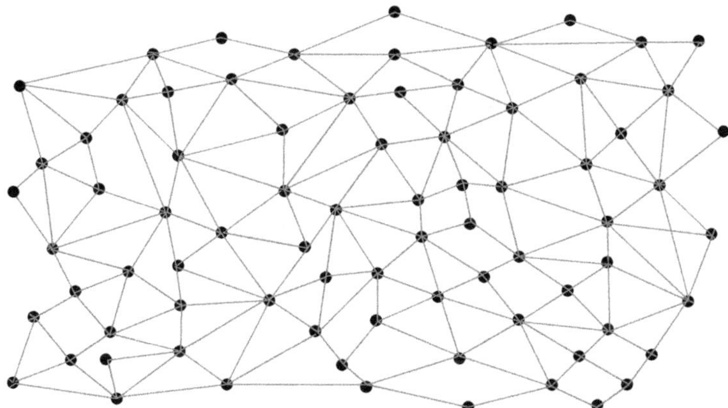

Figure 3.5   Distributed method

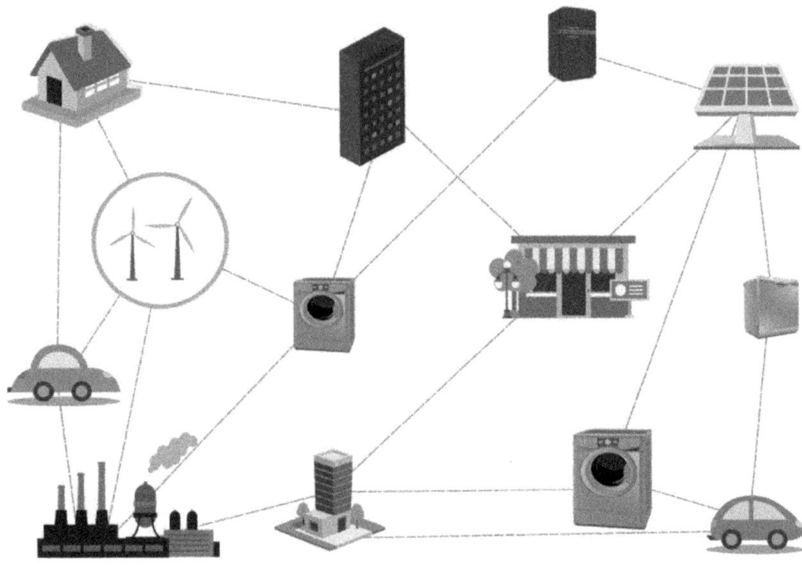

Figure 3.6   Microgrid equipped with distributed approach

Olfati-Saber *et al.* [28] wrote a paper providing a theoretical framework for the analysis of a consensus-based algorithm applied to a multiagent system. Ren *et al.* [29] carried out a comprehensive survey of distributed consensus-based problems. The authors concentrate on the application of consensus problems in cooperative control. Deng *et al.* [8] proposed a real-time DR algorithm to find an economic solution through the participation of both users and producers. They have tried to model a real-time interaction in a smart grid given a multibuyer–multiseller system.

In addition, this algorithm promises to be a privacy protector. Rahbari-Asr *et al.* [30] worked on an incremental welfare consensus (IWC)–based algorithm among responsive smart load and dynamic distributed generators (DGs) inside a smart grid for the purpose of energy management. The IWC algorithm can converge on a global optimum social welfare without having a central controller. The authors are not concerned with scalability, which is verified by a Monte Carlo simulation. Furthermore, this paper states that its proposed algorithm does not need to disclose any private information regarding the utility and cost function. Mudumbai *et al.* [31] represented a distributed algorithm having dual-application. This algorithm can independently control generators' output power in response to frequency deviation, while considering the ED of generators in a microgrid. Some key features of distributed algorithms have been emphasized for the proposed algorithm, such as scalability, dynamic response, and model independence. The researchers also benchmarked their algorithm's performance against the centralized approach using numerical results.

Table 3.1 provides a comprehensive and meaningful comparison of the three methods to clearly highlight the pros and cons of each approach [30,32,33].

Table 3.1   Comparison of centralized, decentralized, and distributed control method

Type	Pros	Cons
Centralized control	• Easy to implement • Easy maintenance in the case of single point failure	• Computational burden • Not easy to expand (so it is not suitable for smart grids) • Single point of failure (highly unstable) • Requires a high level of connectivity
Decentralized control	• Local information only • No need for a comprehensive two-way high-speed communication • Without leaders, system still includes some control island-area • Parallel computation	• Absence of communication links between agents restricts performance • Moderate scalability
Distributed control	• Easy to expand (high scalability) • Low computational cost (parallel computation) • Avoids single point of failure • Suitable for large-scale systems • Not affected by changes in system topology • practical solution for plug-and-play characteristic of smart grid	• Needs synchronization • May be time-consuming for local agents to reach consensus • Convergence rates may be affected by the communication network topology • Needs a two-way communication infrastructure • Cost to upgrade on the existing control and communication infrastructure

3.3   Energy management aspects of microgrids

In the microgrid application scenario, one of the challenging tasks is reducing large energy imbalances due to the uncertainty in power supply from intermittent renewable energy source based DGs and the dynamic nature of electricity consumption [34]. Fortunately, advances in information and communication technologies (ICT) along with more and more heterogeneous flexible loads, such as plug-in electric vehicles (PEVs), thermostatically controlled loads and distributed energy storage (DES), enable a great opportunity to develop the DR and DSM in smart grid applications. These technologies provide a lot of energy management approaches to ensure that the power demand can be rescheduled according to the power supply from utilities or local microgrids through directly or indirectly load control strategy [35]. Within the context of various nondispatchable renewable resources-based microgrid, many DR programs supported by HEMS or SCADA can further promote the participation of active energy customers into power distribution network to provide a way that they can contribute to the optimization of the value chain through directly controlling the self-generated power and electric devices. The potential demand elasticity offered by end-users (e.g., household demand) can postpone or defer grid investments and promote the efficient exploitation of the renewable electricity produced at or close to the consumption level [36]. The implementation of these opportunities require us developing new operational strategies, value mechanism, and ICT tools for enabling the coordination between demand scheduling and microgrid with the objective of supporting the entire power distribution network through providing ancillary services. The feasibility of combined optimal operation of microgrids can also be improved by embedding various DR or DSM strategies into the operation.

Both DR and DSM are mainly aimed at settling down the energy imbalances caused by irrational energy consumption or optimizing the consumption strategies by aligning the energy consumption to the supply and responding immediately to the electricity price signal. Majority of the DR/DSM strategies are designed to reduce the peak demand by shifting the energy demand from peak hours to lean hours, namely peak shaving or valley filling.

3.4   Demand response and demand-side management

In the last few years, there have been more and more retailers and utilities investing in DR programs, utilizing changes in end-users' electricity demand as one of the ways to increase electricity demand elasticity. Usually, most DR actions may be either responses to changes in the electricity prices over time or incentives from utilities that result in peak shaving or even the relief of congested networks incentive agreement [37]. With the development of networked microgrids, those incentives also include local power supply situations and relevant generation forecast. Generally, there are two DR mechanisms, namely incentive-based and price-based. Each DR mechanism comprises a number of DR alternatives that can be adopted, which are shown in Table 3.2.

More specifically, the incentive-based DR mechanism has two subcategories. One of them is the conventional mechanism that is widely used in many applications, including direct control and an interruptible/curtailment program. The other one is innovative market-based mechanism includes emergency actions, demand bidding, reserving market, and various kinds of ancillary services markets.

On the other hand, there are also several alternatives for the priced based DR mechanism. It includes time-of-use (TOU) pricing, critical peak pricing (CPP), extreme day CPP, real-time pricing, and so on. From economic perspective, the benefits of DR actions may be significant for both utilities and customers, if those electricity price mechanisms are introduced in a proper way. The cost reduction for both retailers and end-users may be significant reaching up to 18% when 40% of the controllable devices are considered [36]. As the difference

Table 3.2   Different demand response mechanism

DR program	Time-of-use (ToU)	Critical peak pricing (CPP)	Real-time pricing (RTP)	Direct load control (DLC)	Interruptible	Bidding	Emergency
Mechanism type	Price based	Price based	Price based	Incentive based	Incentive based	Incentive based	Incentive based
Rule	Non-dispatchable	Both	Non-dispatchable	Dispatchable	Dispatchable	Dispatchable	Dispatchable
Response type	Customer side	Customer side	Customer side	Utility side	Customer side	Customer side	Utility side
Advantages	Low price rate during off peak, user can shift load with min. cost	Customer response for a short time period to get discount offers	The customer can minimize the cost with respect to price change in a day, month, or season	The utility offers good discount for limited load reduction or shifting	Customers respond for a short period to get discount rates	The utility offers good discount for limited load reduction or shifting	Customer can get credit or discount rate for the short response
Disadvantages	One price rate for all customers' consumption levels, user should follow the price change with respect to time	The customer should shift or curtail home resource for certain time	Customers need to instantaneously respond to minimize cost	The customer should give the utility a level of authority to shift or curtail certain load to balance energy used	The customer should shift or curtail home resource for certain time	The customer should shift or curtail home resource for certain time	The customer should shift or curtail home resource for certain time

between peak and lean TOU rates increases, demand elasticity increases as well. These mechanisms can be implemented through the main architecture of DSM framework, which is shown in Figure 3.7. It is noteworthy that DSM techniques depend heavily on two-way communication techniques including wide area network (WAN) and home area network (HAN). The realization of the necessary DR actions usually requires frequent communication between customers and utilities or local microgrids, especially considering real-time pervasive uncertainty of the highly dynamic intermittent renewable sources, caused by weather conditions.

However, the main barriers for wide rollout of DR programs as identified by different stakeholders are low consumer interest and ineffective program design. There is also a high correlation between these two barriers because if some more effective program designs were proposed, some of them would possibly encourage customers to actively participate. Otherwise, the benefit will not be big enough to improve customers' interest. So far, the majority of the applied DR mechanisms are based on highly centralized control concepts. Thus, they require the acquisition and processing of a very large amount of local information and configuration data from a central point, conflicting with the popular distributed control approaches for microgrids [38].

This exhibits considerable complexity and burden on the only control center, which affects the scalability of aforementioned various DR mechanisms. Most of the manageable demand actions in DR program implementations just concern large commercial or industrial customers, failing to incorporate a considerable share of small residential customers even with self-generation capability. Thus, some low-level EMSs are needed to deal with energy consumption of residential customers.

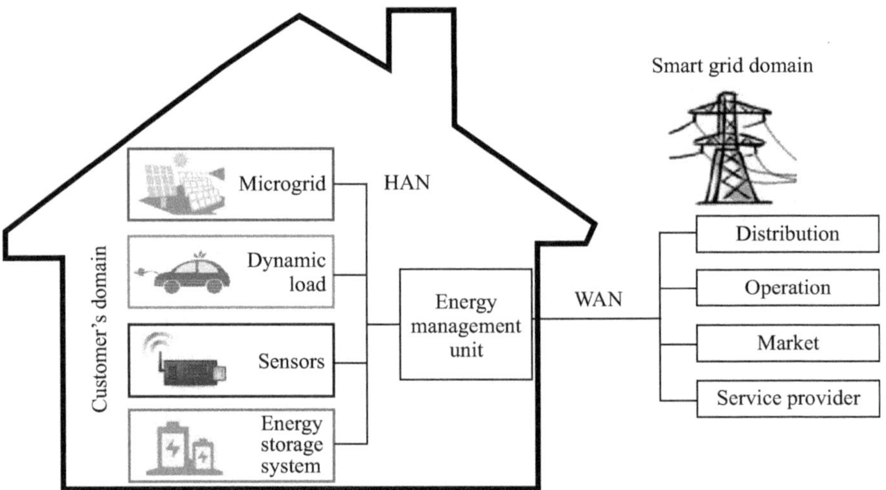

Figure 3.7   Architecture of DSM framework

3.5   Home energy management system

Associated with the DR action and local microgrid installation of many residential customers, there are many researchers proposing the idea of a zero-energy building or a smart greenhouse assisted by a HEMS. To some extent, a HEMS is actually a part of the smart grid on the consumption side, which analyses the microscopic parameters of appliances. It is employed to collect data from home appliances (e.g., solar panel, electric vehicle, geo thermal, LED lamp) using smart meters and pervasive sensors, and then to optimize power demand and supply based on this collected local information. A typical HEMS usually focuses on power consumption monitoring and standby power reduction. With the increasing demand for intelligent and personalized services, the so-called context-aware systems have been implemented in a smart home to support these personalized services with machine learning and reasoning mechanisms. These innovative designed systems have advantages that can offer adaptive energy service prediction with respect to the power consumption patterns and the related human activities. According to the users' activities and predefined requirements, HEMS along with context-aware systems can reason through the adaptive DR actions by analyzing incentives and power management policies [37].

In recent years, HEMS gradually combines context-aware systems to improve energy consumption efficiency and resident satisfaction and comfortable level. However, since conventional HEMSs need excessive resource consumption and long-term pattern analysis to have energy consumption pattern generation, these systems usually have a lot of time delay and energy mismatch. In contrast, modern HEMS usually exploit various embedded sensors (e.g., smart meters or intelligent monitoring sensors) and advanced ICT infrastructure (e.g., cloud computing platform or fog computing platform) to support managing complex applications and services. Large-scale usage of embedded sensors and Internet of Things technology will lead to a great rise in machine-to-machine (M2M) communications over wired and wireless links, which also requires enormous computing resources (Figure 3.8).

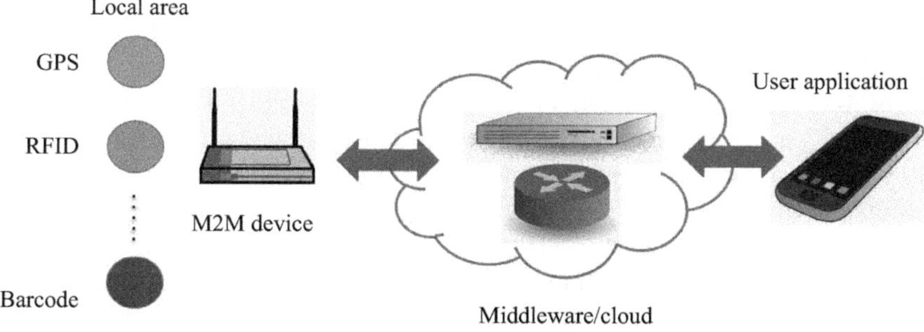

Figure 3.8   M2M communication in HEMS

On the other hand, along with the ability to control home appliances intelligently and efficiently using HEMS, integrating local microgrids with renewable energy systems becomes increasingly important. Compared with the past applications that are usually limited to heating water or heating a room through single energy source, the current applications also include remotely operating home appliances or adjusting lighting system, as well as linking the existing appliances directly with renewable energy generation systems and energy storage systems. The applicability of various renewable energy resources is continuously increasing in recent years.

Future zero energy buildings or smart green homes (which produce and provide the electricity themselves through local microgrid without the external supply of electricity, maximizing energy efficiency) could be realized by context-awareness technology and the M2M technique based intelligent HEMS. More importantly, through the utilization of two-way communication means, it is possible to use HAN to connect different smart devices and measurement units to an EMS and manage the operation of electric appliances in an economic way. With a focus on residential EMSs, a large number of research and demonstration projects have been done recently and related findings have been published in different scientific papers [39–41].

As an example, authors of [39] proposed a residential energy system to provide grid support services and manage different distributed energy resources (DERs), considering the minimum operation cost. Authors of [41] developed the energy scheduling strategy and domestic energy management for a residential building by taking technical and operational issues into account. An innovative single-objective energy management algorithm for domestic load scheduling has been outlined in [40] with the similar goal to minimize energy consumption cost. Other authors have also investigated such residential energy management problem in multiple ways with taking into account a time-domain simulation, incentive-based DR actions, and price-elastic load shifting [42].

As can be observed from the related literature, there is a large and growing body of research addressing the home energy management problem within smart residential micro-grids considering different objectives and related constraints. Although some literature tries to cover the extent of these problems of home energy scheduling in future smart grids with networked microgrid, several challenges are still associated with intelligent energy management production and energy consumption units. In [43], the author proposes a multiobjective dispatching model of a residential smart EMS to coordinate different DERs and smart household devices.

DR is generally performed in the residential district through HEMS, since residential districts will be aware of and more sensitive to the electricity price with "shiftable," controllable, flexible, interruptible, deferrable, elastic, and dispatchable appliances, e.g., PHEV, washer, and dryer. For these appliances, users are only concerned about the results of whether their tasks are finished within a certain time period without referring to the particular intermediate steps. This fact implies that their aggregate energy consumption should not be less than a threshold before a particular deadline. Based on two-way communications in HEMS, smart metering or AMI could gather detailed information regarding users' electricity usage patterns

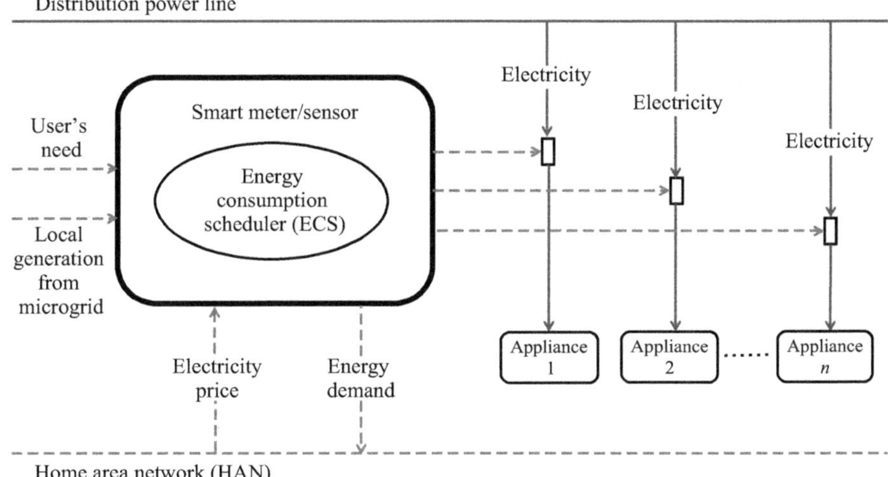

Figure 3.9   Home energy management system

and provide automatic control to household appliances, which forms the core functionality of HEMS.

As illustrated in Figure 3.9, there is an energy consumption scheduler (ECS) embedded in the smart meter at each household, whose role is to control the ON/OFF switch and operating mode of each appliance. The dynamic electricity price signal can be obtained from the power utility and the user's energy demand exchanging information via HAN. The smart meter acts as a controller that coordinates all appliances to satisfy the user's requests. After the DR, the smart meter will send ON/OFF or UP/DOWN control commands with specified operating modes to all appliances, according to the optimized ECS [10].

The electricity price and local energy demand are exchanged via the HAN in the household. On the other hand, the communication between the household and the power utility is based on WAN. In this way, the smart meter or AMI is able to automatically coordinate all electric devices via ON/OFF control commands with specified operating modes.

3.6   Energy management with SCADA

SCADA is a computer system for gathering and analyzing the real-time data of power system. SCADA systems have been widely used to monitor and control a plant or equipment in industries since the 1980s. The supervision and management of a microgrid with SCADA are shown in Figure 3.10 and, [44] based on SCADA systems through web services. A human machine interface (HMI) is developed with the objective of facilitating the interaction between users and systems.

To manage the two-type microgrid, namely, in island mode and connected mode, two different architectures are needed and proposed in [44]. In the hierarchical

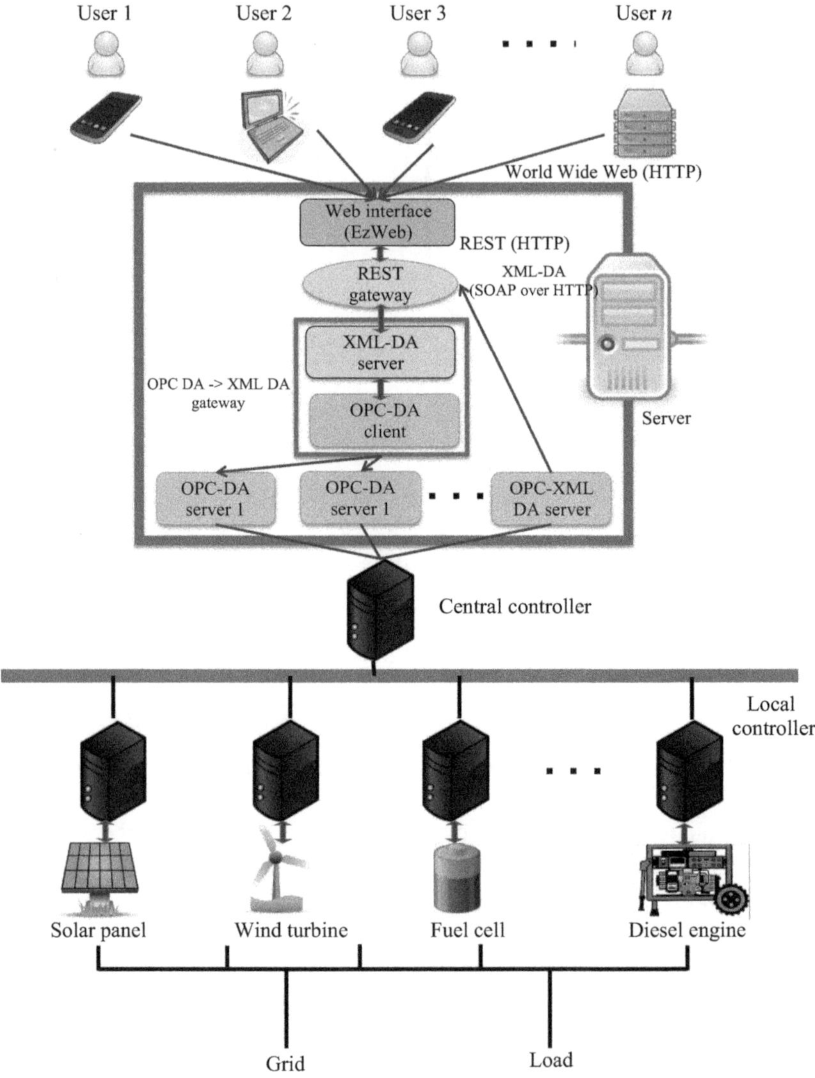

Figure 3.10   General system architecture of SCADA for microgrid management

architecture, there is a central controller element that governs the local controllers of the energy sources. In the decentralized architecture, the control is replicated in the local controllers of the energy sources. Nowadays, the utilization of a SCADA makes configuration and supervision of microgrids much easier than conventional field-test. Any layman can easily make necessary modifications to introduce new energy sources in the entire system or setting the scheduled response sensitivity. The end-users can adapt the system by simply receiving useful information, such as the cost of fuel or the efficiency curve, by means of intuitive graphical user interface and simple screens, which the user can even personalize. The user can access total knowledge of the functioning of every element to supervise the system at all times and therefore to make decision to resolve unforeseen situations quickly. The system also allows users to control different functions such as starting or stopping sources of energy in real-time. As illustrated in Figure 3.10, this SCADA has a series of layers that give the user access to different field variables.

The open platform communications (OPC) client accesses the OPC server and can obtain the variables from the general controller. The information is sent to the XML-DA server, which allows Representational State Transfer requests, making it possible to visualize the information using a web interface [44]. In addition, along with the increasing interactions between customers who have local microgrids and the capability to provide energy for the neighborhood community, an S-SCADA (social SCADA) concept is presented in some literatures [45]. This approach that consists of various computational tools is capable of linking the physical world with the social or cyber world to support community development. The traditional SCADA system requires expertise to gather and analyze real-time data; professional interfaces (e.g., HMI) are needed for system operators, who are expected to have technical knowledge of the plant that is being controlled. The S-SCADA is designed to create dummy interfaces that can easily present processed information to users who have no technical knowledge of the system.

3.7 Supporting infrastructure

The improvement of power equipment technologies has been a great contribution to the development of smart grids. In this section, we mainly focus on smart meter measurements, which are considered as the evolution of the existing grid structure. Besides that, with the advancement of computer communication technologies, smart meters can likely enhance the operation efficiency and reliability of power systems.

3.7.1 Smart meters systems

The smart meter is widely used as a smart grid technology. It is an electronic device for recording electric energy consumption. It also obtains information and communicates it back to the utility for monitoring and billing. The initial implementation of the smart meter technology was in the fields of commercial and industrial because most customers have the need for more sophisticated rates and more granular billing data requirements. For over 15 years, electronic meters have been used effectively by utilities in delivering accurate billing data for at least a portion of their customer bases [46]. The previous technology for collecting energy consumption data from meters is called automated meter reading (AMR). It provides only one-way communication, from the home to the utility. Evolved from the foundations of AMR, the AMI was developed in around 2005. The AMI technology differs from the traditional AMR by providing two-way communication between meters and the central system. The development from the AMR to the AMI, and their functionalities, are summarized in Figure 3.11.

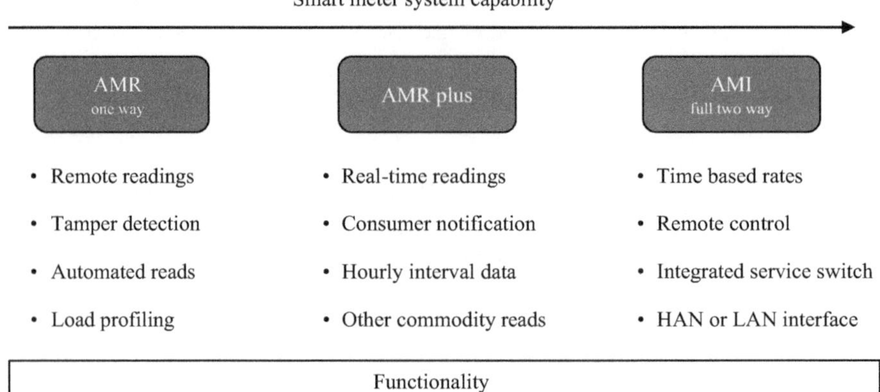

Figure 3.11   Smart meter technology development

In essence, the smart grid is an innovative reconstruction from the aspects of transmission and distribution and the smart meter system is an important integrant of the smart grid infrastructure in data collection and communication. There are three components in the smart meter system: the advanced metering device, communication network management, and data processing system.

Smart meters operate on a two-way communication, so an internal memory component is needed. Smart meters also allow consumers to track their own energy use via the Internet and/or with third-party computer programs. The two-way nature of smart meter systems allows for sending commands to operate grid infrastructure devices, such as distribution switches and recloses, to provide a more reliable energy delivery system. This is known as distribution automation [46]. Smart meter systems have many benefits, no matter who the consumer or electrical company is. From the consumers' point of view, the smart meter enables the delivery of a rapid report to the central system when tampering happens. This can effectively help reduce the rate of theft and improve security. Besides that, everyday billing information is available for every customer so that each one can manage his/her own usage of appliances and, consequently, lower their bill. From the electrical companies' point of view, the management of power consumption data from every meter can be easily gathered and processed, and naturally, the procedure of billing can be made fast with the help of the two-way communication.

3.7.2   Advanced metering infrastructure

AMI is developed from AMR. It is a technology that provides a connection between system operators and consumers. On the one hand, the information is available for consumers so they are more aware of and can make adjustments to energy usage. On the other hand, system operators can improve the service and billing process based on the data provided by AMI. The AMI infrastructure consists of home network systems, including thermostats and other in-home controls, smart

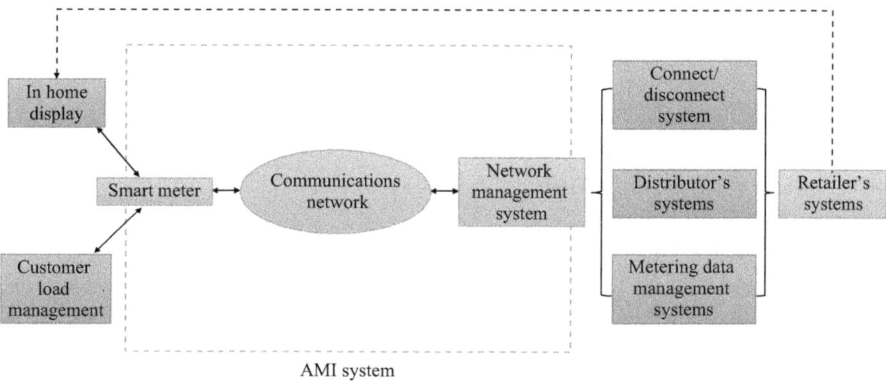

Figure 3.12   Overview of AMI

meters, communication networks from the meters to local data concentrators, backhaul communications networks to corporate data centers, meter data management systems, and finally data integration into existing and new software application platforms [47]. Figure 3.12 illustrates the overview of the AMI system and how it connects with in-home controls and communication networks.

3.7.2.1   Smart meters

For the residents, meters simply record the total energy consumption over a period of time. However, there are also other functions, such as power quality monitoring, net metering, and load limiting, as well. This helps emissions and carbon reductions and eventually improves energy efficiency. Meters show this information to every customer and are likely to cause a reduction of energy usage.

3.7.2.2   Communication infrastructure

The communication infrastructure in AMI builds a platform between consumers, the utility, and the electrical load. In terms of security, the infrastructure must employ bidirectional communication standards. Local concentrators that used to collect data from different meters are commonly implemented. They also transmit the data to the central server, the bandwidth of which should be considered, based on consumer services and other requirements.

3.7.2.3   Home area networks

HANs provide every consumer with a portal that connects smart meters to electrical devices. They also act as the consumer's agent. The consumer can check their energy usage and its cost through the in-home display and set limits for the utility to control the loads from wasting energy.

3.7.3   Privacy and security of smart meters

Cyber security is regarded as one of the biggest challenges in smart grids. Vulnerabilities may allow an attacker to penetrate a system, obtain private user information, gain access to control of the software, and alter load conditions to destabilize the grid in unpredictable ways. We must realize that the advanced infrastructure implemented in smart grids, on one hand, empowers us with more powerful mechanisms to defend against attacks from outside, but on the other hand, exposes many new vulnerabilities [48]. In this part, we mainly discuss some security and privacy issues due to the deployment of smart meters.

3.7.3.1 Security in smart metering

Attacks on smart meters can be classified as physical (external tampering, neutral bypass, missing neutral, etc.), electrical (over/under voltage, circuit probing, etc.), and software and data [49]. Smart meters raise several serious security issues:

1. The risk of widespread fraud: when smart meters are widely implemented, meter readings could be manipulated. The industry is concerned with the reliability of the returned read data. This could ruin the service provider's reputation and lead to unpredictable losses.
2. Excessive technical regulation: equipment suppliers argue that equipment costs have been pushed up and that there is nearly no benefit of continuing to move forward with smart grids. This attitude can be harmful to the prospect of fixing security problems.
3. Strategic vulnerability: a remote off switch exists in all electricity meters. A potential adversary could switch off devices by using unpredictable cyberattacks.
4. Conflict of interest: while it is the governments that prefer to cut energy use, in most countries, the meters will be controlled by energy retailers, whose goal is to maximize sales. Meanwhile, the competition authorities should worry about whether giving energy retailers vast amounts of data on customers will adversely impact competition via increased lock-in [50].
5. Lack of universal standards: communication between meters and appliances is important. An authorized standard can improve the interoperability and management in the central system, eventually relieving the competition. However, many countries cannot even decide on the architecture for connecting appliances to meters.

3.7.3.2 Privacy in smart metering

Smart meters also have potential problems for customer privacy. Retailers have the ability to obtain huge amount of data from meters or other electric devices. Not only would this reveal energy usage information, but personal habits, behaviors, and preferences could also be disclosed to some interested parties. Smart meter data, which consists of granular, fine-grained, and high-frequency type energy usage measurements, can be used by others either maliciously or inadvertently using existing or developing technology to infer types of activities or occupancies of a home for specific periods of time. Analysis of granular smart meter energy data may result in [51]:

1. Invasion of privacy and intrusion of solitude.
2. Near real-time surveillance.
3. Behavior profiling.

4. Endangering the physical security of life, family and property.
5. Unwanted publicity and embarrassment (e.g., public disclosure of private facts or the publication of facts which place a person in a false light).
6. Determine how many people are home and at what times.
7. Determine what appliances you use when, e.g., washer, dryer, toaster, furnace, A/C, microwave, medical devices ... the list is almost endless, depending on the granularity of the data.
8. Determine when a home is vacant (for planning a burglary), who has high priced appliances, and who has a security system.
9. Law enforcement can obtain information to identify suspicious or illegal behavior or later determine whether you were home on the night of an alleged crime.
10. Landlords can spy on tenants through an online utility account portal.
11. For consumers with PEVs, charging data can be used to identify travel routines and history.
12. Utilities can promote targeted energy management services and products.
13. Marketers could obtain information for targeted advertising.

To address the privacy problems related to smart meters, some approaches have been discussed or proposed:

1. Compress the meter readings and use random sequences in the compressed sensing to enhance the privacy and integrity of the meter readings [52].
2. For billing or performing calculations, improve the existing protocol and ensure the data is accurate without disclosing it.
3. Encrypt all consumption data, make most personal information anonymized, and collect and directly send both of them to the central system.

3.8   Conclusion and future research trends

As a cutting-edge technology, microgrids feature intelligent EMSs and sophisticated control, which will dramatically change our energy infrastructure. The modern microgrids are a relatively recent development with high potential to bring distributed generation, DES devices, controllable loads, communication infrastructure, and many new technologies into the mainstream. As a more controllable and intelligent entity, a microgrid has more growth potential than ever before. However, there are still many open questions, such as the future business models and economics. What is the cost-benefit to the end-user? How should we systematically evaluate the potential benefits and costs of control and energy management in a microgrid?

References

[1] S&C Electric Company, "Microgrids: an old idea with new potential," in *Tech. Rep.* Chicago, IL, USA: S&C Electric Company, 2013.

[2] P. F. Schewe, *The Grid: A Journey Through the Heart of Our Electrified World.* Washington, DC: The National Academies Press, Feb 2007.

[3]   W. Su, "The role of customers in the U.S. electricity market: past, present and future," *The Electricity Journal*, vol. 27(Aug), pp. 112–125, 2014.

[4]   W. Su and J. Wang, "Energy management systems in microgrid operations," *The Electricity Journal*, vol. 25(Oct), pp. 45–60, 2012.

[5]   *Market Data: Microgrids, Campus/Institutional, Commercial & Industrial, Community, Community Resilience, Military, Utility Distribution, and Remote Microgrid Deployments: Global Capacity and Revenue Forecasts*, Navigant Research, 2016.

[6]   H. Pourbabak and A. Kazemi, "A new technique for islanding detection using voltage phase angle of inverter-based DGs," *International Journal of Electrical Power & Energy Systems*, vol. 57(May), pp. 198–205, 2014.

[7]   S. Kar, G. Hug, J. Mohammadi, and J. M. F. Moura, "Distributed state estimation and energy management in smart grids: a consensus + innovations approach," *IEEE Journal of Selected Topics in Signal Processing*, vol. 8(Dec), pp. 1022–1038, 2014.

[8]   R. Deng, Z. Yang, F. Hou, M.-Y. Chow, and J. Chen, "Distributed real-time demand response in multiseller–multibuyer smart distribution grid," *IEEE Transactions on Power Systems*, vol. 30(Sep), pp. 2364–2374, 2015.

[9]   W. Saad, Z. Han, H. Poor, and T. Basar, "Game-theoretic methods for the smart grid: an overview of microgrid systems, demand-side management, and smart grid communications," *IEEE Signal Processing Magazine*, vol. 29(Sep), pp. 86–105, 2012.

[10] R. Deng, Z. Yang, M.-Y. Chow, and J. Chen, "A survey on demand response in smart grids: mathematical models and approaches," *IEEE Transactions on Industrial Informatics*, vol. 11, no. 3, pp. 1–1, 2015.

[11] A. Kazemi and H. Pourbabak, "Islanding detection method based on a new approach to voltage phase angle of constant power inverters," *IET Generation, Transmission & Distribution*, vol. 10(Apr), pp. 1190–1198, 2016.

[12] H. K. Karegar and B. Sobhani, "Wavelet transform method for islanding detection of wind turbines," *Renewable Energy*, vol. 38(Feb), pp. 94–106, 2012.

[13] A. Cardenas, K. Agbossou, and M. L. Doumbia, "An active anti-islanding algorithm for inverter based multi-source DER systems," in *Asia-Pacific Power and Energy Engineering Conference, APPEEC*, 2009.

[14] J. Sadeh and E. Kamyab, "Islanding detection method for photovoltaic distributed generation based on voltage drifting," *IET Generation, Transmission & Distribution*, vol. 7(Jun), pp. 584–592, 2013.

[15] W.-J. Chiang, H.-L. Jou, and J.-C. Wu, "Active islanding detection method for inverter-based distribution generation power system," *International Journal of Electrical Power & Energy Systems*, vol. 42(Nov), pp. 158–166, 2012.

[16] X. Fang, S. Misra, G. Xue, and D. Yang, "Smart grid—the new and improved power grid: a survey," *IEEE Communications Surveys & Tutorials*, vol. 14(Jan), pp. 944–980, 2012.

[17] J. E. Tate and T. J. Overbye, "Line outage detection using phasor angle measurements," *IEEE Transactions on Power Systems*, vol. 23(Nov), pp. 1644–1652, 2008.

[18] J. Zhu and A. Abur, "Improvements in network parameter error identification via synchronized phasors," *IEEE Transactions on Power Systems*, vol. 25, no. 1, pp. 44–50, 2010.

[19] J. Li, C. C. Liu, and K. P. Schneider, "Controlled partitioning of a power network considering real and reactive power balance," *IEEE Transactions on Smart Grid*, vol. 1, no. 3, pp. 261–269, 2010.

[20] W. Su, J. Wang, and D. Ton, "Smart grid impact on operation and planning of electric energy systems," in *Handbook of Clean Energy Systems* (J. Yan, ed.), Chichester: John Wiley & Sons, Ltd., Jul 2015, pp. 1–13.

[21] M. Higgs, "Electrical SCADA systems from the operator's perspective," *IEE Seminar Condition Monitoring for Rail Transport Systems*, vol. 1998, pp. 3/1–3/4, IEE, 1998.

[22] N. Rahbari-Asr and M.-Y. Chow, "Cooperative distributed demand management for community charging of PHEV/PEVs based on KKT conditions and consensus networks," *IEEE Transactions on Industrial Informatics*, pp. 1907–1916, 2014.

[23] J. Cao, Consensus-based distributed control for economic dispatch problem with comprehensive constraints in a smart grid. *Electronic Theses, Electronic Theses, Treatises and Dissertations*, Tallahassee, FL, USA: Florida State University, 2014.

[24] L. Gan, U. Topcu, and S. H. Low, "Optimal decentralized protocol for electric vehicle charging," *IEEE Transactions on Power Systems*, vol. 28(May), pp. 940–951, 2013.

[25] Y. Guo, J. Xiong, S. Xu, and W. Su, "*Two-Stage Economic Operation of Microgrid-Like Electric Vehicle Parking Deck*," IEEE, 2015.

[26] Y. He, B. Venkatesh, and L. Guan, "Optimal scheduling for charging and discharging of electric vehicles," *IEEE Transactions on Smart Grid*, vol. 3, no. 3, pp. 1095–1105, 2012.

[27] T. Senjyu, R. Kuninaka, N. Urasaki, H. Fujita, and T. Funabashi, "Power system stabilization based on robust centralized and decentralized controllers," in *2005 International Power Engineering Conference*, vol. 2, pp. 905–910, 2005.

[28] R. Olfati-Saber, J. A. Fax, and R. M. Murray, "Consensus and cooperation in networked multi-agent systems," *Proceedings of the IEEE*, vol. 95(Jan), pp. 215–233, 2007.

[29] W. Ren, R. Beard, and E. Atkins, "A survey of consensus problems in multiagent coordination," in *Proceedings of the 2005, American Control Conference, 2005*, pp. 1859–1864, IEEE, 2005.

[30] N. Rahbari-Asr, U. Ojha, Z. Zhang, and M.-Y. Chow, "Incremental welfare consensus algorithm for cooperative distributed generation/demand response in smart grid," *IEEE Transactions on Smart Grid*, vol. 5(Nov), pp. 2836–2845, 2014.

[31] R. Mudumbai, S. Dasgupta, and B. B. Cho, "Distributed control for optimal economic dispatch of a network of heterogeneous power generators," *IEEE Transactions on Power Systems*, vol. 27(Nov), pp. 1750–1760, 2012.

[32] Z. Zhang and M. Y. Chow, "Incremental cost consensus algorithm in a smart grid environment," in *IEEE Power and Energy Society General Meeting*, 2011.

[33] V. Loia and A. Vaccaro, "Decentralized economic dispatch in smart grids by self-organizing dynamic agents," *IEEE Transactions on Systems, Man, and Cybernetics: Systems*, vol. 44, no. 4, pp. 397–408, 2014.

[34] H. K. Nunna, A. M. Saklani, A. Sesetti, S. Battula, S. Doolla, and D. Srinivasan, "Multi-agent based demand response management system for combined operation of smart microgrids," *Sustainable Energy, Grids and Networks*, vol. 6(Jun), pp. 25–34, 2016.

[35] H. T. Haider, O. H. See, and W. Elmenreich, "A review of residential demand response of smart grid," *Renewable and Sustainable Energy Reviews*, vol. 59(Jun), pp. 166–178, 2016.

[36] E. Karfopoulos, L. Tena, A. Torres, *et al.*, "A multi-agent system providing demand response services from residential consumers," *Electric Power Systems Research*, vol. 120(Mar), pp. 163–176, 2015.

[37] J. Byun, I. Hong, and S. Park, "Intelligent cloud home energy management system using household appliance priority based scheduling based on prediction of renewable energy capability," *IEEE Transactions on Consumer Electronics*, vol. 58(Nov), pp. 1194–1201, 2012.

[38] F. Katiraei and M. Iravani, "Power management strategies for a microgrid with multiple distributed generation units," *IEEE Transactions on Power Systems*, vol. 21(Nov), pp. 1821–1831, 2006.

[39] H. Karami, M. J. Sanjari, S. H. Hosseinian, and G. B. Gharehpetian, "An optimal dispatch algorithm for managing residential distributed energy resources," *IEEE Transactions on Smart Grid*, vol. 5(Sep), pp. 2360–2367, 2014.

[40] A. Barbato, A. Capone, G. Carello, M. Delfanti, M. Merlo, and A. Zaminga, "House energy demand optimization in single and multi-user scenarios," in *2011 IEEE International Conference on Smart Grid Communications (SmartGridComm)*, pp. 345–350, IEEE, Oct 2011.

[41] M. Tasdighi, H. Ghasemi, and A. Rahimi-Kian, "Residential microgrid scheduling based on smart meters data and temperature dependent thermal load modeling," *IEEE Transactions on Smart Grid*, vol. 5(Jan), pp. 349–357, 2014.

[42] M. Parvizimosaed, F. Farmani, and A. Anvari-Moghaddam, "Optimal energy management of a micro-grid with renewable energy resources and demand response," *Journal of Renewable and Sustainable Energy*, vol. 5, p. 053148, Oct 2013.

[43] A. Anvari-Moghaddam, H. Monsef, A. Rahimi-Kian, J. M. Guerrero, and J. C. Vasquez, "Optimized energy management of a single-house residential microgrid with automated demand response," in *2015 IEEE Eindhoven PowerTech*, pp. 1–6, IEEE, Jun 2015.

[44] E. Álvarez, A. Campos, and R. García, "Scalable and usable web based supervisory and control system for micro-grid management," *Energies and Power*, vol. 1, no. 8, pp. 763–768, 2010.

[45] R. Palma-Behnke, D. Ortiz, L. Reyes, G. Jimenez-Estevez, and N. Garrido, "A social SCADA approach for a renewable based microgrid – The Huatacondo project," in *2011 IEEE Power and Energy Society General Meeting*, pp. 1–7, IEEE, Jul 2011.

[46] Edison Electric Institute (EEI), "Smart meter systems: a metering industry perspective," *A Joint Project of the EEI and AEIC Meter Committees*, Edison Electric Institute (EEI), 2011.

[47] National Energy Technology Laboratory (NETL), "Advanced metering infrastructure," *Tech. Rep.*, US Department of Energy, Office of Electricity and Energy Reliability, 2008.

[48] National Institute of Standard and Technology (NIST), "Introduction to NISTIR 7628 guidelines for smart grid cyber security," *Guideline*, National Institute of Standard and Technology (NIST), Sep 2010.

[49] P. McDaniel and S. McLaughlin, "Security and privacy challenges in the smart grid," *IEEE Security & Privacy*, pp. 763–768, 2009.

[50] W. Wang and Z. Lu, "Cyber security in the smart grid: survey and challenges," *Computer Networks*, vol. 57, pp. 1344–1371, 2013.

[51] E. Quinn, "Smart metering and privacy: existing laws and competing policies," Available at *SSRN 1462285*, pp. 1–29, 2009.

[52] H. Li, R. Mao, L. Lai, and R. Qiu, "Compressed meter reading for delay-sensitive and secure load report in smart grid," *Smart Grid Communications*, pp. 114–119, IEEE, 2010.

Chapter 4

Storage systems for microgrids

Shin'ya Obara

4.1   Introduction: storage systems for microgrids

Storage-of-electricity technology strongly influences the electric-power quality of an independent microgrid, reliability and economic efficiency. The performance required of the battery for electric power is efficiency, a charge and discharge rate, and economic efficiency. In this chapter, the following two storage-of-electricity technologies are introduced. The technology for the operation of a battery based on online weather prediction is described first. Next, control of electric-power fluctuation of the microgrid with a sodium–sulphur battery (NaS battery) and with a hydrogen career using organic chemical hydride methylcyclohexane (OCHM) are described.

- **First investigation**

A fuel-cell microgrid with photovoltaics effectively reduces greenhouse gas emission. A system-operation-optimisation technique is important for microgrid with photovoltaics and unstable power. In this chapter, the optimal operation algorithm of this compound microgrid is developed using numerical weather information (NWI) that is freely available. A GA (genetic algorithm) was developed to minimise system fuel consumption. Furthermore, the relation between the NWI error characteristics and the operation results of the system was investigated. As a result, the optimised operation algorithm using NWI reduced the energy cost of the system.

- **Second investigation**

The present study uses numerical analysis to investigate the operating methods and costs of an independent microgrid incorporating a NaS battery or an energy-storage system using organic hydrides. Details of the operation of the system and its installed capacity and cost were obtained, assuming an independent microgrid in Kitami City, a cold region in Japan. Analysis results indicate that energy-storage technology using the organic hydride system is economically inferior to the NaS battery owing to greater losses associated with the water electrolyser and dehydration reactor. Therefore, for the widespread use of the organic hydride system, it is necessary to improve the efficiency of these components.

4.2 Operation planning for a compound microgrid containing a PEFC and photovoltaics with prediction of electricity production using GA and NWI

4.2.1 Introduction

An energy-supply system using a microgrid provides the optimal system for energy demand. Therefore, its use as a clean energy-supply technique is expected to spread [1–3]. A microgrid using a PEFC (proton exchange membrane fuel cell) may become the mainstream of future distributed energy. In addition, the application of green energy to a microgrid is desirable. Accordingly, this chapter examines a PEFC and photovoltaics compound system. Power can be supplied to a grid from both the PEFC and photovoltaic components in the system. The hydrogen-supply method to the PEFC assumes that the steam reforms the LPG (liquefied petroleum gas). However, the power-generation-output characteristics and PEFC exhaust heat with a steam reformer are non-linear with a load factor [4]. Furthermore, although the power and exhaust heat of the proposed system are utilised effectively, battery installation and a heat-storage tank are planned. Consequently, the operation plan of the proposed microgrid must be optimised as a non-linear system considering electricity and heat storage. We have summarised the use of a GA on the operation optimisation of a non-linear system with heat storage [5–7]. In addition, it is necessary to predict unstable photovoltaic electricity production for every sample time while optimising operation of a compound microgrid with a PEFC and photovoltaics. Accordingly, NWI is used to predict photovoltaic electricity production [8,9]. Anyone can obtain NWI in Japan through the Internet. However, there is an error in the photovoltaic electricity production calculated using NWI compared to using the actual meteorological data. Consequently, the operation plan of the system using the NWI differs from operation under actual weather conditions. The cause of this difference in operation is not addressed in this section. Instead, the relationship between the NWI error and the operation results of the system is investigated. It is shown that the operation optimisation algorithm using NWI is important for operation of a PEFC microgrid with photovoltaics. The objective of this study is to develop an analysis algorithm to optimise operation of a PEFC microgrid with green energy.

4.2.2 System configurations

4.2.2.1 PEFC and photovoltaics compound microgrid

Figure 4.1 shows a scheme of a compound microgrid with PEFC and photovoltaics. The compound microgrid consists of a power system and a heat system. Here, the power system is not connected to a commercial power system. The power from a PEFC and a solar cell can be supplied simultaneously to the microgrid. Moreover, these power sources can accumulate electricity using a battery. The hydrogen (reformed gas) supplied to a PEFC is produced from LPG using a steam reformer.

Each piece of equipment of the power system and the heat system is operated by a system controller. The photovoltaic electricity production for every sample

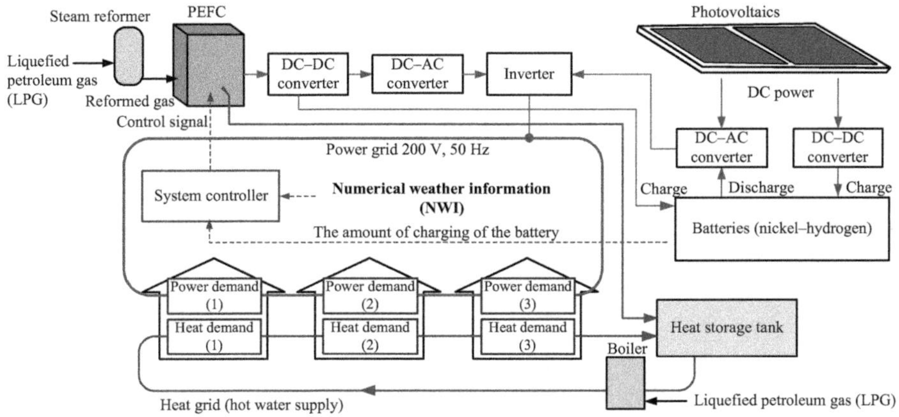

Figure 4.1   System scheme

time in a target day is predicted using the NWI (the amount of solar radiation and outdoor air temperature) obtained by the system controller from 0:00 on the target day. Based on this prediction, the optimal system operation on the target day is planned by the system controller. The objective given to the system controller is to minimise fuel (LPG) consumption. As Figure 4.1 shows, fuel is consumed by the PEFC and a boiler in the proposed system. The optimisation analysis of the operation plan in this section considers operation of a power system and a heat system. The NWI used for analysis is the information obtained at 0:00 on the target day. Therefore, the NWI does not match actual meteorological data. If a system is operated according to the first optimisation plan (the plan made at 0:00 on the target day), then depending on the magnitude of this error, the fuel consumption may get worse. For example, the operation hours of a PEFC and the boiler may be extended under the actual weather conditions. Investigating the relation between the NWI error and system fuel consumption evaluates the operation optimisation algorithm using NWI.

4.2.2.2   System operation

Figure 4.2 shows power demand (a) on a representative day, heat demand (d), electricity production and exhaust heat of PEFC (b), (e), the operation model of a battery (c) and the operation model of heat storage and the boiler (f). Predicted photovoltaic electricity production based on the NWI obtained at 23:00 on a representative day is shown in Figure 4.2(a) and (d). Furthermore, the photovoltaic electricity production obtained under actual weather conditions at each time is shown in this figure.

The relation between the load factor and power generation efficiency of a PEFC with a reformer, load factor and heat generation efficiency is non-linear. Figure 4.3 shows the relation of the load factor and power generation efficiency of a home fuel cell with an LPG reformer released by Tokyo Gas Co. Ltd. [10]. As shown in Figure 4.3(a), power generation efficiency also falls with load decreases.

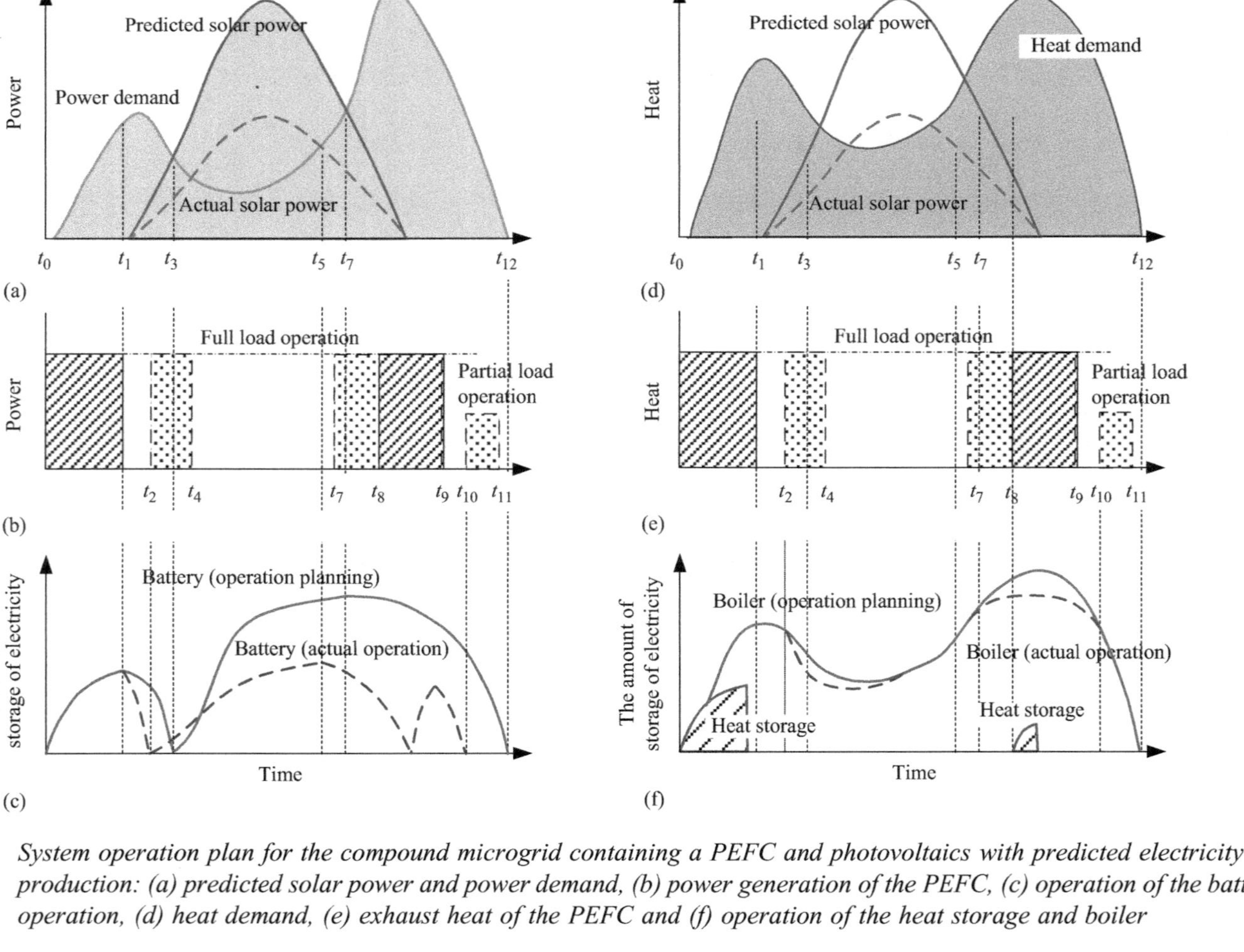

Figure 4.2   System operation plan for the compound microgrid containing a PEFC and photovoltaics with predicted electricity production: (a) predicted solar power and power demand, (b) power generation of the PEFC, (c) operation of the battery operation, (d) heat demand, (e) exhaust heat of the PEFC and (f) operation of the heat storage and boiler

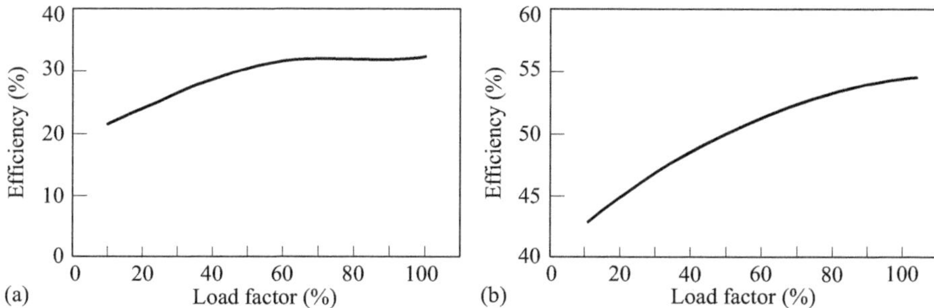

Figure 4.3 Efficiency characteristics of the PEFC with steam reformer: (a) power generation efficiency and (b) heat generation efficiency

Therefore, the PEFC operates well under high load. Accordingly, as shown in Figure 4.2(b), it is desirable to operate the PEFC near its maximum efficiency point. When the PEFC is operated near the maximum efficiency point, exhaust heat of the PEFC has an output characteristic shown in Figure 4.2(e). Operations of the PEFC in the periods from t_0 to t_1 and t_8 to t_9 in Figure 4.2(b) and (e) are the optimal operation plan based on predicted solar power shown in Figure 4.2(a) and (d). On the other hand, operation of PEFC in the periods from t_2 to t_4, t_7 to t_8 and t_{10} to t_{11} in Figure 4.2(b) and (e) covers the photovoltaic shortage compared to the actual solar power (the actual solar power is smaller than the predicted solar power), as shown in Figure 4.2(a) and (d). When there is little actual solar power compared to the predicted solar power, additional PEFC operation is required. In this case, as shown in Figure 4.2(c) and (f), battery, heat storage tank and boiler operations change. Accordingly, the relation between the magnitude of the difference of the predicted solar power and the actual solar power, and the fuel consumption of the system is investigated. By considering this result, the influence of the NWI error on the system operation plan can be identified.

4.2.3   Analysis method

4.2.3.1   Power system

Photovoltaics

In this section, installation of the polycrystalline silicon solar module of area S_s is assumed. The average production of electricity $P_{s,t}$ of the solar module from sample time t to $t + 1$ on a representative day is calculated as shown in Figure 4.1. R_T in (4.1) is the temperature coefficient, and when the temperature $T_{c,t}$ of the solar cell rises, power generation efficiency will fall. T_o is a reference temperature, and η_s is the power generation efficiency under T_o. The temperature $T_{c,t}$ of the solar cell is calculated from the specific heat of the polycrystalline silicon and the amount of solar radiation at sampling time t. When the intensities of direct solar and sky solar radiation are expressed by $H_{D,t}$ and $H_{M,t}$, respectively, as the solar radiation input into the acceptance surface, $P_{s,t}$ will be calculated by (4.1).

Direct insolation and sky solar radiation are used for power generation in a flat solar cell. Global-solar-radiation intensity, direct solar radiation intensity and horizontal sky solar radiation intensity at time t ($t = 0.1, 2, \ldots, 23$) are expressed with $I_{H,t}$, $I_{D,t}$ and $I_{M,t}$, respectively. $I_{H,t}$ and $I_{D,t}$ can be determined from the NWI. Moreover, $I_{M,t}$ can be calculated using $I_{H,t}$ and $I_{D,t}$. The incidence angle θ to the acceptance surface of sunlight is calculated using (4.2). Here, φ, δ and ω show the latitude of a setting point, the solar celestial declination and hour angle, respectively, whereas (4.3) is a calculation formula for the sky solar radiation component $H_{D,t}$.

$$P_{s,t} = S_s \cdot \eta_s \cdot (H_{D,t} + H_{M,t}) \cdot \left\{ 1 - (T_{c,t} - T_o) \cdot \left(\frac{R_T}{100} \right) \right\} \tag{4.1}$$

$$\sin \theta = \cos \varphi \cdot \sin \delta - \sin \varphi \cdot \cos \omega \cdot \cos \delta \tag{4.2}$$

$$H_{D,t} = I_{D,t} \cdot \cos \theta \tag{4.3}$$

Equation (4.4) calculates the incidence sky solar radiation component $H_{M,t}$ of the solar cell. The first term on the right-hand side of (4.4) is the air solar radiation component; the second term is the reflective solar radiation component; β is the angle of gradient of the acceptance surface by (4.5); and ρ is the reflection factor of the ground.

$$H_{M,t} = I_{M,t} \cdot \frac{1 + \cos \beta}{2} + \rho \cdot I_{H,t} \cdot \frac{1 - \cos \beta}{2} \tag{4.4}$$

$$\cot \beta = \cos \varphi \cdot \cot \omega + \sin \varphi \cdot \operatorname{cosec} \omega \cdot \tan \delta \tag{4.5}$$

Power balance
Equation (4.6) is a power balance equation. $P_{fc,t}$, $P_{pv,t}$ and $P_{bt,t}$ on the left-hand side in the equation are the PEFC power, photovoltaic power and battery power, respectively. $P_{need,t}$, $P_{btc,t}$, $P_{loss,t}$ on the right-hand side in the equation represent power demand, the amount of battery charge and loss of power, respectively. Charge-and-discharge loss of a battery is included in the power loss $P_{loss,t}$.

$$P_{fc,t} + P_{pv,t} + P_{bt,t} = P_{need,t} + P_{btc,t} + P_{loss,t} \tag{4.6}$$

4.2.3.2   Heat balance

Equation (4.7) is a heat-balance equation.

$H_{fc,t}$, $H_{bl,t}$ and $H_{st,t}$ on the left-hand side in the equation are the heat power of a fuel cell, a boiler, and a heat storage tank, respectively. $H_{need,t}$, $H_{sts,t}$ and $H_{loss,t}$ on the right-hand side of the equation are heat demand, the amount of heat storage and the heat loss, respectively. Heat-storage loss is included in the heat loss $H_{loss,t}$ on the right-hand side of the equation.

$$H_{fc,t} + H_{bl,t} + H_{st,t} = H_{need,t} + H_{sts,t} + H_{loss,t} \tag{4.7}$$

4.2.3.3   Optimal analysis using GA

Objective function

If $P_{\mathrm{fc},t}$ in (4.6) and $H_{\mathrm{bl},t}$ in (4.7) are determined, the heating value of LPG $Q_{\mathrm{fuel},t}$ consumed by a compound microgrid is calculable. Here, the amount of fuel with the output power of $P_{\mathrm{fc},t}$ and $H_{\mathrm{bl},t}$ is decided by the PEFC power generation efficiency and the thermal efficiency of the boiler. Equation (4.8) defines the objective function in this study. The objective function minimises the system fuel consumption $Q_{\mathrm{system,day}}$ on one day. The fuel consumption $Q_{\mathrm{fuel},t}$ of the system from sample time t to $t+1$ is the sum of the fuel consumption $Q_{\mathrm{fc},t}$ of a fuel cell, and the fuel consumption $Q_{\mathrm{bl},t}$ of a boiler.

$$Q_{\mathrm{system,day}} = \sum_{t=0}^{23} Q_{\mathrm{fuel},t} = \sum_{t=0}^{23} \left(Q_{\mathrm{fc},t} + Q_{\mathrm{bl},t} \right) \tag{4.8}$$

Optimal operation planning algorithm

In this study, the optimal operation plan for the proposed compound microgrid is analysed using a GA. Figure 4.4 shows the operation optimisation algorithm developed in this section, and the flow is explained below.

1. The energy-demand pattern data, equipment specifications, GA parameters, numerical weather data, efficiencies, initial conditions and system loss are used as input into a computer (system controller) in Calculation (A) in Figure 4.4.
2. In Calculation (B), many initial generation chromosome models are generated at random. One individual of the chromosome model expresses PEFC operation and power. The PEFC operation is represented with a 1-bit binary number, and the PEFC power is represented by a 14 bit binary number.
3. In Calculation (F), the PEFC power is determined by decoding the chromosome model. Furthermore, in Calculation (G), the production of electricity of photovoltaics is calculated using NWI (Section 3.1.1).
4. In Calculations (H) through (K), battery, heat storage tank and boiler operations are planned on the basis of the power-balance and heat-balance equations in (Sections 3.1.2 and 3.2).
5. The fuel consumption is calculated from the amount of PEFC and boiler power. In Calculation (L), these values are totalled, and the fuel consumption of the system in the sampling time t is determined.
6. Calculations (E) through (M) are repeated from sampling time 0 to 23 for one chromosome model. In Calculation (N), the adaptive value [namely, the objective function shown in (4.8)] of the chromosome model is obtained from this result.
7. The adaptive value of all the chromosome models is decided by repeating Calculations (D) through (O). The ranking of the chromosome models is decided according to the magnitude of the adaptive value of each chromosome [Calculation (P)].

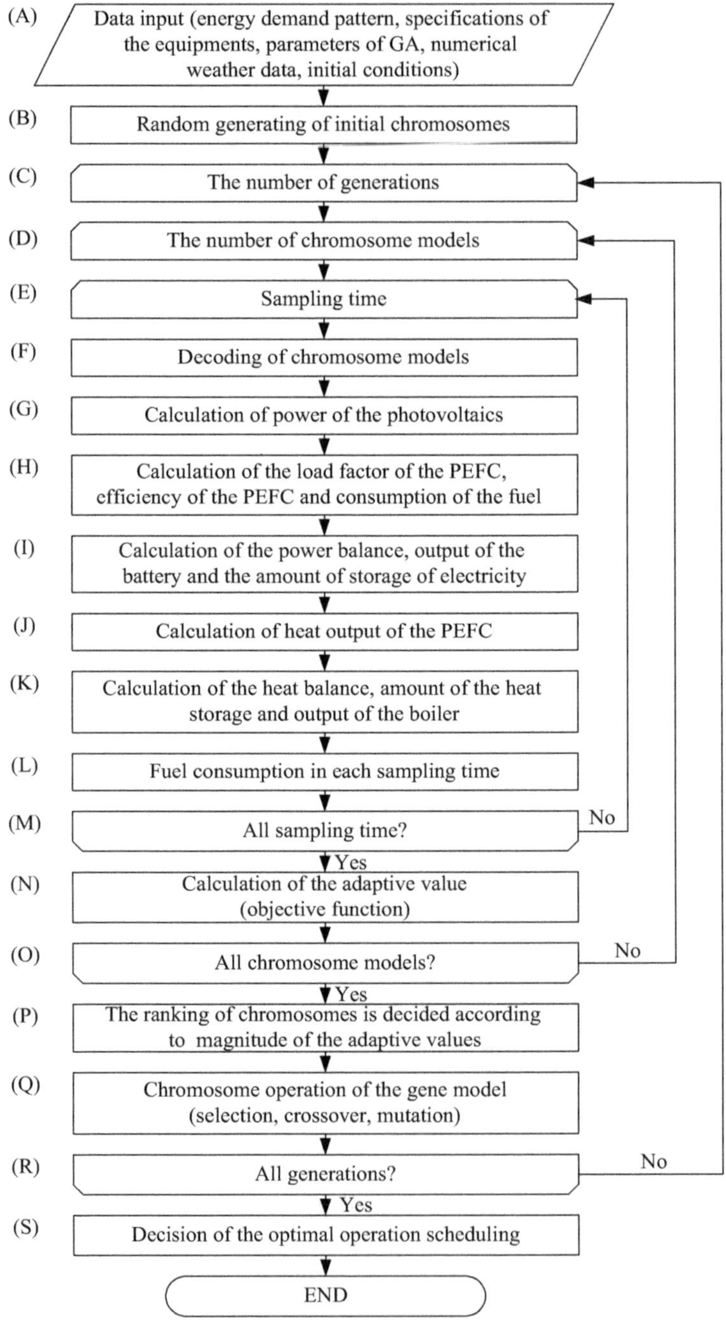

Figure 4.4   Optimal operation algorithm

8. The chromosome models with low adaptive value are selected, and they are exchanged for a new randomly generated model. Moreover, the genetic manipulation of crossover and mutation is added on the basis of the probability given in Calculation (A) on the chromosome models with high adaptive value [Calculation (Q)].
9. Calculations (D) through (Q) [repeated calculation of Calculations (C) to (R)] are repeated for a defined number of generations. In the last generation's chromosome group, the solution with the highest adaptive value is chosen to be the optimal system operation plan [Calculation (S)].

4.2.4   Case analysis

4.2.4.1   Equipment specifications

The equipment specifications for the case analysis of the PEFC and photovoltaics compound microgrid are shown in Table 4.1. The analysis assumes that the microgrid equipment is installed in Sapporo, Japan (latitude 43.062-degree north and longitude 141.354-degree east, a cold and snowy area).

PEFC with a reformer

The maximum PEFC power with a reformer is 3 kW, and this performance is shown Figure 4.3.

Photovoltaics

The maximum efficiency and photovoltaic temperature coefficients are 16.4% and 0.4%/K, respectively. These values are general facility values used in Japan. The solar panel is installed on the roof, with a slope of 30-degrees facing south and the solar cell area is set to 60.0 m^2.

The area of the general solar cell installed on individual houses in Japan is usually 25–40 m^2 (for a solar cell with a 3–5 kW capacity).

Table 4.1   Equipment specifications

PEFC maximum power	3 kW
PEFC performance	Figure 4.3
Solar cell type	Multicrystalline silicon
Area of solar cell	60.0 m^2
Maximum generation efficiency of the solar cell	16.4%
Temperature coefficient of the solar cell	0.4%/K
Battery type	Nickel–hydrogen
Amount of battery self-discharge	10%/hour
Efficiency of converter	95%
Efficiency of inverter	95%
Loss of heat storage	5%/hour
Efficiency of boiler	90%

Table 4.2   GA parameters

The number of chromosome models	10,000
Generation number	100
Probability of mutation	10%
Selection	50% of a low rank is replaced

Battery, converter and inverter
The self-discharge of a battery is set to 10%/h. The converter and inverter efficiencies are both set to 95%.

Heat-storage tank and boiler
Heat dissipation loss of a heat-storage tank is set to 5%/h, and the boiler efficiency is set to 90%.

4.2.4.2   GA parameters

The GA parameters in the proposed algorithm are shown in Table 4.2. These values were chosen by repeating trial and error approach so that the convergence solution was as stable as possible. Since the convergence solution (analysis result) has dispersion for every analysis, the optimal solution is obtained by repeating the same analysis.

4.2.4.3   Energy-demand pattern

Power and heat are supplied to three individual houses in Sapporo, Japan using the proposed microgrid. Figure 4.5 shows the power and heat demand on a representative day every month [11]. There is no cooling load and heating is included in the heat load. Therefore, the power-load pattern on a representative day of every month does not vary significantly throughout the year. On the other hand, the magnitude of demanded heat varies greatly between the summer season and winter season.

4.2.4.4   Error of the NWI

Various error characteristics of the NWI can be considered. However, in the investigated case, various NWI errors were not found. Accordingly, the error pattern of the following two types is used in this study. Figure 4.6 shows the error pattern of the two types used to analyse the proposed algorithm (Figure 4.4). As shown in Figure 4.6(a), a linear error and a quadratic error are used as error patterns. Target day operation plans are determined by the NWI at 0:00 on the target day. In this section, the error that is proportional ($error_t = const1 \cdot t$) to time is defined as the linear error. On the other hand, the error that follows ($error_t = const2 \cdot t^2$) a secondary curve relative to time is defined as the quadratic error. Here, the integrated values of the two error types are set equal to each other. Therefore, const1 and const2 were decided so that Area A and Area B [shown in Figure 4.6(a)] might become equal [in Figure 4.6(a), Area A = Area B = 1.0]. The common characteristic of these error patterns is the increase of the NWI error as

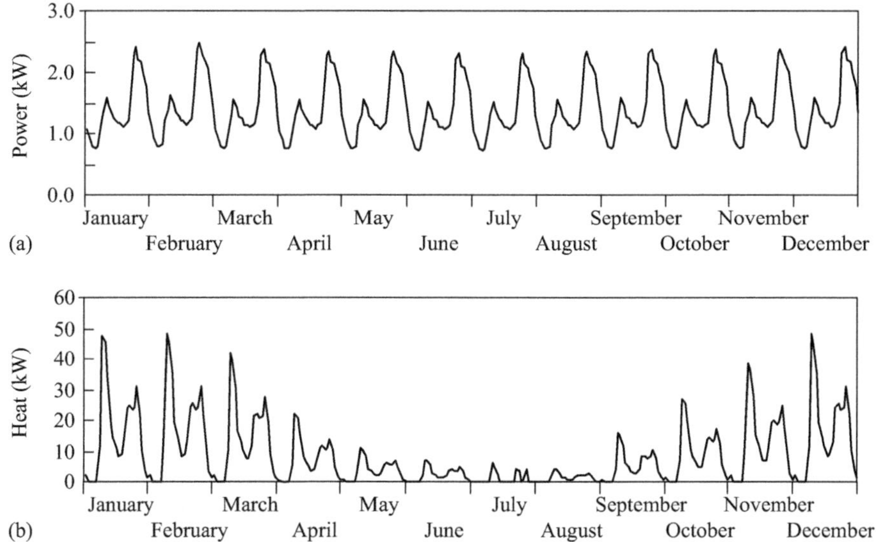

Figure 4.5   Energy-demand model. Load patterns for three individual houses on representative days in Sapporo, Japan: (a) power demand model and (b) heat demand model

time increases. Moreover, as shown in Figure 4.6(b) and (c), fluctuation errors of ±20% and ±40% at random are added to the two error types. These fluctuation errors simulate the instability of the solar insolation data.

4.2.5   Results and discussion

4.2.5.1   Operation planning

Figure 4.7 shows the results of the system operation plan optimisation analysis on representative February days (winter). Figure 4.7(a) and (b) is the optimal operation plans for a power system and a heat system, respectively. Moreover, Figure 4.7(c) shows the fuel consumption plan in this case. The fuel consumption is the sum total of each value of the PEFC with a reformer and a boiler. Figure 4.8 shows the operation results of the system at the time of the linear error and quadratic error of the NWI. Similarly, Figure 4.9 shows the results of the system operating plan optimisation analysis on representative August days (summer season). Figure 4.10 shows the operation result with the two error types on the NWI.

The battery operation plans shown in Figures 4.7(a) and 4.9(a) differ greatly for each month. Accordingly, the maximum electricity storage in August is clearly large compared with that in February. This is because of the difference in the photovoltaic power generation in February and August. Moreover, when Figure 4.7(b) is compared with Figure 4.9(b), the ratios of the PEFC exhaust heat-to-heat demand vary greatly for each month. The PEFC exhaust heat to the heat

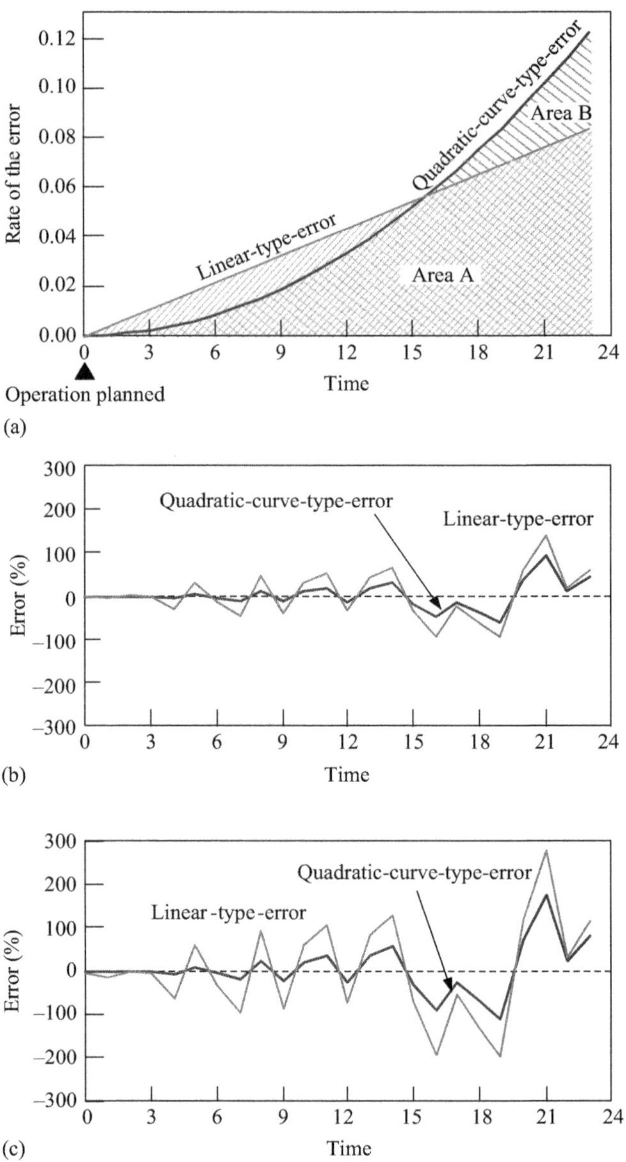

Figure 4.6   *Error function and error function with random error in the numerical weather information: (a) error function, (b) error of the numerical weather information ±20% and (c) error of the numerical weather information ±40%*

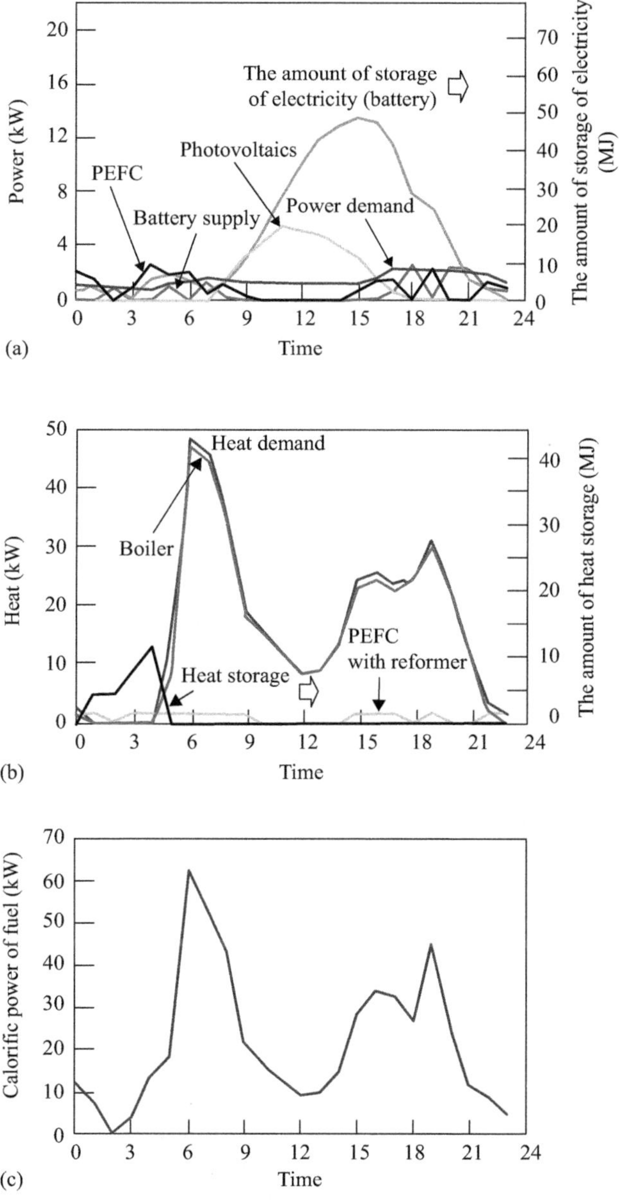

Figure 4.7   Analysis results of the proposed microgrid operation plan in February: (a) operating planning of the power system, (b) operating planning of the heat system and (c) fuel consumption

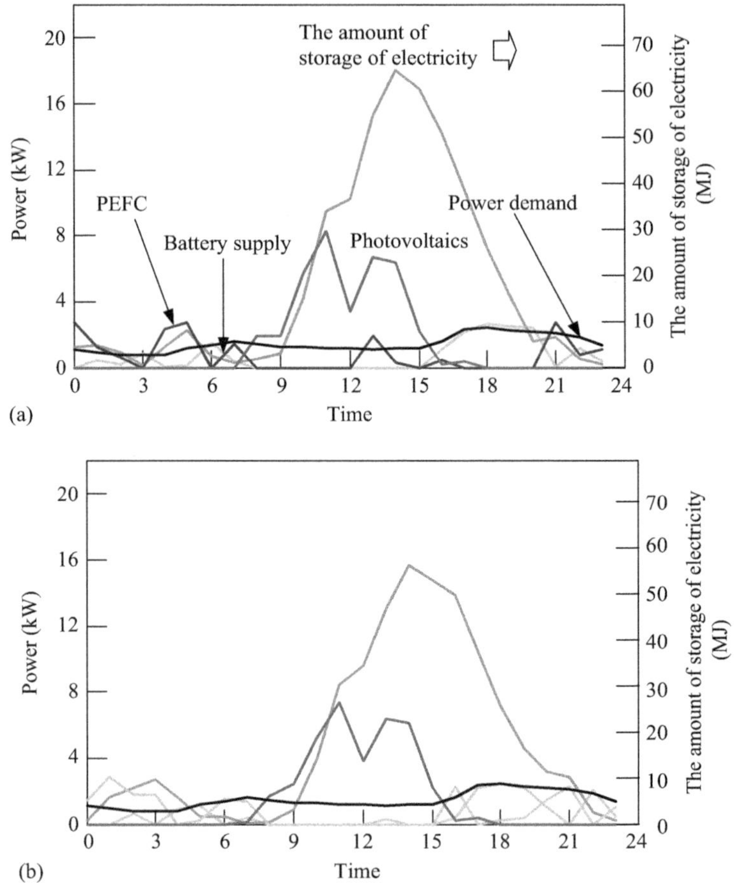

Figure 4.8   Operation planning in the case of numerical weather information with two types of error; a power system in February: (a) linear-type-error and (b) quadratic-curve-type-error

demand ratio is very low in February. As a result, heat supply on February representative days is mainly boiler heat. The summer season has little system fuel consumption based on the difference in the heat power of a boiler [Figure 4.7(c) and 4.9(c)]. Therefore, if the proposed compound microgrid is optimised on the basis of the objective function in (4.8), electric power should generally be optimised in the summer, and heat should be generally optimised in the winter.

4.2.5.2   Influence of the numerical weather information error

The relation between the NWI error and the power system operation results is investigated (Figures 4.8 and 4.10). If an error is included in the NWI, storage of

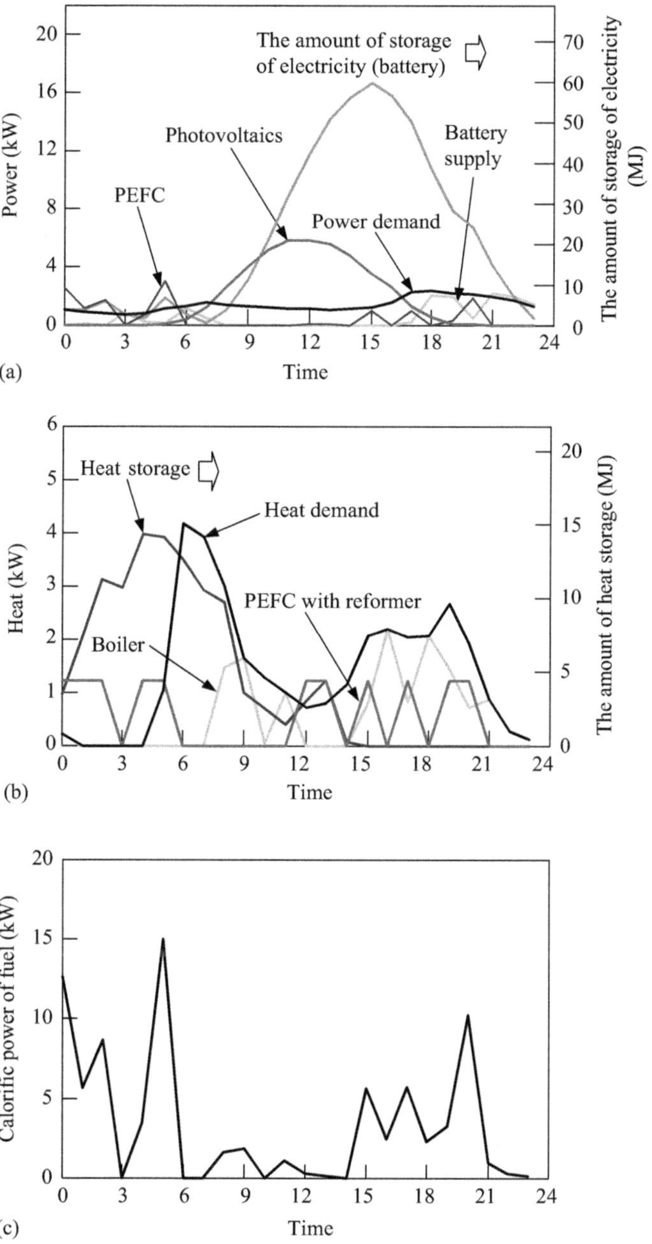

Figure 4.9   Analysis results of the operation planning of the proposed microgrid in August: (a) operation planning of the power system, (b) operation planning of the heat system and (c) fuel consumption

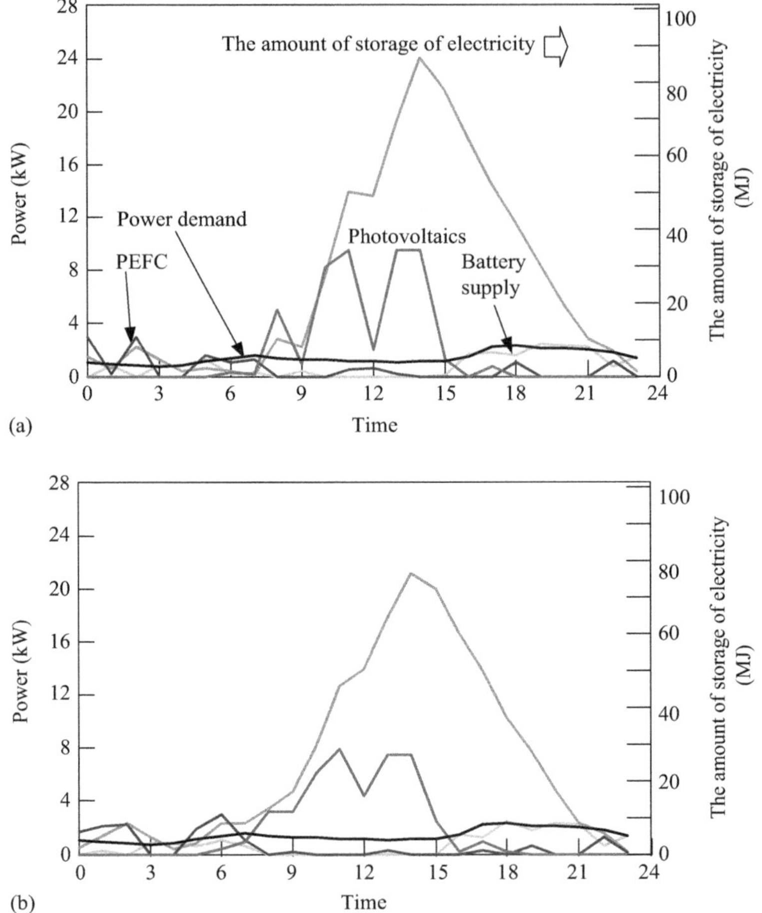

*Figure 4.10   Operation planning in the case of numerical weather information
with two error types; power system in August: (a) linear-type-error
and (b) quadratic-curve-type-error*

electricity will increase sharply for any month. From this result, the time shift of power is inferred and used to perform an important role of optimising system operation with NWI error. Accordingly, because the operation plan is strongly influenced by battery capacity setup, it is assumed that the fuel consumption of the system changes with battery capacity.

When adding the linear error to the NWI, large battery capacity is required compared with the quadratic error. Therefore, the system-operation method changes with the error characteristics of the NWI. To minimise the battery capacity, the NWI quadratic error is desirable.

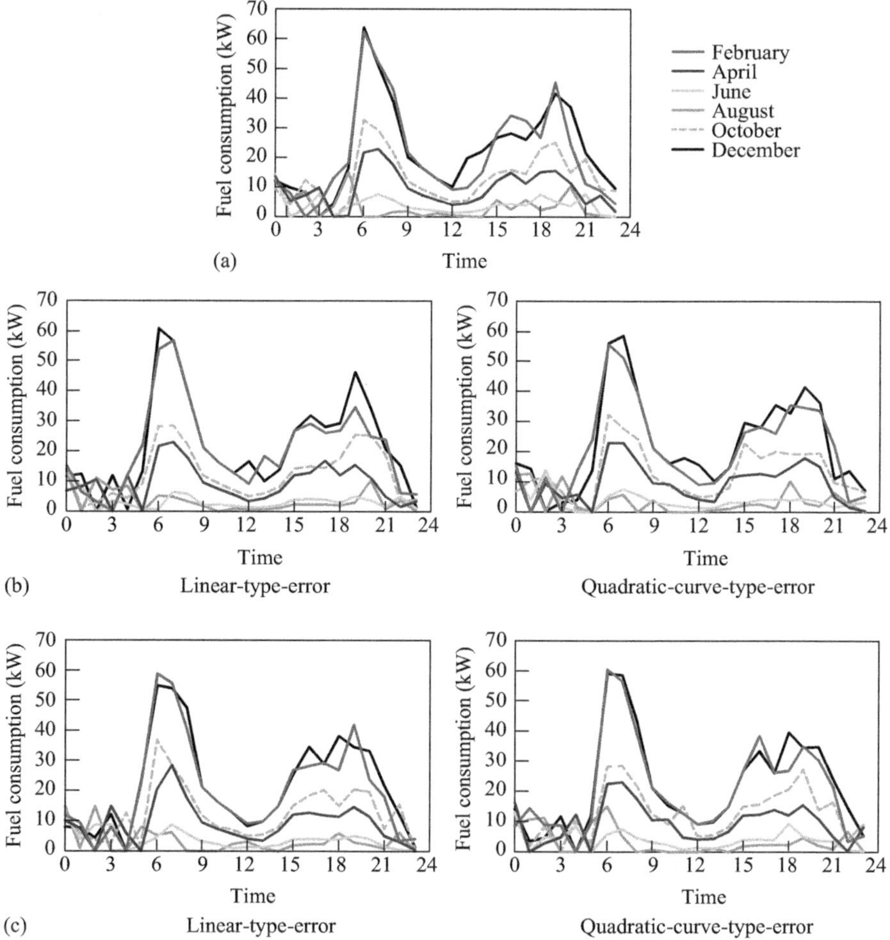

Figure 4.11   Analysis results of the fuel consumption: (a) optimal operation planning, (b) in the case of a numerical weather information with ±20% error and (c) in the case of a numerical weather information with ±40% error

4.2.5.3   Fuel consumption

Figure 4.11(a) shows the fuel-consumption plan in the case of the optimal system-operation plan. Moreover, Figure 4.11(b) and (c) shows the operation results of the fuel consumption with NWI error. The winter season (February and December) with the high heat output of the boiler requires significant fuel consumption. Moreover, when Figure 4.11(b) and (c) is compared with the fuel-consumption pattern for every month, shown in Figure 4.11(a), there is a clear difference. Accordingly, the total fuel consumption on the representative day was

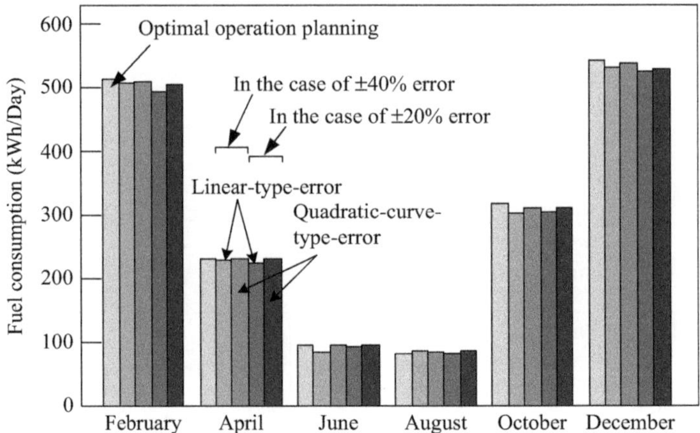

Figure 4.12   Analysis results of the fuel consumption on a representative day

calculated about every month (Figure 4.12). As shown in Figure 4.12, the results of the fuel-consumption plan with the optimal system operation plan and the fuel consumption when operating the system with NWI error were small in value. From this result, it is concluded that operation optimisation using the NWI in the fuel-cell microgrid with unstable photovoltaics achieves good operation. Even if it includes error in the NWI, the system maintains good operation and hardly suffers from the error.

4.2.6   Conclusion

In this section, the photovoltaic electricity production was predicted using NWI, and a system operation optimisation algorithm based on NWI was proposed. The proposed algorithm uses GA to optimise the system operation plan. However, since error exists between NWI and meteorological data in real time, the operation of an actual system differs from the optimal operation plan defined beforehand. Accordingly, in this section, the relation between the error characteristic of the NWI and fuel consumption of the system was investigated. Moreover, it was concluded as follows. First, when the proposed compound microgrid is installed in a cold region and optimised, power is mainly optimised in summer, and heat is primarily optimised in winter. Second, for system operation with NWI error, the power-time shift has an important role. Accordingly, the operation plan changes greatly with the magnitude of the battery capacity. As a result, system fuel consumption varies greatly from month to month. Lastly, high-performance operation can be achieved by using the operation optimisation method based on the NWI on the fuel-cell microgrid with unstable photovoltaics. Even if the error shown in this section is included in the NWI, the influence on the system fuel consumption is small.

4.3 Economic efficiency of a renewable energy-independent microgrid with energy storage using a sodium–sulphur battery or organic chemical hydride

4.3.1 Introduction

It is necessary to control unpredictable fluctuations of renewable energy to allow the development of an independent microgrid utilising renewable energy. Introduction of a battery, which can allow high-speed control of output, is effective in this regard. In the present section, two means of controlling power fluctuations are investigated: a NaS battery and energy storage using a hydrogen medium incorporating the OCHM, hereafter referred to as OCHM. NaS batteries are already produced commercially and have been proposed for controlling fluctuations of systems with large-scale renewable energy sources [12,13]. Moreover, NaS batteries have been investigated extensively to assess their possible introduction to power networks [14–16]. The OCHM technology investigated here obtains hydrogen from water electrolysis, providing electric power from renewable energy, and produces methylcyclohexane through a hydrogenation reaction of this hydrogen with toluene. Because abundant hydrogen is included, the energy density of methylcyclohexane is high. Furthermore, because methylcyclohexane is a liquid, its performance in terms of conveyance and storage is excellent. Hydrogen supplied to a fuel cell or engine generator can be obtained by dehydration of methylcyclohexane, and renewable energy generated by the OCHM does not require power transmission lines for transport by land [17–19]. Moreover, although the energy density of the OCHM is about seven times that of NaS batteries, the economic advantages of the OCHM are offset by transport costs and loss of energy from the water electrolyser, hydrogenation reactor, and dehydration reactor. Furthermore, although direct current power can be obtained directly from NaS batteries, power-generation systems must be installed to convert hydrogen obtained by dehydration of methylcyclohexane into power. Therefore, the economic advantages of addressing power fluctuations using the OCHM in an independent microgrid remain unclear.

Therefore, this study aims to establish the equipment capacity, operating methods, and costs of an independent microgrid incorporating renewable energy and energy storage using NaS batteries or the OCHM. To this end, citywide residential energy supply is investigated in Kitami City, Japan by assuming the installation of an independent microgrid comprising a large-scale solar power system and a wind farm. The operating method of the microgrid is determined by analysing the balance between supply and demand and energy storage such that the annual energy balance of the entire city may be satisfied. Based on the results of this analysis, the capacities, operating methods and costs of the large-scale solar power system, wind farm, NaS battery, OCHM system and other auxiliary equipment are obtained and compared in economic terms.

4.3.2   Proposed system

4.3.2.1   Independent microgrid

In the present study, renewable energy equipment, an electrical power grid, energy storage equipment and users are assumed to be located within approximately 100 km of a microgrid. For example, it is proposed that the wind farm must be installed in a coastal area, where wind speeds are high, and the large-scale solar power system must be installed in a suburb that receives sufficient solar radiation. Power fluctuations of the proposed microgrid are controlled by the introduction of a NaS battery or an energy storage system incorporating the OCHM. Heat for space heating and hot water can be generated by supplying a hot antifreeze solution from multiple heat pumps.

Table 4.3 presents the characteristics of NaS batteries and OCHM systems reported in previous studies; the information is obtained from NGK Insulators Ltd. [20] and examination of existing systems [21], respectively. Although the theoretical energy density of a NaS battery is greater than that of an OCHM, the actual operational energy density of an OCHM is about seven times that of a NaS battery. Moreover, the charge and discharge efficiencies of an OCHM and NaS battery are 50% and 75%, respectively. However, although the charge and discharge efficiencies of a NaS battery are based on the charge and discharge (AC output) of electric power, those of the OCHM depend on energy storage and release based on the heating value of hydrogen. Although specific details of the design life of the OCHM are unknown, it is expected that periodic replacement of the catalyst layer will be required because the dehydration reactor will be maintained at temperatures of 300 °C or more. The facilities cost for the OCHM is much higher than that for NaS batteries. However, since the operating costs of the OCHM are very small compared to those of NaS batteries, the economic efficiency of the OCHM may become advantageous for large outputs and long operating times.

Table 4.3   Specifications of the NaS battery and OCHM

		Japanese Yen (1 USD = 94 JPY)
	NaS battery	**Organic chemical hydride (methylcyclohexane)**
Energy density (Wh/kg)	784 (theoretical), 100–160 (real)	684
Efficiency of discharge and charge (%)	75 (AC)	50 (Heating value of hydrogen)[*]
Design life of equipment (year)	15 (4,500 cycle)	–
Self-discharge (%/h)	None	None
Equipment cost (JPY/kWh)	25,000	170,000
Operation cost (JPY/kW)	200,000	18.8
Application scale (kW)	2000–tens of thousands	No limit

*112,000 kg/day, transportation distance of 50 km.

Seasonal energy variations are possible with both NaS batteries and the OCHM. Therefore, the installed capacity is determined on the basis of annual energy supply and demand balances. Accordingly, surplus renewable energy generated in the summer season, when little heat is required, is stored by NaS batteries or by hydrogenation of toluene (liquid methylcyclohexane). Then, the energy stored in summer is supplied to the demand side in winter through discharge of the NaS batteries or dehydration of methylcyclohexane in order to meet the higher heat demand. The amount of renewable energy supplied can be controlled by introducing seasonal energy variations into the microgrid.

4.3.2.2 Independent microgrid with NaS batteries

Figure 4.13(a) illustrates the proposed independent microgrid system with the NaS battery. Electric power is supplied to a power transmission line from renewable energy sources (a large-scale solar power system and a wind farm) at 6.6 kV three-phase, three-wire alternating current (AC) system. The electric power supplied to the demand side is converted into 200 V AC by a transformer that is part of the power distribution network. Voltage fluctuations of $\pm20\%$ can be compensated by the NaS battery via charge and discharge cycles. The rated temperature of the NaS battery is 300 °C, and the battery is heated by an electric heater. Electric power generated by renewable methods is supplied to the NaS battery through DC/AC conversion, to the heat pump and to the power distribution network. The NaS battery plays two roles: stabilisation of the supply and demand balance and introduction of seasonal energy variations. The microgrid illustrated in Figure 4.13(a) is referred to as the NaS battery type.

4.3.2.3 Independent microgrid with energy storage by an organic hydride

When incorporating energy OCHM storage into a microgrid, the operations involved are similar to those for a storage battery [Figure 4.13(b)], and the OCHM is transported instead of power being transmitted through power lines [Figure 4.13(c)]. When there is a deficit in the supply of renewable energy, as shown in Figure 4.13(b), hydrogen obtained by dehydration of methylcyclohexane is supplied to a power generator. However, the stored methylcyclohexane is supplied to a generator installed near the power distribution network through land transportation, for example by a pipeline, as shown in Figure 4.13(c). In the case of the system illustrated in Figure 4.13(c), existing infrastructure (such as conventional tankers and gas stations) can be used. In the present study, the systems illustrated in Figure 4.13 (b) and (c) are referred to as OCHM-A and OCHM-B systems, respectively.

4.3.3 System components

4.3.3.1 Renewable energy

Wind farm
Wind generator system used for the NaS battery system
The wind generator system can be classified as a direct drive duplex feeding induction generator (DFIG), and DFIGs have been adopted extensively for use as

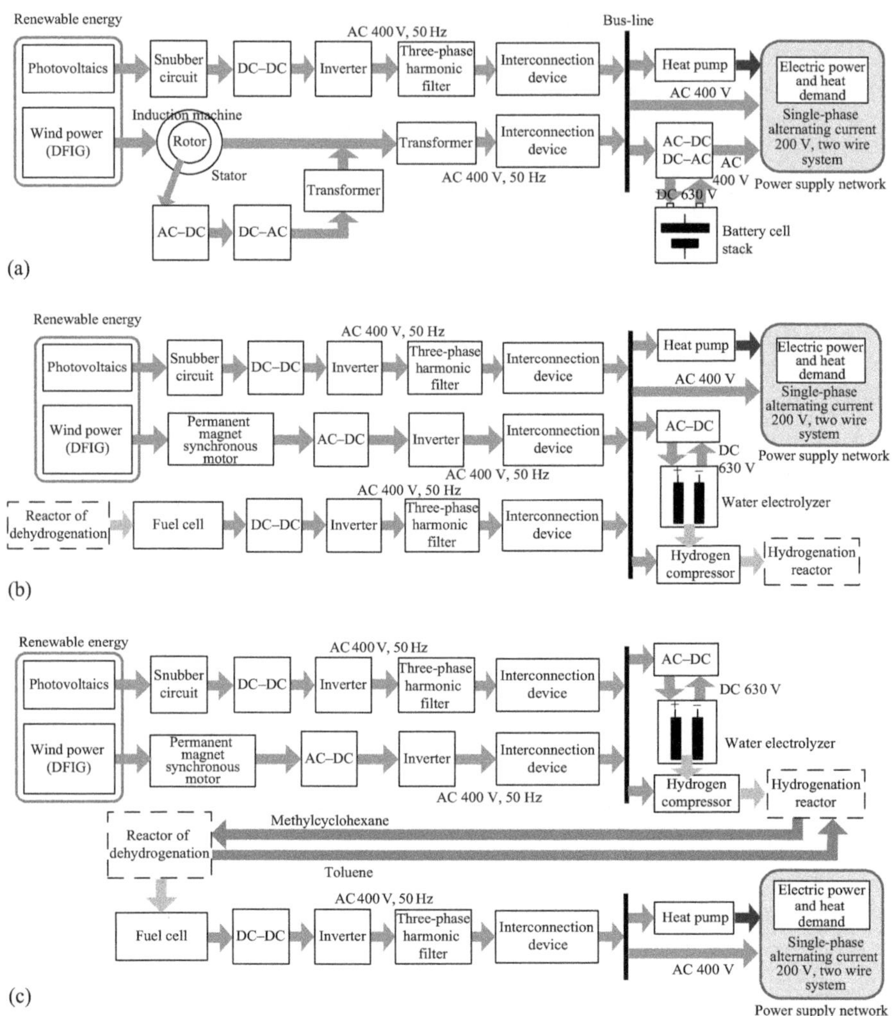

Figure 4.13   Independent microgrid with NaS battery or OCHM system: (a) NaS battery type, (b) OCHM-A type and (c) OCHM-B type

power generators in combination with large adjustable-speed wind turbines. All wind generator systems included in the wind farm installed in the NaS battery system are assumed to be of the DFIG type. Figure 4.14(a) is a block diagram illustrating the independent microgrid of the NaS battery system. The DFIG wind generator is shown in the upper part of Figure 4.14(a), and a photovoltaic is illustrated in the lower part. The wind farm and large-scale solar power system included in the NaS battery microgrid require the installation of two or more sets of the wind generator and photovoltaic systems illustrated in Figure 4.14(a).

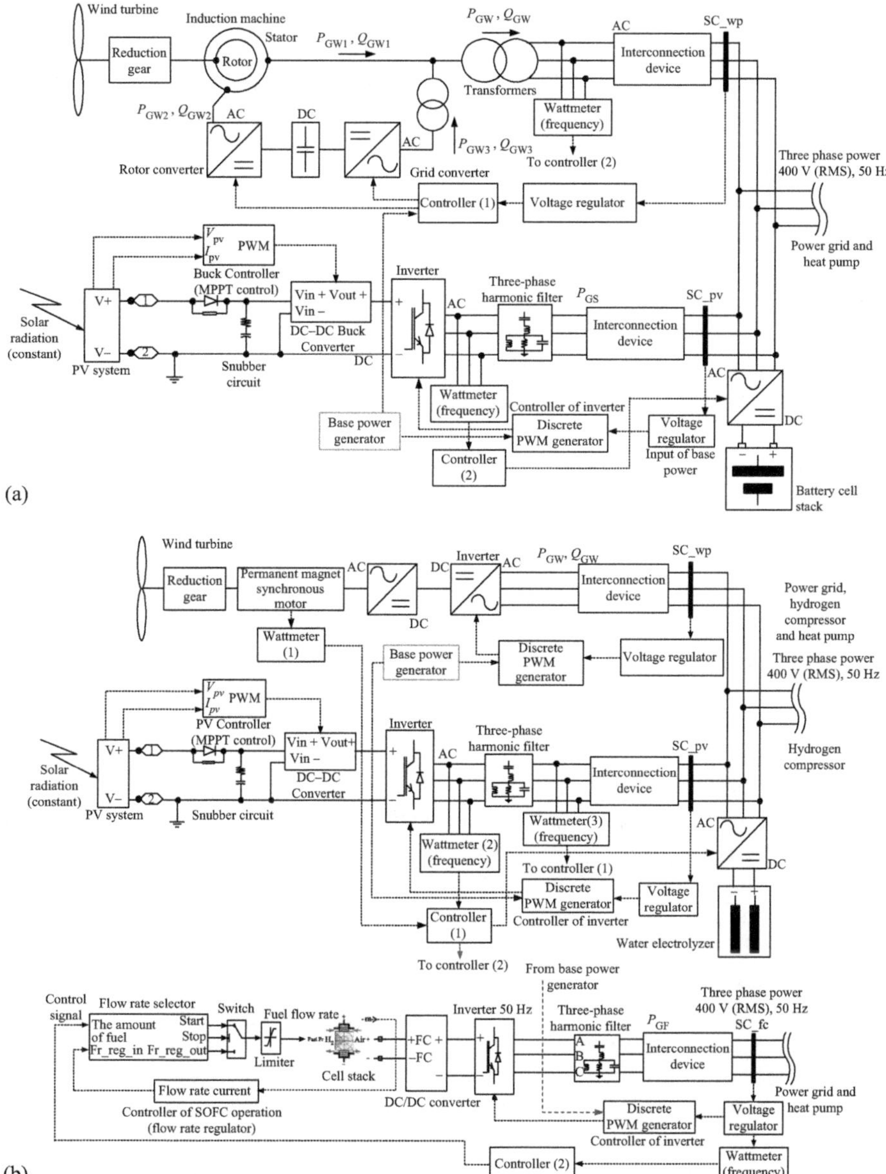

Figure 4.14   Block diagram of the proposed system: (a) NaS battery type and (b) OCHM type

In the DFIG type, the primary stator coil is connected to the electric power system, and AC with a frequency determined by the difference between the rotational frequency of the rotor and the system frequency is supplied to the secondary coil of the rotor, which is connected to the wind turbine. By magnetising this secondary coil, output generated at the system frequency can be obtained at the output of the primary stator coil. The mechanical and electrical rotor frequencies are separated, and the rotor frequency can be synchronised with the system frequency. DFIG wind generator systems have high power generation efficiency and low facilities cost because of the small capacity inverters. The effective power output from the inverter P_{GW3} is defined by (4.9). Here, s is the sliding variable calculated using (4.10), n_0 is synchronous speed and n is the engine speed of the power generator. The effective power supplied to the demand side P_{GW} is derived using (4.11). Moreover, the generated electric power is supplied to the power distribution network, NaS battery and heat pump via the bus line SC_wp in Figure 4.14(a).

$$P_{GW3} = -s \cdot P_{GW1} \tag{4.9}$$

$$s = \frac{(n_0 - n)}{n_0} \tag{4.10}$$

$$P_{GW} = P_{GW1} + P_{GW3} = (1 - s) \cdot P_{GW1} \tag{4.11}$$

Wind-generator system used for the OCHM system

For the wind farm connected to the OCHM microgrid illustrated in Figure 4.13(b) and (c), wind power is generated by a permanent magnet type synchronous generator. The OCHM system with a wind farm and large-scale solar power system supplies electric power to the power distribution network, water electrolyser, heat pump and hydrogen compressor. Figure 4.14(b) is a block diagram illustrating the microgrid of the OCHM type. Power transmitted to reduction gears from the turbine of the wind generator system is passed to the synchronous generator, as shown in the upper part of the figure, and the output of the power generator is supplied to bus line SC_wp through AC/DC, DC/AC and system interconnection equipment. Electric power is supplied to a distribution network, water electrolyser, heat pump and hydrogen compressor from bus line SC_wp in the OCHM-A system, whereas electric power is supplied to a water electrolyser and hydrogen compressor in the OCHM-B system. The measured value of the electric power generated from wind power in Figure 4.14(b) is sent from Wattmeter (1) to Controller (1), and as a high-speed command signal from Controller (1) to the AC/DC converter next to the water electrolyser, which corresponds to electric power generation. The current used for cold-water electrolysis corresponds to the difference between the electric power generated and the load (i.e. the total power consumed owing to the electricity demands of the hydrogen compressor and heat pump). Although the output of wind power generation can change sharply within tens of seconds, it has been confirmed [22] that hydrogen production by water electrolysis can cope with such fluctuations.

Large-scale solar power system

The use of photovoltaics in a large-scale solar power system requires the same equipment as that used for the NaS battery and OCHM systems, assuming that bifacial photovoltaics are used [23]. The electric power output from the solar module is supplied to bus line SC_pv through a snubber circuit (protection circuit), DC/DC converter, inverter, three-phase AC harmonic filter and system inter-connection equipment. Electric power is distributed to the NaS battery, power distribution network and heat pump from the bus line SC_pv in the NaS battery system, whereas electric power is distributed to a power distribution network, water electrolyser, hydrogen compressor and heat pump from bus line SC_pv in the OCHM-A system. Electric power is supplied to the water electrolyser and hydrogen compressor from bus line SC_pv in the OCHM-B system.

4.3.3.2 Fuel cell

The lower part of Figure 4.14(b) is a block diagram of a fuel cell introduced into the microgrid of the OCHM-A and OCHM-B systems. The fuel cell is assumed to be a solid oxide fuel cell (SOFC), and the power output is controlled by adjusting the fuel supply using an electric energy signal transmitted by Controller (2) via bus line SC_fc. Furthermore, to interconnect the fuel cell with the electric power system, it is necessary to synchronise the supply capability with the electric power system. Therefore, the synchronisation of the fuel cell and other electric power systems is adjusted by measuring the voltage of bus line SC_fc and sending a feedback signal to the inverter from the PWM generator.

4.3.3.3 NaS battery

Figure 4.15(a) outlines the NaS battery connected to the proposed system. The NaS battery must maintain all active materials in a molten state, and the ionic conduction properties of beta alumina are enhanced at high temperatures (300–350 °C). Equations (4.12) and (4.13) describe the reactions within the NaS battery. Although the anode is a two-component region, with neutral sulphur and Na_2S_5 in the initial discharge (i.e., the last stage of charge), it has only one component (Na_2S_x; $x < 5$) at advanced stages of discharge.

$$2Na + xS \leftrightarrow Na_2S_5 + (x-5)S, \quad \text{for } x > 5 \tag{4.12}$$

$$2Na + xS \leftrightarrow Na_2S_x, \quad \text{for } x < 5 \tag{4.13}$$

The theoretical specific energy of sulphur Na_2S_x is 755 W h/kg, whereas the total theoretical specific energy is 784 W h/kg or 1,005 W h/l. Because energy consumption by auxiliary equipment, loss of reaction heat and overvoltage occur, the actual specific energy is only 100–160 W h/kg. As the characteristics of the NaS battery are strongly dependent on battery temperature, a complete analysis of transient characteristics must consider the effects of battery temperature. Therefore,

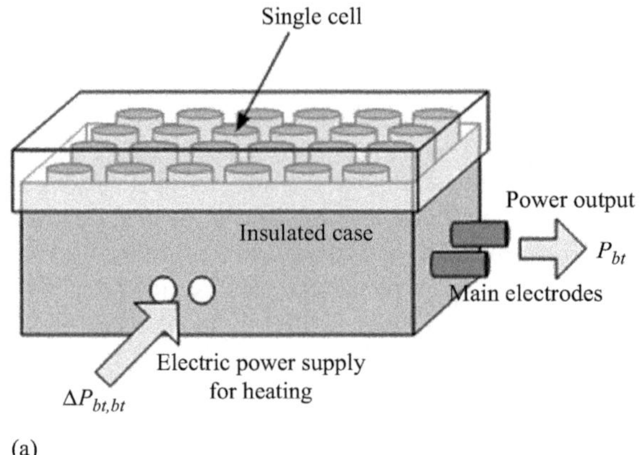

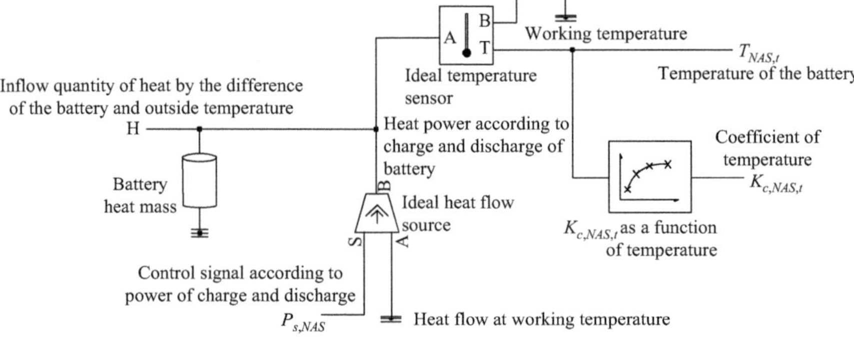

Figure 4.15   NaS battery: (a) schematic of a cell stack of a NaS battery, (b) block diagram of a NaS battery and (c) characteristics of temperature dependence of the battery (battery sell stack)

the amount of heat radiation emitted by the battery is calculated, as shown in Figure 4.15(b), from the difference between the outside air temperature and the reference temperature. Then, the operating temperature of the battery is calculated from its heat capacity. This operating temperature is controlled as illustrated in the block diagram of Figure 4.15(c) in order to modify the difference between the working temperature and battery temperature. Moreover, the charge and discharge efficiencies and energy density, which are affected by changes in battery temperature, are derived from the relationship between battery temperature and the battery temperature characteristics described above.

4.3.3.4 Energy storage in a hydrogen medium using an organic chemical hydride

Water electrolysis

Equation (4.14) describes the water electrolysis reaction; the standard enthalpy of water formation at 1 atm and 25 °C is $\Delta H° = -285.8$ kJ/mol. Typically, the energy efficiency of a water electrolyser using alkaline type water electrolysis is 70%–75%. The electric power supplied to the water electrolyser and the quantity of hydrogen produced depend on the standard enthalpy of water formation and the energy efficiency of the water electrolyser.

$$H_2O \rightarrow H_2 + 1/2O_2 - 285.8 \text{ kJ/mol} \tag{4.14}$$

Hydrogenation and dehydration reactions

Equation (4.15) describes the hydrogenation of toluene (C_7H_8) and (4.16) describes the dehydration of methylcyclohexane (C_7H_{14}). For optimum reaction temperatures and pressures, the ratios of chemical reaction of the hydrogenation and dehydration reactions are 100% and 99%, respectively. Catalysts such as Ru–CeO$_2$ are used for the hydrogenation reaction under conditions of 1 MPa and 70 °C. On the other hand, catalysts such as those containing Pt are used for the dehydration reaction under conditions of 1.1 MPa and 300–350 °C. Because the dehydration reaction requires heat supply, its energy efficiency is typically around 85%.

$$C_7H_8 + 3H_2 \rightarrow C_7H_{14} + 204.8 \text{ kJ/mol} \tag{4.15}$$

$$C_7H_{14} \rightarrow C_7H_8 + 3H_2 - 204.8 \text{ kJ/mol} \tag{4.16}$$

Water electrolyser and hydrogenation reactor

Figure 4.16 presents a block diagram of the hydrogenation reactor adopted in this study. Hydrogen is produced by supplying the water electrolyser with electric power derived from renewable energy. The resulting hydrogen is further pressurised to 1.0 MPa by a compressor. The work of the compressor is assumed to be the work required for the compression of an ideal gas and is calculated as in (4.17). φ_c in this equation represents the overall efficiency of the compressor and incorporates several components: power consumption of the inverter control equipment

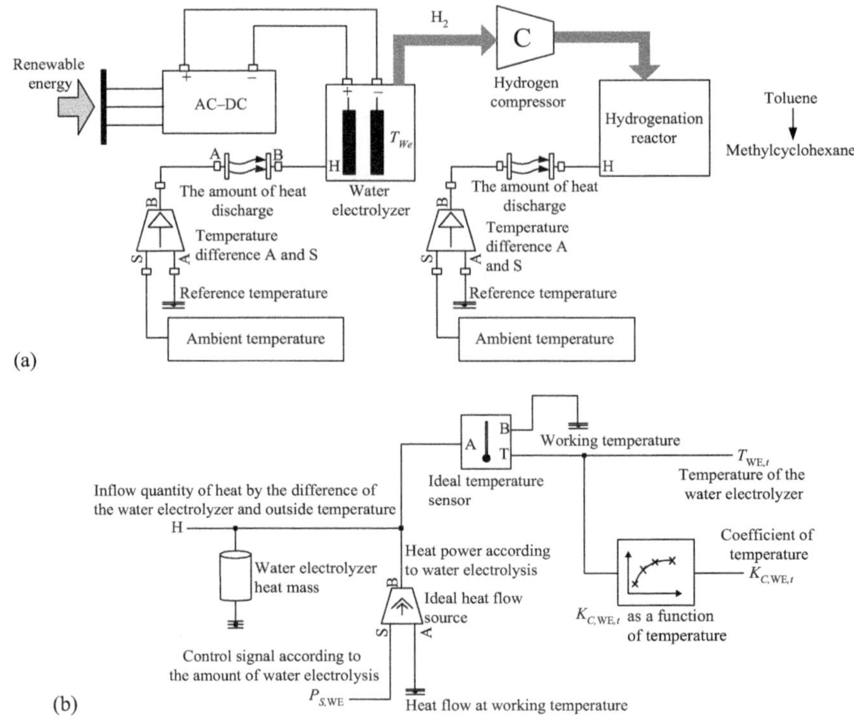

Figure 4.16   Hydrogenation system: (a) configuration of hydrogenation and (b) characteristics of temperature dependence of the water electrolyser and hydrogenation reactor

and electric motor, transfer loss of power, loss through air leaks, insufficient cooling and other mechanical loss. Moreover, $L_{c,H_2,t}$, U_∞ and C represent the work of the compressor, flow of hydrogen and pressure, respectively.

$$L_{c,H_2,t} = \frac{C_\infty \cdot U_{\infty,t} \cdot \ln\left(C_{Comp,H_2}/C_\infty\right)}{\varphi_c} \tag{4.17}$$

The performance of the water electrolyser and hydrogenation reactor depends on temperature. Therefore, the amount of heat radiated from these components is calculated as the difference between the outside air temperature and the reference temperature, as shown in the block diagram of Figure 4.16(a). Furthermore, the present operating temperature is calculated from the heat capacity of the equipment, and the reaction temperature is controlled by the difference between the operating temperature and working temperature, as shown in the block diagram of Figure 4.16(b). The reaction efficiency, which is based on the equipment temperature, is obtained with reference to the relationships between the reaction

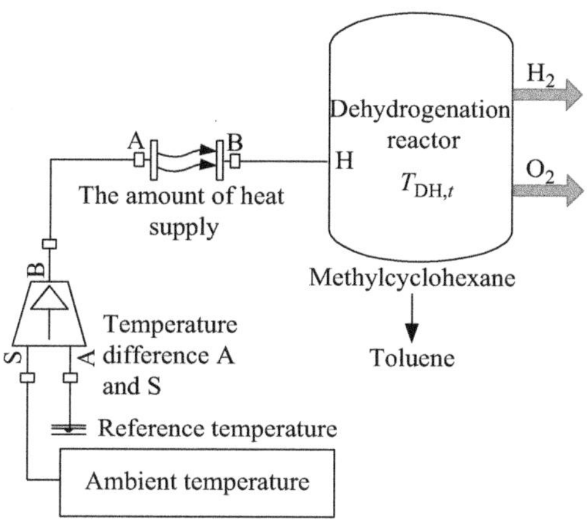

(a)

(b)

*Figure 4.17 Dehydrogenation system: (a) configuration of the dehydrogenation
system and (b) characteristics of temperature dependence of
dehydrogenation*

temperature and output characteristics of each component of the system [$K_{c,t}$ in
Figure 4.16(b)].

Dehydration reaction equipment
Figure 4.17 presents a block diagram of a reactor incorporating the dehydration
reaction described by (4.16). The rate of the dehydration reaction is dependent on
temperature. Therefore, reaction temperature must be considered when assessing
the heat capacity of the equipment shown in Figure 4.17(b), and the amount of heat
radiated owing to the difference in temperature between the outside air and the

reactor must be calculated. Furthermore, the rate of dehydration (for a given sampling time) is calculated according to the block diagram of Figure 4.17(a).

4.3.4   Control method

4.3.4.1   Microgrid of the NaS battery type

The control method for the NaS battery type microgrid is illustrated in Figures 4.13(a) and 4.14(a) and is as described below. Electric power from the DFIG wind farm [P_{GW} in Figure 4.14(a)] is supplied to bus line SC_wp. On the other hand, electric power from a large-scale solar power system [P_{GS} in Figure 4.14(a)] is supplied to bus line SC_pv and connects electric power of SC_pv and SC_wp to an electric power system. The electric power of SC_wp and SC_pv synchronises with the output power of the base power generator of Figure 4.14(a) through transmission of a control signal to the inverter control equipment of the wind farm and solar power system. Furthermore, the maximum output point control is introduced to the photovoltaics and the operating point of modular current and voltage is controlled by a common solar cell at a maximum output point.

Charge and discharge of the NaS battery are set with reference to the frequency measured by wattmeters installed at the wind farm and solar power system and are controlled by commands issued to the AC/DC converter of the NaS battery by Controller (2) in Figure 4.14(a). Because surplus electricity is generated when the measured frequency exceeds the rated value, Controller (2) commands charge operation the AC/DC converter of NaS battery. On the other hand, because the supply of electric power will be insufficient when the measured frequency is less than the rated value, Controller (2) commands discharge operation the AC/DC converter of NaS battery.

4.3.4.2   Microgrid of the OCHM type

The control method for the microgrid with energy storage by the OCHM is illustrated in Figure 4.13(b) and (c). When managing the OCHM in addition to the NaS battery [OCHM-A, Figure 4.13(b)], power generated by renewable means is supplied to an electrical power grid through a power transmission line, and the surplus power is supplied to the water electrolyser. The power supply and demand balance for the wind farm and solar power system can be determined by Controller (1) measuring the frequency at each wattmeter, as illustrated in the upper part of Figure 4.14(b). The shortfalls and excesses of electric power are assessed on the basis of this frequency, and Controller (1) commands hydrogen production through the AC/DC converter of water electrolyser when surplus power is supplied. On the other hand, when the power supply is insufficient, the controller outputs a command to increase production of electricity by the SOFC, as shown in the lower part of Figure 4.14(b).

Power generated by the wind farm and large-scale solar power system is supplied to the water electrolyser, power distribution grid, hydrogen compressor and heat pump in the OCHM-A microgrid. Moreover, bus lines SC_wp, SC_pv and SC_fc in Figure 4.14(b) are interconnected with the electric power system.

However, although the power generated by the wind farm and solar power system in the OCHM-B system is supplied to the water electrolyser and hydrogen compressor through bus lines SC_wp and SC_pv, bus line SC_fc operates independent of SC_wp and SC_pv.

Hydrogen generated by the water electrolyser is supplied to the hydrogenation reactor, which is filled with toluene, and methylcyclohexane is generated in the OCHM system. This methylcyclohexane is supplied to a catalyst layer in the dehydration reactor and desorption of hydrogen is promoted by the application of heat. The inverter is controlled by the fuel-cell system illustrated in the lower part of Figure 4.14(b), such that the frequencies of the electric power system and the power P_{GF} become synchronised. Moreover, shortfalls and excesses of electricity are assessed by Controller (2) based on the frequency of the electric power system. The supply of H_2 to the fuel cell is changed accordingly, allowing the generation of electricity power to be adjusted. Furthermore, the power of the fuel cell is stabilised by a flow-rate regulator.

4.3.5 Analysis example

4.3.5.1 Proposed system

A microgrid is installed at Kitami City in Hokkaido, Japan for this analysis. Kitami is the largest city in the Sea of Okhotsk area of Hokkaido and has a population of about 125,000 in around 61,214 residences. The installation of wind power generation equipment and a large-scale solar power system (in the coastal area and inland, respectively) is planned for Kitami City. The city experiences a maximum air temperature of 35 °C in summer and a minimum temperature of −20 °C in winter; and the temperature range is larger inland than at the coast. The annual mean air temperature in Kitami is about 6 °C, and the city experiences the lowest winter temperatures among all urban areas in Japan. However, little rainfall or snowfall occurs in the city. Figure 4.18 outlines the microgrid installed at Kitami. The detailed climate condition is released by the Japanese Meteorological Agency [24].

The NaS battery is installed at the energy supply base in the NaS battery system. Moreover, a water electrolyser, hydrogenation reactor, dehydration reactor and fuel cell are installed at the energy supply base in the OCHM-A system. The energy supply and generation bases are connected by a power transmission line in both systems. Moreover, electric power generated by renewable methods is supplied to the energy generation base (Figure 4.18), and electricity is transmitted to the energy supply base through the power transmission line. On the other hand, in the OCHM-B system, hydrogen produced at the energy generation base is incorporated into methylcyclohexane in the hydrogenation reactor. Then, methylcyclohexane is transported to the energy supply base by land. The dehydration reactor and fuel cell are installed at the energy supply base, from which electric power is supplied to the distribution network. No transmission line is installed between the energy generation and supply bases in the OCHM-B system.

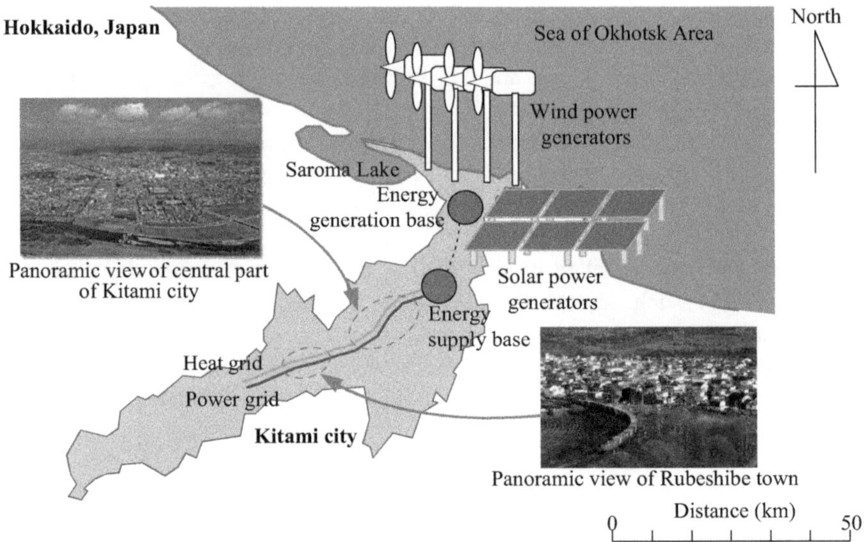

Figure 4.18   Kitami city microgrid

Table 4.4   Performance of equipment

Equipment	Efficiency
Inverter	0.95
DC–DC converter	0.97
AC–DC converter	0.97
SOFC (electric power)	0.5
SOFC (heat)	0.35
COP of heat pump	Figure 4.19
Wind power generator	0.335
Photovoltaics (bifacial solar cell)	0.213
Temperature loss of solar cell	0.4%/K
Water electrolysis	0.71
Reactor of hydrogenation	0.99
Reactor of dehydrogenation	0.85
Charge–discharge efficiency of NaS	0.75
Hydrogen compressor	0.5

4.3.5.2   Performance and cost of equipment

The performance of the equipment constituting the proposed system is shown in
Table 4.4. Figure 4.19 shows the characteristics of a heat pump for cold regions that
uses a CO_2 coolant that is currently commercially available. These characteristics
were chosen because the coefficient of performance (COP) of the heat pump is
dependent on the outside air temperature [25]. The power generation efficiency and
the efficiency of thermal power for the SOFC are maintained at constant values and

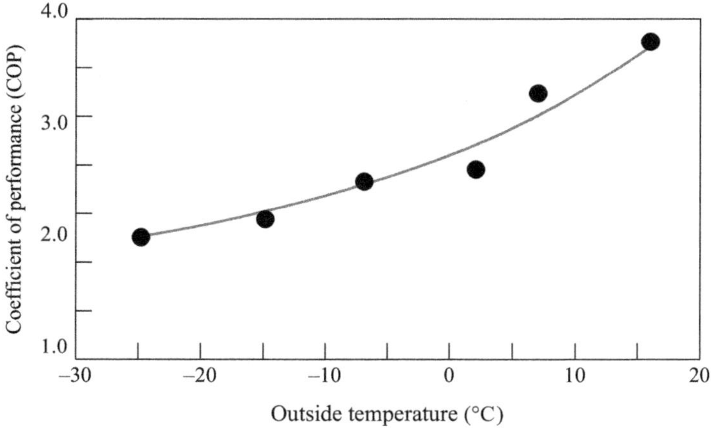

Figure 4.19   Performance of the heat pump

Table 4.5   Unit price for the setup

	Japanese Yen (1 USD = 94 JPY)
Photovoltaics	325,000 JPY/kW
Rental expense of land	150 JPY/m^2
Maintenance	15%/year
Personnel expenses	3,000,000 JPY/year
Wind turbine	210,000 JPY/kW
Maintenance	6,000 JPY/kW
SOFC	500,000 JPY/kW
Heat pump	100,000 JPY/kW
NaS battery	25,000 JPY/kW
Water electrolysis, hydrogenation, dehydrogenation equipment	4,820 JPY/(m^3/h)

are related to the performance and cost of hydrogenation, dehydration and water electrolysis [21]. The reason for using the heat-to-power ratio of fixed SOFC is that analysis becomes easy. Table 4.5 presents information relating to the cost of the equipment. Furthermore, values of Table 4.5 were inferred from present product price. Capacity of wind power generation assumed to be 2,400 kW, whereas the large-scale solar power system assumes that bifacial solar cells are used [23]. The cost of the solar and wind power generation equipment is considered in conjunction with maintenance costs obtained from records of previous operations.

4.3.5.3   Analysis method

Equations (4.18)–(4.21) describe the electric power balance of the microgrid for the NaS battery, OCHM-A and OCHM-B types of systems, respectively. Equation (4.22) is a heat balance equation common to all systems. The subscript t indicates sampling time; the left-hand sides of (4.18)–(4.22) represent energy supply terms, whereas the right-hand sides represent energy consumption. The OCHM-B type is divided into a hydrogen generation system driven by renewable energy [see (4.20)] and a power-supply system that is linked to the distribution network and heat pump by SOFCs [see (4.21)]. Although P_{GW} is the output of a wind-power generator, the wind farm consists of a set of N_{GW} generators, each with a maximum output of 2,400 kW; these wind-power generators are installed in areas where wind speed is high, that is the coastal area or the ridges of the mountains surrounding Kitami. Moreover, although P_{GS} represents the output of one double-sided solar cell, the large-scale solar power system consists of a set of N_{GS} systems. The value of P_{GS} is calculated from the power generation area, power generation efficiency (Table 4.4) and slope face solar insolation at Kitami. ΔP_{need} on the right-hand side of (4.18), (4.19) and (4.21) represents the power demand of Kitami, which is obtained from the electricity demand characteristics of average individual houses and the number of such houses present. Loss of electric power ΔP_{loss} represents the total power loss related to the equipment listed in Table 4.4. ΔP_{hp} is a power consumption term for the heat pump. The COP of the heat pump can be obtained when the outside air temperature is introduced into Figure 4.19. Then, ΔP_{hp} can be obtained by dividing the heat demand of the entire Kitami region by this COP.

The supply and demand balance of electric power with the NaS battery system [see (4.18)] is adjusted by varying the charge and discharge of P_{NaS}, that is the power of the NaS battery. The power consumption terms P_{WE} and P_{HC} are included in (4.19) and (4.20) to represent the water electrolyser and hydrogen compressor, respectively, in the OCHM systems. P_{WE} is calculated from the efficiency of the water electrolyser (Table 4.4) and the amount of power supplied, whereas P_{HC} is obtained from (4.17). When the electricity supplied by renewable energy exceeds the demand (excluding that of the water electrolyser and hydrogen compressor), as defined by the right-hand side of (4.19), the surplus power is supplied to the water electrolyser, and hydrogen is produced in the OCHM-A system. Then, methylcyclohexane is produced from toluene by supplying this hydrogen to the hydrogenation reactor, and the methylcyclohexane produced is stored in a pressure tank. When the power supply generated from renewable energy is insufficient to meet the power demands, as defined by the right-hand side of (4.19), the SOFC is operated using hydrogen obtained by dehydration of stored methylcyclohexane. All electric power derived from renewable energy (P_{GW} and P_{GS}) is given to the water electrolyser and hydrogen compressor in the OCHM-B system [see (4.20)], and methylcyclohexane is produced by hydrogen through water electrolysis and the hydrogenation of toluene. Hydrogen can subsequently be produced by supplying this methylcyclohexane to the dehydration reactor installed near the SOFC, supplying hydrogen to the SOFC to provide electric power at the demand side [see (4.21)].

The amount of renewable energy introduced into each type of microgrid is set as a parameter and the capacity of the energy storage equipment (NaS battery, water electrolyser and so on) is decided by the annual energy balance. The cost of each type of microgrid depends primarily on the unit price of equipment (Table 4.5) and equipment capacity.

Power balance of the NaS battery system

$$\sum_{l=1}^{N_{\mathrm{GW}}} P_{\mathrm{GW},l,t} + \sum_{m=1}^{N_{\mathrm{GS}}} P_{\mathrm{GS},m,t} + P_{\mathrm{NaS},t} = \sum_{n=1}^{N_{\mathrm{need}}} \Delta P_{\mathrm{need},n,t} + \sum_{o=1}^{N_{\mathrm{hp}}} \Delta P_{\mathrm{hp},o,t} + \sum_{p=1}^{N_{\mathrm{loss}}} \Delta P_{\mathrm{loss},p,t}$$

$$(4.18)$$

Power balance of the OCHM-A system

$$\sum_{l=1}^{N_{\mathrm{GW}}} P_{\mathrm{GW},l,t} + \sum_{m=1}^{N_{\mathrm{GS}}} P_{\mathrm{GS},m,t} + \sum_{q=1}^{N_{\mathrm{SOFC}}} P_{\mathrm{SOFC},q,t}$$

$$= \sum_{n=1}^{N_{\mathrm{need}}} \Delta P_{\mathrm{need},n,t} + \sum_{o=1}^{N_{\mathrm{hp}}} \Delta P_{\mathrm{hp},o,t} + \sum_{r=1}^{N_{\mathrm{WE}}} \Delta P_{\mathrm{WE},r,t} + \sum_{s=1}^{N_{\mathrm{HC}}} \Delta P_{\mathrm{HC},s,t} + \sum_{p=1}^{N_{\mathrm{loss}}} \Delta P_{\mathrm{loss},p,t}$$

$$(4.19)$$

Power balance of the OCHM-B system

$$\sum_{l=1}^{N_{\mathrm{GW}}} P_{\mathrm{GW},l,t} + \sum_{m=1}^{N_{\mathrm{GS}}} P_{\mathrm{GS},m,t} = \sum_{r=1}^{N_{\mathrm{WE}}} \Delta P_{\mathrm{WE},r,t} + \sum_{s=1}^{N_{\mathrm{HC}}} \Delta P_{\mathrm{HC},s,t} + \sum_{p=1}^{N_{\mathrm{loss}}} \Delta P_{\mathrm{loss},p,t} \qquad (4.20)$$

$$\sum_{q=1}^{N_{\mathrm{SOFC}}} P_{\mathrm{SOFC},q,t} = \sum_{n=1}^{N_{\mathrm{need}}} \Delta P_{\mathrm{need},n,t} + \sum_{o=1}^{N_{\mathrm{hp}}} \Delta P_{\mathrm{hp},o,t} + \sum_{p=1}^{N_{\mathrm{loss}}} \Delta P_{\mathrm{loss},p,t} \qquad (4.21)$$

Heat balance of all types of microgrid

$$\sum_{o=1}^{N_{\mathrm{hp}}} H_{\mathrm{hp},o,t} = \sum_{n=1}^{N_{\mathrm{need}}} \Delta H_{\mathrm{need},n} + \Delta H_{\mathrm{loss},t} \qquad (4.22)$$

4.3.6 Results and discussion

4.3.6.1 Operational results for the NaS battery

Figure 4.20 illustrates the results of the operational analysis of the microgrid incorporating a NaS battery. Figure 4.20(a) presents the power consumption [right-hand side of (4.18)] for a representative day and the monthly electricity supply derived from renewable sources, whereas Figure 4.20(b) presents the operational results for the NaS battery. The capacity of the NaS battery is set to balance the annual supply and demand of the system, as described by (4.18). In addition, much charging (discharging) occurs from March to November (December to February) owing to the power consumption of the heat pump.

4.3.6.2 Installed capacity of the OCHM type microgrid

Figure 4.21(a) and (b) illustrates the supply capability (including that of the wind farm, large-scale solar power system, and SOFC) of the microgrid for the OCHM-A and OCHM-B systems and the power consumed by the heat pump. The labels A1–A9

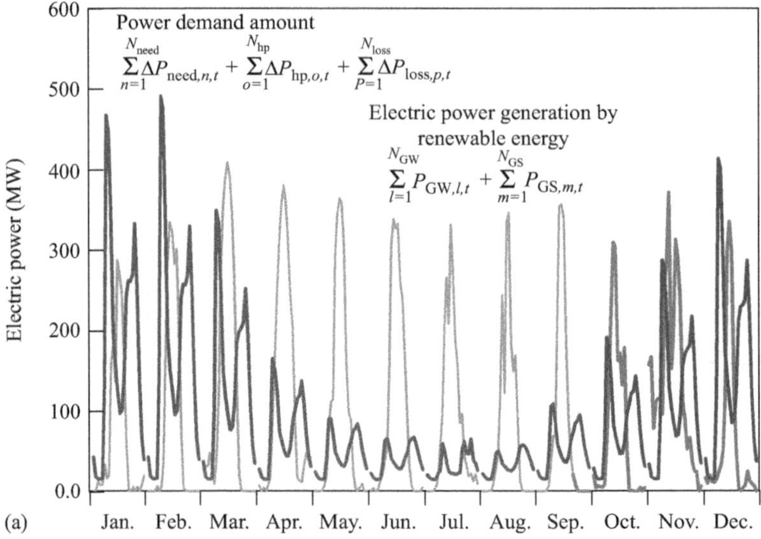

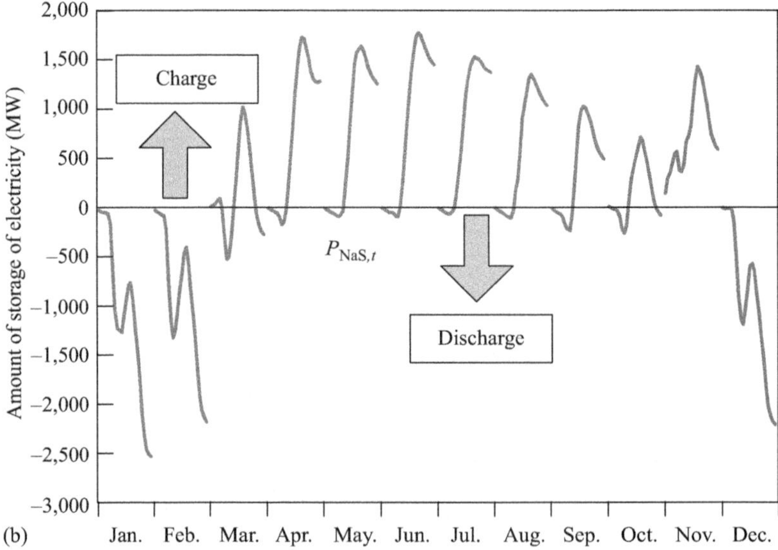

*Figure 4.20   Operational analysis results for the NaS battery system: (a) analysis
results for electric power supply and demand and (b) operational
analysis results for charge and discharge of the NaS battery*

represent combinations of the power generation area of bifacial solar cells and
the number of 2400-kW wind power generators installed. To satisfy the energy
balance equations [see (4.19–4.3.14)], the power generated by photovoltaics will
increase as that generated by wind power decreases. However, possible increases in

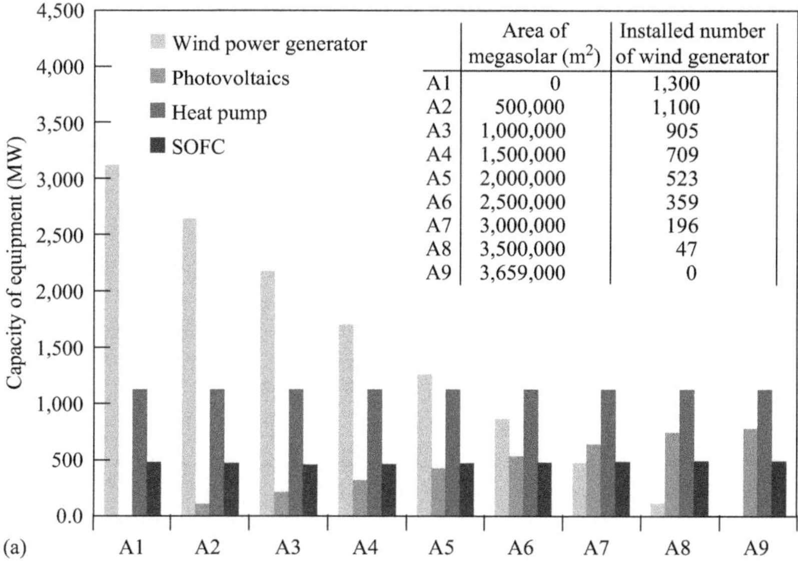

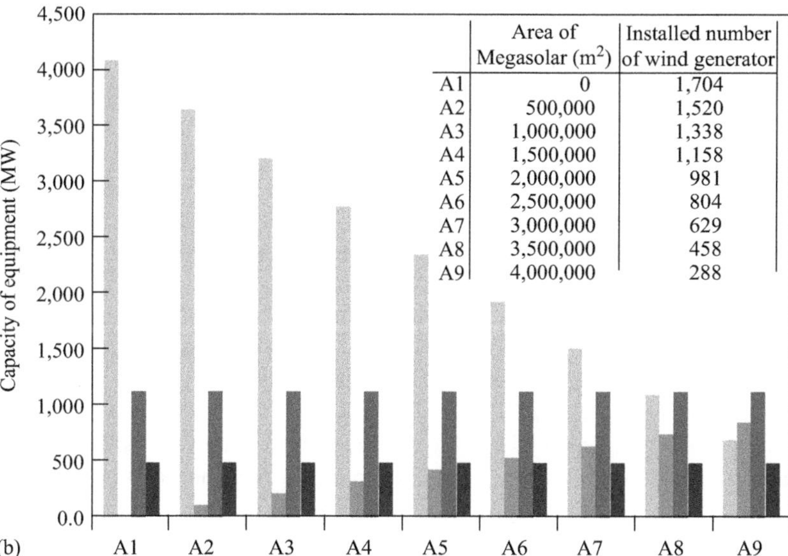

Figure 4.21   Capacity analysis results for the OCHM systems: (a) OCHM-A system and (b) OCHM-B system

the capacity of photovoltaics are small compared to the possible reduction of wind power generation, as shown in Figure 4.21(a) and (b). Moreover, wind speed is low in the region where the wind farm is located, meaning that production of electricity does not increase with increasing capacity for wind power generation. Therefore, the introductory capacity of the SOFC is not related to the amount of renewable energy introduced from both sources.

4.3.6.3 Equipment cost

The cost of each microgrid investigated in this study represents the sum total of equipment and maintenance costs for a 20-year period. The results of the cost analysis for all types of microgrids are shown in Figure 4.22. However, the facilities cost for the power transmission and distribution lines were not included in the microgrids. Moreover, the cost of land transportation was not considered for the OCHM-B system.

The combinations of renewable energy types (A1–A9) shown in Figure 4.22(b) and (c) for the OCHM system are the same as those illustrated in Figure 4.21(a) and (b), respectively. Similarly, the combinations for the NaS battery system (A1–A7) are shown in Figure 4.22(a). A reduction in the amount of wind power generated results in a lower facilities cost, which decreases as more photovoltaics are introduced in the cost analysis for the NaS battery system [Figure 4.22(a)]. Wind conditions are poor at the introductory point of the wind farm, and there are more annual productions of electricity from large-scale solar power systems than wind farm. The relationship between the energy combination and the facilities cost exhibits a similar trend for both the OCHM and NaS battery systems.

Figure 4.22(b) and (c) presents the facilities cost of the microgrid for the OCHM-A and OCHM-B systems, respectively. For the microgrid of the OCHM-A system, equipment cost comprises 74% of the total cost, with maintenance and personnel costs account for the remainder. Similarly, equipment cost represents 79% of the facilities cost for the OCHM-B system. The equipment cost for hydrogenation and dehydration and the cost of the water electrolyser are comparatively small. Energy combination A5 represents the minimum cost option for the microgrid of the OCHM-A system, whereas option A6 represents the minimum cost for the OCHM-B system. Moreover, with energy combination A5 for OCHM-A, the cost of the system is 12% less than that with energy combination A6 for OCHM-B. The microgrid of the OCHM-A system has cost advantage over the OCHM-B system. This is because there are many losses in hydrogen production and energy storage by organic hydride. Furthermore, the OCHM-A system has few quantities to be stored such as the organic hydride for the OCHM-B system. The efficiencies of dehydration of the OCHM and hydrogen production are 85% and 71%, respectively, as shown in Table 4.4. Therefore, the installed capacity of renewable energy and the equipment costs of the entire system will increase when large amounts of energy are stored using the OCHM. Moreover, the exhaust heat of the SOFC can be considered part of the heat demand for the microgrid of the OCHM-A system.

Figure 4.23 illustrates the results of cost analysis for a 20-year period for both the NaS battery and OCHM systems. The NaS battery system costs less than the OCHM systems for the conditions analysed in the present study. Therefore, an energy storage system incorporating the OCHM can be considered economically disadvantageous. The minimum equipment cost of the NaS battery system is 16%

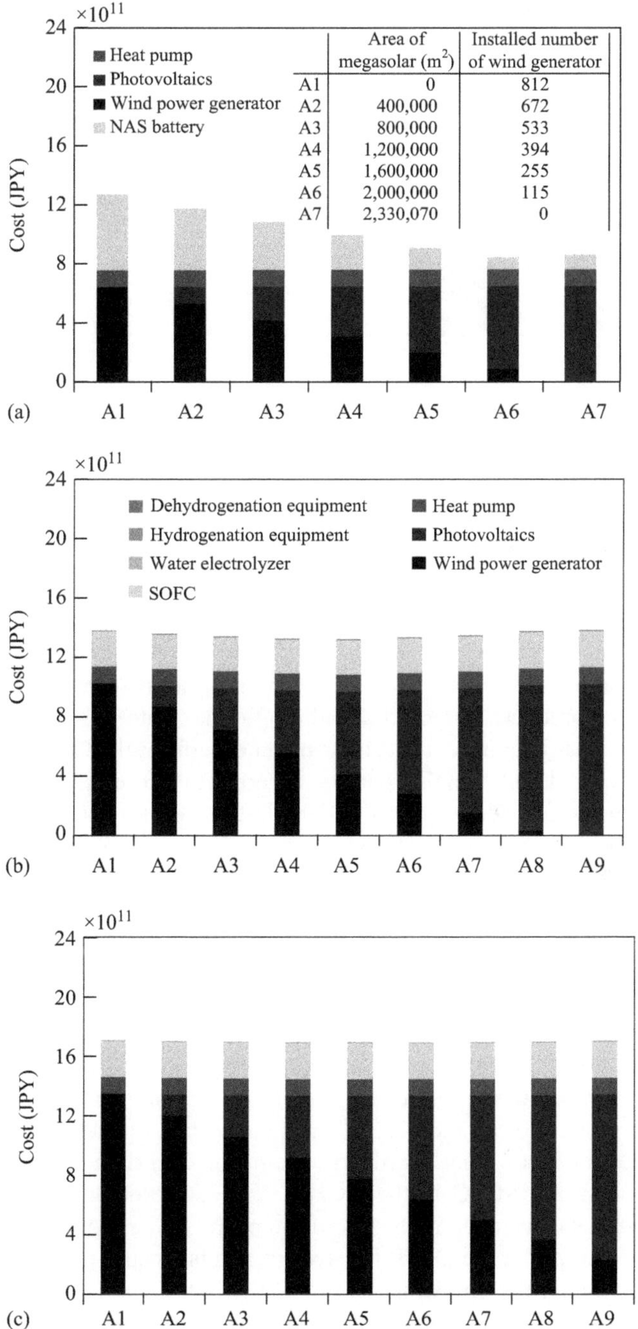

Figure 4.22   Cost analysis results for each system for a 20-year period: (a) NaS battery system, (b) OCHM-A system and (c) OCHM-B system

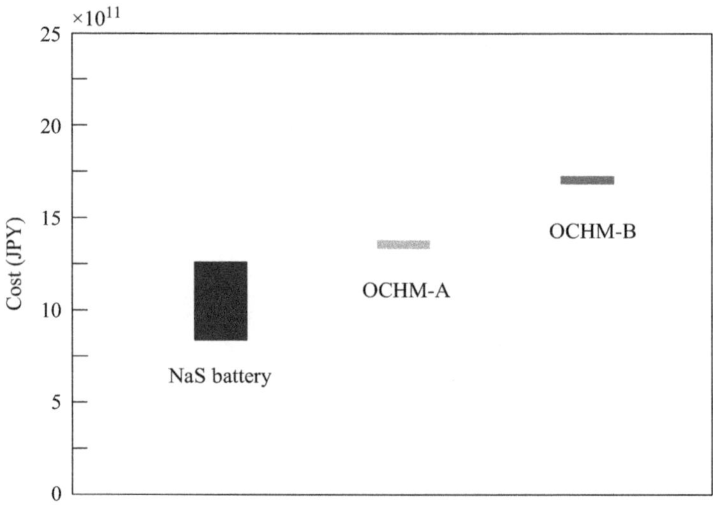

*Figure 4.23   Analysis results for operation and equipment cost for a 20-year
period*

cheaper than that of the OCHM-A system. The microgrid with energy storage by
organic hydride produces large losses through the water electrolyser and the
dehydration reactor. Therefore, the renewable energy from the OCHM-A system is
large scale compared with the NaS battery system. Therefore, improvements in the
efficiency of the water electrolyser and dehydration reactor must be achieved for
implementation of a microgrid of OCHM type.

4.3.6.4   Operational results for OCHM systems

Figure 4.24 illustrates the results of operational analysis of a representative day for
each month in order to assess minimum equipment cost [A5 and A6 in Figure 4.21
(a) and (b), respectively] for the OCHM-A and OCHM-B systems. As wind velocity is high along the coast of Kitami in November, the wind farm can produce
considerable amounts of electricity at this time [Figure 4.24(a) and (b)]. The
amount of energy stored by a hydrogen medium incorporating OCHM is greater for
the OCHM-B system for all months. When the electricity generated by renewable
energy is insufficient to meet the energy demand, electricity is supplied to the
demand side from the SOFC in the OCHM-A system microgrid. Therefore, the
capacity of the SOFC installed in the microgrid of the OCHM-A system is expected
to be smaller than that of the OCHM-B system and is usually generated according
to the load. However, the operating pattern of the SOFC in Figure 4.15(a) and (b)
barely changed so that there is little difference in the capacity of the SOFC between
the OCHM-A and OCHM-B systems.

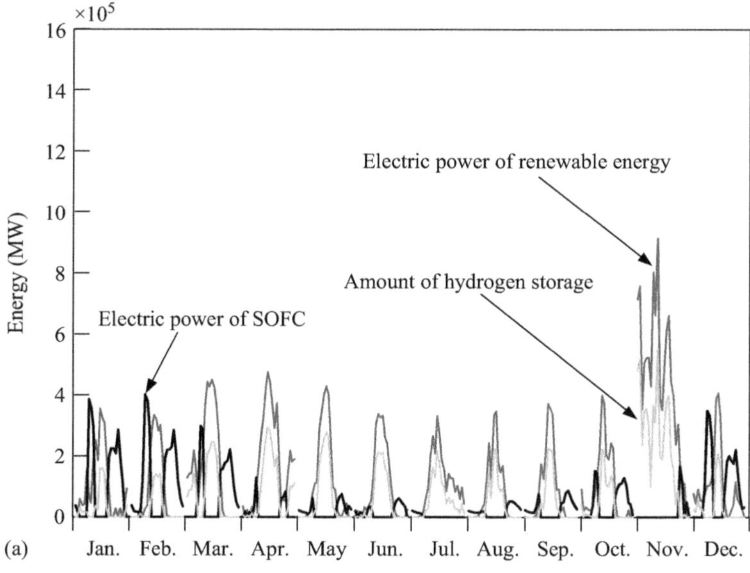

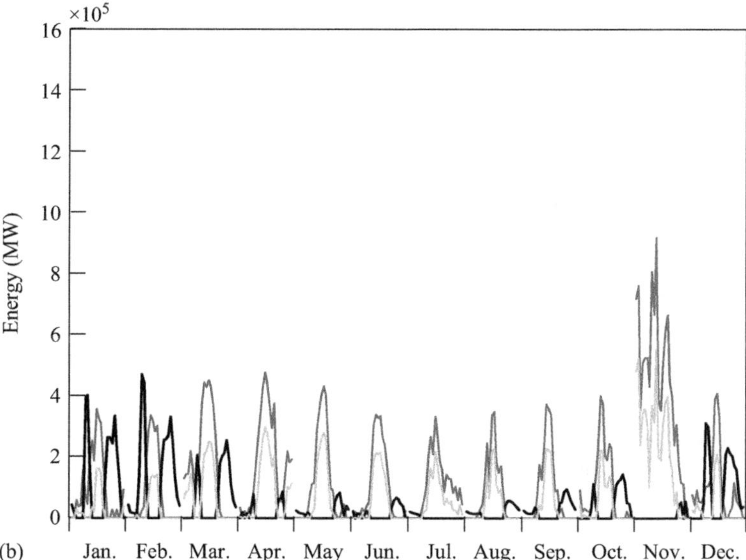

Figure 4.24   Operational analysis results for the OCHM-type system: (a) OCHM-A system for A5 and (b) OCHM-B system for A6

4.3.7   Conclusions

Change in the output of renewable energy considerably influences the quality of electric power in independent microgrids designed to realise local supply and consumption of energy. Accordingly, operating methods and costs were analysed for a microgrid incorporating two types of energy storage equipment: a sodium–sulphur (NaS) battery, and a hydrogen medium incorporating the OCHM. The installed capacity and cost were calculated, assuming an independent microgrid in Kitami City, a cold region in Japan. The OCHM-A system was designed to simulate energy storage similar to that of a battery, whereas OCHM-B was simulated to substitute power transmission or transportation of the OCHM. The conclusions of the investigation can be summarised as follows.

1. The facilities cost of the microgrid decreased with reduction of the amount of wind power generated and an associated increase in the amount of photovoltaics introduced. For this reason, storage capacity is determined according to the maximum output of the wind farm. Furthermore, wind conditions at the introductory point were found to be poor, suggesting that the annual production of electricity by the large-scale solar power system should be large.
2. Equipment costs for the OCHM-A and OCHM-B systems represented 74% and 79% of their respective facilities costs (where the cost of the electrical power grid is not included). The costs of the water electrolyser, hydrogenation reactor and dehydration reactors are extremely small relative to the cost of generating the energy. In addition, the microgrid for the OCHM-A system was found to be 12% cheaper than that for the OCHM-B system.
3. Because energy storage by the OCHM is inefficient owing to the inefficiency of the water electrolyser and dehydration reactor, an OCHM system requires renewable energy to be produced on a much larger scale compared with a NaS battery system. Therefore, the cost of the system tends to increase with reduction in the installed capacity required to make the OCHM system viable. Improvements in the efficiency of the water electrolyser and dehydration reactor are important for realisation of OCHM-type microgrids.
4. The capacity of the SOFC introduced into the microgrids of the OCHM-A and OCHM-B systems is unrelated to the energy combination adopted.

Nomenclature

H	heat (W)
I_D	direct solar radiation intensity (W/m^2)
I_H	global-solar-radiation intensity (W/m^2)
I_M	horizontal sky solar radiation intensity (W/m^2)
H_D	intensity of direct solar (W/m^2)
H_M	intensity of sky solar radiation (W/m^2)
P	power (W)
Q	fuel consumption (W)
R_T	temperature coefficient (%/K)

S_s	area of the solar cell (m^2)
T	temperature (K)
T_c	temperature of the solar cell (K)
T_o	reference temperature (K)
t	sample time
AC	alternating current
C_{Comp,H_2}	compression pressure of hydrogen (MPa)
C_∞	atmospheric pressure (MPa)
DC	direct current
DH	dehydrogenation equipment
H	heat (kW h)
ΔH	consumption of heat power (kW)
I	current (A)
$K_{c,t}$	temperature coefficient
L_{c,H_2}	work of hydrogen compressor (Nm)
N	number
NaS	sodium–sulphur battery
n	number of revolutions of power generator (rpm)
n_o	synchronous speed (rpm)
OCHM	organic chemical hydride
P	effective electric power (kW)
ΔP	consumption of electric power (kW)
P_s	signal of pressure gauge
PWM	pulse width modulation
Q	reactive power (var)
s	sliding
SC	bus line
SOFC	solid oxide fuel cell
T	temperature (K)
t	sampling time (h)
U_∞	flow of hydrogen (m^3/s)
V	voltage (V)

Greek symbols

β	angle of the acceptance surface gradient
δ	the solar celestial declination
φ	the latitude of a setting point
η_s	photovoltaics efficiency at T_o (%)
θ	incident angle to the sunlight acceptance surface

ρ	ground reflection factor
ω	hour angle
φ_c	overall efficiency of compressor

Subscripts

bl	boiler
bt	battery
btc	battery discharge
fc	PEFC with reformer
loss	energy loss
need	energy demand
pv	photovoltaics
s	solar module
st	heat storage
sts	heat-storage output
fc	fuel cell
GS	photovoltaics
GW	wind power generator
HC	hydrogen compressor
hp	heat pump
loss	loss
need	demand
pv	photovoltaics
WE	water electrolysis
wp	wind power generation

References

[1]   S. Obara, "Equipment plan of compound interconnection micro-grid composed from diesel power plants and solid polymer membrane-type fuel cell", *International Journal of Hydrogen Energy*, 2008;33(1):179–188.

[2]   H. Aki, S. Yamamoto, J. Kondoh, *et al.*, "Fuel cells and energy networks of electricity, heat, and hydrogen in residential areas", *International Journal of Hydrogen Energy*, 2006;31(8):967–980.

[3]   H. Jiayi, J. Chuanwen and X. Rong, "A review on distributed energy resources and MicroGrid", *Renewable and Sustainable Energy Reviews*, 2008;12(9):2472–2483.

[4]   S. Obara, "Load response characteristics of a fuel cell micro-grid with control of number of units", *International Journal of Hydrogen Energy*, 2006;31(13):1819–1830.

[5] S. Obara, "Operating schedule of a combined energy network system with fuel cell", *International Journal of Energy Research*, 2006;30(13):1055–1073.

[6] S. Obara, "Equipment arrangement planning of a fuel cell energy network optimized for cost minimization", *Renewable Energy*, 2007;32(3):382–406.

[7] S. Obara, "The exhaust heat use plan when connecting solar modules to a fuel cell energy network", *Transactions of the ASME, Journal of Energy Resources Technology*, 2007;129(1):18–28.

[8] Online Data Service, GPV/GSM (Grid Point Value/GSM (Global Spectral Model)), http://www.jmbsc.or.jp/hp/online/f-online0a.html, Japan meteorological business support center, 2009.

[9] Data of Japan Meteorological Agency, http://database.rish.kyoto-u.ac.jp/arch/jmadata/gpv-original.html, Kyoto University, 2009.

[10] Development of Home Fuel Cell Cogeneration System in Tokyo Gas, ALIA News Vol. 89, http://www.alianet.org/homedock/15kinen/5-2.html, 2009.

[11] K. Narita, "Research on unused energy of cold region cities and utilization for district heat and cooling", Ph.D. Thesis, Department of Socio-Environmental Engineering, Faculty of Engineering, Hokkaido University, Sapporo, Japan, 1996.

[12] K. M. June and H. L. Chang, "Numerical study on the thermal management system of a molten sodium–sulphur battery module", *Journal of Power Sources*, 2012;210:101–109.

[13] W. Zhaoyin, C. Jiadi, G. Zhonghu, C. Xiaohe, Z. Fuli and L. Zuxiang, "Research on sodium sulphur battery for energy storage", *Solid State Ionics*, 2008;179:1697–1701.

[14] W. P. Cheol, S. R. Ho, W. K. Ki, H. A. Jou, Y. L. Jai and J. A. Hy, "Discharge properties of all-solid sodium–sulphur battery using poly (ethylene oxide) electrolyte", *Journal of Power Sources*, 2007;165:450–454.

[15] A. Sone, T. Kato, T. Shimakage and Y. Suzuoki, "Influence of forecast accuracy of photovoltaic power output on capacity optimization of microgrid composition under 30-minute power balancing control", *Electrical Engineering in Japan*, 2013;182:20–29.

[16] J. R. Carl and A. S. Björn, "Energy analysis of batteries in photovoltaic systems. Part I: Performance and energy requirements", *Energy Conversion and Management*, 2005;46:1957–1979.

[17] Y. Okada, E. Sasaki, E. Watanabe, S. Hyodo and H. Nishijima, "Development of dehydrogenation catalyst for hydrogen generation in organic chemical hydride method", *International Journal of Hydrogen Energy*, 2006;31:1348–1356.

[18] B. B. Rajesh, S. Rayalu, S. Devotta, and M. Ichikawa, "Chemical hydrides: A solution to high capacity hydrogen storage and supply", *International Journal of Hydrogen Energy*, 2008;33:360–365.

[19] S. Anshu, K. Shilpi and B. B. Rajesh, "Hydrogen delivery through liquid organic hydrides: Considerations for a potential technology", *International Journal of Hydrogen Energy*, 2012;37:3719–3726.

[20] NGK Insulators, Ltd., http://www.ngk.co.jp/english/products/power/nas/index.html, 2012.

[21] Hrein Energy Inc., http://www.hrein.jp/english/index.htm, 2012.

[22] Section 5.3 Introductory concept for hydrogen supply base by green energy, Kumejima-cho new energy vision, Kumejima town office, http://www.town.kumejima.okinawa.jp/industry/new_enevision.html, 2013 (in Japanese).

[23] PVG Solutions Inc., http://www.pvgs.jp/files/ja/PVGS_Catalogue_Oct2011.pdf, 2013.

[24] Japanese Meteorological Agency, http://www.jma.go.jp/jma/index.html, 2013.

[25] Mitsubishi heavy industries, The development of the CO_2 heat pump hot water supply unit for business use that can be operated down to -25 degrees Celsius of outside air temperature, Technical report of Mitsubishi heavy industries, Ltd., 2011;48(4):86–88.

Chapter 5

Reliability and power quality

Jorge Morel

5.1 Introduction

5.1.1 Overview

The Fukushima Daiichi nuclear disaster of 2011 contributed to establish an increasing awareness for the need of a safer energy supply system for Japan and other countries mainly supplied by nuclear generation. In the case of Japan, the government is now increasing its effort in the promotion of non-CO_2 emitting renewable generation, such as solar and wind power, as well as in the adoption of a distributed generation architecture for specific regions in the country. This new strategy will also make the system less vulnerable to the effects of shutdowns of the conventional centralised systems in the case of natural disasters and other large disturbances [1,2].

Microgrids, which can be considered the building blocks for the smart grid, have, as one of their most important roles, to improve the reliability and power quality of the entire system in the presence of renewable-based power generation or large disturbances. The inherently variable and uncertain power generated by renewable generators in microgrids and smart grids, together with the possible flow of current in two directions in a grid line, do not exist in the traditional centralised power systems. This can affect the quality of the electricity, at the frequency, voltage and stability levels, constituting challenges to be solved in the development of smart grids and microgrids.

Even though the challenges are large, the improvement in reliability, power quality and resilience, especially under conditions of catastrophic event or shut downs in a main power supply, will represent a revolutionary advance for the power grid industry.

5.1.2 Chapter's aim and scope

This chapter aims to present different aspects of reliability and power quality in microgrids from a perspective of its interconnection with a larger system, such a smart grid or a centralised power system.

After definitions and key concepts are presented, a case study is included in order to highlight how vulnerable a microgrid can be under disturbances, and the important role of control and storage systems to keep the grid stable and working properly.

5.2   Power quality

5.2.1   *What is power quality?*

The quality of the electricity supplied to customers in a power system has been one of the most important fields of study in the power industry since its beginning, several decades ago, because low power-quality has the potential to negatively affect the operation and life of electric equipment forming the power grid, for example power transformers and customer's devices.

Today good electricity quality continues to be a very important requirement to be satisfied for operators and customers, not only in centralised power grids but also in modern smart grids and their building blocks: the microgrids.

According to Fuchs and Masoum [3] there is no generally accepted definition of power quality, but the term has always referred to the quality of voltage and current based on arbitrary tolerance levels. What it is agreed is that electricity quality has a key impact on three of the most important characteristics of a power grid: reliability, security and efficiency. A definition can be given as 'degree to which near sinusoidal waveforms of bus voltages and currents are maintained at rated magnitude and frequency'.

The presence of variable output renewable generation and the large amounts of power converters used for interconnection can have a negative impact on power quality of smart grids if proper countermeasures are not implemented.

The different types of disturbance affecting power quality in a given electric grid can be classified into seven categories. The concepts presented here regarding disturbances give an overview of the different types of disturbances presented in [3].

1.   Transient events, which are fast and short-duration events that produce distortion and can be impulsive or oscillatory.
2.   Short-duration voltage variations, which are classified as interruption, sag and swell. According to the duration, each of them can be (a) instantaneous (from 0.5 to 30 cycles), (b) momentary (from cycles to 3 s) or (c) temporary (from 3 s to 1 min).
3.   Long-duration voltage variation, which is defined as the deviation of the RMS voltage from nominal value for more than 1 min. It can be classified as sustained interruption, under-voltage or over-voltage.
4.   Voltage imbalance, defined as the situation when the voltages of a three-phase system are different in magnitude or the phases angles are not 120 degrees.
5.   Waveform distortion, defined as the steady-state deviation of the waveform from a sine wave. It can be grouped in the following:
    i.   DC offset, or the presence of DC current or DC voltage component in an AC system.
    ii.   Harmonics, defined as AC components of voltage or current with integer multiples of the fundamental frequency of the system (FREQ).
    iii.   Interharmonics, which are similar to harmonics but their frequency are not integer multiples of the fundamental.
    iv.   Notching, or the 'periodic voltage disturbance caused by line-commuted thyristor circuits'.

  v. Electric noise, defined as 'unwanted electrical signals with broadband spectral content lower than 200 kHz superimposed on the power system voltage or current in phase conductors, or found in neutral conductors or signal lines'.

6. Voltage fluctuations and flickers. The first defined as 'systemic variation of the voltage envelope or random voltage changes the magnitude of which does not normally exceed specified voltage ranges', and the second as 'continuous and rapid variations in the load current which causes voltage variations'.

7. Frequency variations defined as 'the deviation of the power system fundamental frequency from its specified nominal value'.

5.2.2 Why power quality is important?

Power quality is important because low quality can negatively affect electric equipment in the power grid, from operators' assets to customers' devices, and in this way reduce the reliability of the system.

The negative effects of a poor power-quality on the different components of the power system are, among others [3].

- Possible failure of equipment due to peak voltage caused by harmonics.
- Heating, noise and reduction in lifetime of most power system equipment.
- Need for a reduction of the maximum usable capacity of power transformers in order to avoid failure due to overheating.
- Losses in transmission lines, generators, AC motors and so on, due to harmonics.

Also, the inefficient operation of sensitive electronic equipment, such as TV and computers, is one of these negative effects [4].

5.2.3 Smart grids and power-quality issues

Large interconnection of renewable generation in a given power grid decreases the total inertia of the system. Wind and solar generation, for instance, are connected to the grid by power converters that decouple the dynamics of these renewable generators from that of the grid. Even though the wind turbines (WTs) have rotating parts with large moments of inertia, these are not 'felt' by the grid, there is no positive effect on damping frequency deviations as may be in the case of generators directly connected to the grid, such as conventional coal-based power plants.

Grid stability, for steady state and transient condition, for a grid connected to renewable generation or microgrids, is affected by the reduced inertia and also by voltage profiles caused by possible bidirectional flow of power (compared to the traditional one-way flow) caused by generators at different points in the grid. Therefore, stability must be enhanced for a more reliable and safe grid operation, especially under large disturbances, such as loss of a large generation unit or load.

In the case of small microgrids, they can operate connected to a larger grid or independently. In the first case, the frequency is regulated by the larger generation units. For independent operation, this task is left to, for example, a diesel engine-generator, and due to the reduced inertia, a more stringent frequency control condition is created.

Smart grids, in its broader sense, have the task to better integrate these renewable generators taking into account these potential issues. They can improve power quality and reliability with the deployment of smart meters and management technology to have more reliable information on voltage, frequency and energy consumption at customer level [4]. According to the Department of Energy of the United States of America, smart grids are expected to improve not only reliability, but also flexibility, resiliency and efficiency [5].

5.2.4   *The concept of virtual generator in microgrids*

Virtual synchronous generators (VSGs) are control schemes for storage systems designed to emulate the dynamics of synchronous generators connected to the grid during disturbances and to improve frequency stability of the system. In this section, an overview of this novel concept is presented.

5.2.4.1   Research in the literature

Distributed generation systems possess characteristics that do not match the old established centralised electricity generation structure. In conventional power systems, which have evolved to the current state from more than a century ago, large power plants generate electricity, which is transmitted first by high-voltage transmission lines and then distributed by low-voltage lines to the consumer located generally far away from the large generators. These generators keep the demand and supply balanced by adjusting their output to follow variations in the demand. All generators in the network operate at synchronous speed and are capable of adjusting their outputs to keep the frequency variations to a minimum, with spinning reserves ready to be supplied in the case of shortage of energy. As the generators are composed of heavy rotating parts that can absorb or supply their kinetic energy when the rotating speed changes, the imbalances between supply and demand are buffered by the inertia of these masses. Furthermore, the different loads that constitute the total demand are dispersed over a wide area and are not completely correlated, causing the generators to 'see' only a smoother change in the demand. Due to this, frequency control is a task normally and successfully performed by these synchronous generators [6].

In contrast, the distributed small generation systems, or microgrids, generate power at the point of consumption – that is, local generation for local consumption approach – using photovoltaic (PV) systems, which do not possess rotating parts that can buffer the supply–demand imbalances, or WTs generally connected to the grid through power converters, which decouple their interactions with the rest of the grid. The only sources of inertia in the system are the gas or diesel engine-generators, which have also reduced sizes and moments of inertia, since they are designed to supply only the local small demand at the point of connection. Also the gas or diesel engine-generators only supply the part not covered by the renewable energy generators [7]. Due to this, microgrids have reduced inertia compared to a system of the same rated power and completely supplied by synchronous generators, and their capability for supply–demand balance is drastically reduced, increasing the risk of frequency variations exceeding the standard permitted range.

Several approaches to deal with this problem have been considered in literature. The most common approach is the direct utilisation of fast acting storage systems, such as sodium–sulphur (NaS) batteries, superconducting magnetic energy storage (SMES), supercapacitors and flywheel energy storage to balance supply and demand. The main disadvantage of this approaches is that the storage system must be capable not only of storing energy for future use, but also it must have additional capacity in case a sudden supply-imbalance occurs, and it must be ready to quickly charge or discharge energy, which may increase its cost [8–10].

For instance, in [8], the authors present a control strategy to balance supply and demand in a microgrid containing a WT by utilising SMES including the reduction of the DC–DC converter of the SMES. In [9], the authors analyse the operation of a SMES in a microgrid containing wind, solar and diesel generator for frequency regulation under islanded operation with two types of SMES magnets. A very interesting approach is presented in [10] highlighting the importance of battery cost in the frequency regulation in a microgrid.

The approach includes implementation of a controllable load through the application of robust proportional integral controller for a remote hybrid microgrid as an alternative for storage systems in developing countries.

Another approach to solving the issue of inertia reduction is the concept of virtual inertia (VI) provided by a VSG. With this approach, the dynamics of a synchronous generator are emulated by controlling the output of a battery via a control loop in the voltage source converter (VSC) [11–14].

A complete review of the current technology related to VSGs applied to grids containing generators with few or no rotating mass, covering the fundamentals, recent applications and issues in this area, is presented in [11]. The comparison between two similar approaches for frequency control in microgrids, which have large proportion of their supply through power converters: the concept of VI and the Frequency-Droop-Based control schemes, and the equivalence of both of the schemes under certain conditions are demonstrated in [14]. The operation and dynamics during synchronous machine emulation by VSCs is considered in [12]. The utilisation and dynamics of fast acting batteries to compensate supply–demand imbalances in an isolated grid containing high penetration of solar and wind power, in the form of VI, and the mitigation of the impact of these intermittent generators are considered in [13].

The transient stability improvement in microgrids by the implementation of VI with a controller where the inverters emulates the behaviour of a synchronous generator is considered in [15]. In [16], the concept of virtual machine to provide VI is studied together with the investigation of the stabilisation value of the frequency in a microgrid containing a diesel generator. In [17], the authors studied the frequency stability in a hybrid autonomous microgrid containing a diesel unit and intermittent renewable generation together with the concept of VSG in energy storage systems through digital simulations. In [18], the frequency-droop-based VSG for microgrids by the control of VSCs for grid-connected and islanded operations by emulating the behaviour of synchronous generators through the implementation of swing equation with a primary frequency controller is presented.

The controller is able to change the dynamic response without compromising the steady-state performance, thereby improving the performance compared to the conventional droop control. Finally, in [19], a coordinated distributed control scheme for primary and secondary control based on frequency-bus-signalling method for a low-voltage AC three-phase microgrid under islanded operation is studied using real-time simulations for different scenarios.

Both of the battery-based approaches presented above require reserve capacity with some of the battery stored energy supplied when load exceeds the generation and some additional available capacity for power absorption in the battery when there is excess generation. The addition of a coaxial rotating mass to the existing engine-generator in order to compensate the loss of inertia in the system without depending on batteries is another approach to be considered in the future. However, a point to consider carefully in this approach is the need for an additional mass and the need for mechanical modification of the existing engine-generator, that is flanges, base frame and so on, which will add cost and take up more space.

5.2.4.2   Virtual inertia principle

The VI approach is the utilisation of a VSG implemented by a control algorithm applied to the output of the VSC of a storage system in order to absorb or inject power, emulating the dynamics of a synchronous machine present in a conventional power system.

Figure 5.1 shows the configuration of the VI approach using a VSG. The controller implements an algorithm according to (5.1). Here, the frequency variation in

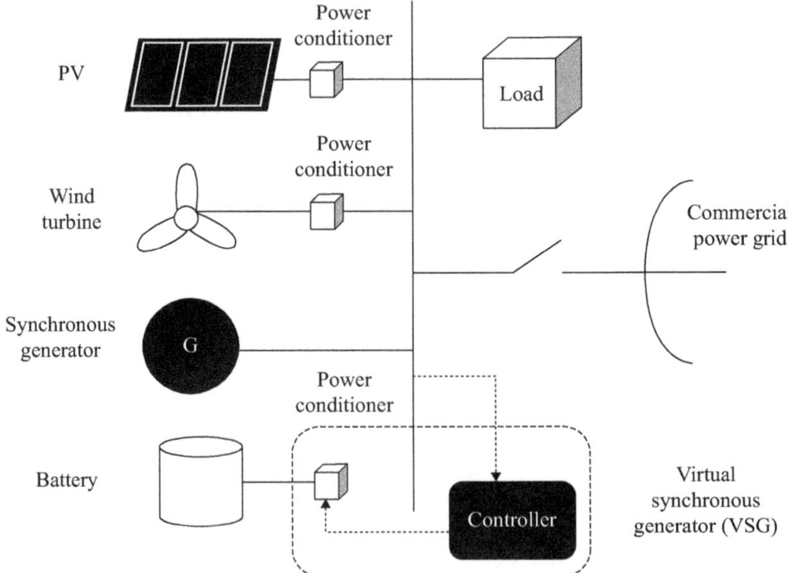

Figure 5.1   Configuration of the virtual inertia approach

the system is measured through the speed deviation $\triangle\omega$, which is utilised as an error signal for the control algorithm to determine the required power output from the storage system by controlling the VSC [11].

$$P_{\text{BAT}} = P^o_{\text{BAT}} + K_i \frac{d\Delta\omega}{dt} + K_p\Delta\omega \tag{5.1}$$

Equation (5.1) has the same form as the swing equation of a synchronous generator, where K_i is the term corresponding to the inertia of the machine, K_p is the term corresponding to the damping constant of the machine and the term P^o_{BAT} is the power to be transmitted by the battery during normal operation. Therefore, the constant K_i is the term which defines the VI of the VSG.

5.2.4.3 Comparison

The VI and real inertia (RI) methods possess the following characteristics:

1. The VI approach encompasses (1) larger battery and VSC sizes, compared to the case when no additional support from the battery is needed and (2) fast charge–discharge operation speed for the battery.
2. The RI approach encompasses (1) the need for modification of the physical structure of the engine-generator, such as enlargement of base frame and rotational axis, requiring additional space and (2) the addition of a rotating mass.

In summary, the main differences are that there is no need for gas-engine generator modification, and no cost of modification or additional space requirements arise in the case of a VI approach. In the case of the RI approach, the battery and the VSC size and type make them cheaper than the VI case. They are smaller (and not necessarily very fast) as there is no need for battery reserve capacity.

5.2.4.4 Discussion

Increasing the moment of inertia of the existing generator by the addition of a rotating mass to the axis of a synchronous generator in a microgrid – the RI approach – can improve the frequency regulation capability of the system but adds additional costs, of the added mass and also of the necessary modifications.

The virtual inertia approach, on the other hand, improves the frequency regulation of the system in a similar way and does not need additional mass (added or manufactured in this way) or modifications in the engine generator. However, it requires additional battery capacity for its implementation.

5.3 Reliability

5.3.1 What is reliability?

According to [20], the term 'high reliability' for a given power system refers to fewer interruptions to supply and high-quality electricity delivered to customers, despite the temporary or permanent unavailability of power plants and transmission lines, or the presence of external large disturbances that may include load fluctuations or severe weather condition. The reliability of power grid is maintained or enhanced under the following situations:

- The components are dependable, replaceable and easily replaceable.
- The system performance is flexible and predictable.
- The information on system condition flows smoothly among actors in the power grid.
- No abrupt failure or extensive damage in the grid under natural or malicious attack.
- Prompt recovery after failure in the system.

Smart grids are expected to increase the level of reliability in a power distribution system. According to [21], three types of outages can be mentioned:

- Major events, affecting a large percentage of the customers,
- Sustained interruptions, with duration of more than 5 min,
- Momentary interruptions, defined as brief loss of power to few customers caused by the operation of interruption devices.

Indexes are used to assess the degree of reliability in a power grid. Four reliability indexes can be identified [21] as follows:

1. System average interruption frequency index
2. System average interruption duration index
3. Customer average interruption duration index
4. Momentary average interruption frequency index

5.3.2   Interoperability

Interoperability is a very important concept and vital for the existence of smart grids. According to [20], interoperability is defined as

> *The capability of systems or units to provide and receive services and information between each other, and to use the services and information exchanged to operate effectively together in predictable ways without significant user intervention. Within the electricity system, interoperability means the seamless, end-to-end connectivity of hardware and software from the customers' appliances all the way thorough the transmission and distribution system to the power source, enhancing the coordination of energy flows with real-time flows of information and analysis.*

Smart grid reliability is improved by interoperability. The main contributions of interoperability to reliability are [20] as follows:

- Fewer outages due to better monitoring, analysis and management of the grid
- Faster system restoration following a failure
- Higher system resistance to attacks
- Better integration of supply and demand-side resources

The ways in which these contributions help the system are [20] as follows:

- Improvement of situational awareness due to better information collection and flow
- Improvement of operator reaction to and understanding of the grid, under certain grid conditions
- Improvement of the required coordinated operation among the different actors on the grid.

As mentioned in [20], there are three dimensions of reliability: Operating reliability, system adequacy and national security. The first one is refers to the capacity of a power grid to withstand sudden disturbances such as short circuits or loss of components. The second one refers to the capacity of a system to supply electric energy at all times. The last one refers, in a broader societal context, to the 'integrity of a nation's citizens, infrastructure, economy, and environment from physical, medical or economic harm', as electricity is vital to the economy and physical security.

5.3.3   Cybersecurity

The evolving smart grid technology calls for an efficient and reliable communication technology, that is information technology and telecommunication, for the realisation of a reliable and secure electric sector. The communication systems are required to quickly and accurately contribute to the solution of any problem on the grid caused by natural disasters, equipment failures and operation error, and of course, the possible deliberate cyber-attacks. Cyber-attacks refer to attacks that can take control of system software and database, accessing private information with the potential to cause severe damage to the operation and infrastructure of the grid [22].

Communication must be also capable of properly addressing the disturbances or attacks in the network in order to avoid cascade failure in the entire system.

An important difference between a smart grid attack and a 'purely' software attack is that it involves tangible machines and infrastructure that can be damaged with the malign operation. Thus, additional considerations and special analysis are required for the prevention of cyber-attacks in smart grids. The National Institute of Standards and Technology (NIST) has published recommendation and standards related to cybersecurity for smart grids [22].

5.3.4   Flexible operation and self-healing

By definition, microgrids must be capable of operating independently from another larger grid under certain events. Reconfiguration of distribution grids under these circumstances is part of this flexible operation capability, in order to quickly restore the service to certain areas and to higher number of customers as possible, which improves system reliability.

A good definition of the concept of self-healing can be found in [23]:

The self-healing grid is a system comprised of sensors, automated controls, and advanced software that utilises real-time distribution data to detect and isolate faults and to reconfigure the distribution network to minimise the customers impacted.

The process involves a reconfiguration of devices, such as sectionalisers and reclosers, to rapidly isolate the faulted area and to continue the supply of power to customers not affected by the failure. The system is reconfigured in 1–5 min, which require proper communication systems. In the past, this operation of isolating and re-establishing the service has been performed for years by power utilities, but with today's severe loading conditions of the grid and the presence of distributed generation, the operation requires a more advanced system for a safer and rapid implementation [23].

5.3.5   Demand response

Demand response programmes in smart grids can also contribute to increasing the reliability of the system. These include a series of pricing strategies according to the time the consumer uses the electricity and programs to allow the power companies to directly control some customers' load during periods of peak demand. By identifying critical loading conditions, the grid operator can better manage and operate the grid optimally to avoid failure in the system [24].

In the longer term, demand side management (DSM) aims to modify or reduce costumers' demand, in order to increase energy utilisation efficiency in a given power grid [25]. The main benefits of DSM are cost reduction of supplying energy demand, reduction of greenhouse gas emissions due to reduced energy consumption, increase reliability in the medium and long term and improved markets by providing a short-term response to certain market conditions, such as periods of peak power prices.

5.3.6   Smart grids and microgrid standards

Several institutions are publishing standards regarding smart grids. The International Electrotechnical Commission and the Institute of Electrical and Electronic Engineers have sets of standards for the different components of the smart grid [26,27].

Also, the NIST has been working on the implementation of standards on smart grid interoperability [28].

5.4   Case study[*]

5.4.1   Introduction

Driven by an urgent need of CO_2 emission reduction and a strong public opinion resisting nuclear power generation following the Fukushima Daiichi nuclear disaster in 2011, Japan now has the challenge to rebuild and adapt its power system to increase its reliability and resiliency in the case of natural disasters. The considered approach is based mainly on the deployment of intermittent and clean renewable

---

[*]This section is entirely based on the paper 'Stability Enhancement of a Power System Containing High-Penetration Intermittent Renewable Generation' by Jorge Morel, Shin'ya Obara, Yuta Morizane, first published in the *Journal of Sustainable Development of Energy, Water and Environment Systems (JSDEWES)*, volume 3, issue 2, June 2015, pages 151–162.

energy generation, with a less centralised structure. The Japanese government has established policies to address this issue [29].

The United States of America and the European Union, in contrast, have been implementing policies to reduce CO_2 emissions by utilising renewable energy sources even before the disaster of 2011 [30,31].

Microgrids have evolved as a type of architecture that makes possible the generation and consumption of energy in limited and well-defined areas. Their flexibility can make microgrids an active part of a larger smart grid system. Another important benefit of constructing a self-sufficient microgrid is the reduction of power to be transferred over a long distance from where centralised power plants are normally located. This reduces the losses and congestion in the transmission lines. Besides the possibility of interconnection of variable output renewable generators, there are other benefits such as increased participation of customers in the reduction of peak demands for the entire system as well as participation in the electricity market [32].

Among the leading research teams in the field of clean energy systems is Aalborg University in Denmark, with the development of an energy system analysis tool called EnergyPLAN [33,34]. Several studies have been performed in the design of energy systems for specific regions in Europe [35,36] where not only electrical aspects were considered but also heat and transportation. This view is much wider than the case of smart grids alone. However, most of this work focuses on long-term operations and planning and do not consider the dynamics of the system for normal operating condition and for the case of transient events in the electrical system.

Despite the slower development of smart energy systems in Japan, great effort has been put in by certain research groups before the nuclear disaster of 2011 [37,38]. Recently, research activities on independent microgrids for local generation and local consumption, containing sustainable renewable energy generation such as wind, solar and tidal power have increased considerably. Operation of microgrids, aiming at the reduction of CO_2 emissions and the safety of energy supply, in cold, urban and remote areas has been studied [39–41].

In microgrids, there is a need to match instantaneous imbalances, not only for transient events in the system but also for the fast oscillations of the power outputs of renewable generators by the utilisation of fast acting batteries, such as the NaS battery. Ohtaka and Iwamoto [42] analysed the utilisation of NaS batteries in suppressing instantaneous or fast oscillations on the grid, as well the possibility of independent active and reactive power control in this type of storage systems.

Storage systems have been also analysed from economic and environmental points of view [43,44]. In [43], NaS batteries are compared to a storage system based on organic chemical hydride. However, the dynamic performance (fast charging/discharging capability) of the NaS battery should also be considered for a complete analysis. In [44], a system with no storage is analysed. Here, good CO_2 reduction is obtained despite the absence of the battery. However, for a 100% renewable supply, batteries may be necessary to shift energy between seasons or to keep frequency balance, as well as to compensate for any transient faults in the system.

For a complete analysis of a power grid, the consideration of dynamic properties of supply and demand, especially for transient events, is vital. This is particularly important because of the reduction in the system inertia as WTs are connected to the power system through converters, which decouple the inertia of the rotating mass from the system. All the energy system designs mentioned above do not consider this aspect.

The objective of this section is to evaluate the effect of fast acting NaS batteries on the transient stability of a power system containing high penetration of renewable sources with highly variable outputs and with potential to exert strong disturbances on the system (e.g. disconnection of WTs due to storms) and reduced system inertia. In order to achieve this, digital simulations are performed using MATLAB®/Simulink®.

Results of this section show that fast acting NaS batteries improve the stability level of the system, keeping the system stable after strong transient disturbances, especially for the case of isolated operation. NaS batteries can also be used for short-term balance between supply and demand during normal operating condition.

5.4.2   Study system

In this section, the location of the study area, Kitami City and the components of the target power system are described.

5.4.2.1   Location

Kitami City is located in a cold region, on the Hokkaido Island in the northern part of Japan, as shown in Figure 5.2. Kitami has an annual demand characteristic with a high heat-to-power demand ratio during winter. The temperature reaches a minimum of 20 °C in winter and a maximum of 35 °C in summer. Despite the low temperature in winter, the city gets little rainfall and snowfall. Kitami has rich natural resources such as wind, solar and tidal power that can be utilised for the generation of clean electrical energy. It is one of the richest areas in solar radiation in Japan. Also, it has open areas with good average wind speeds which can be exploited for the generation of electricity. The currents in the channels connecting the Saroma Lake and the Sea of Okhotsk offer the possibility for tidal generation [45].

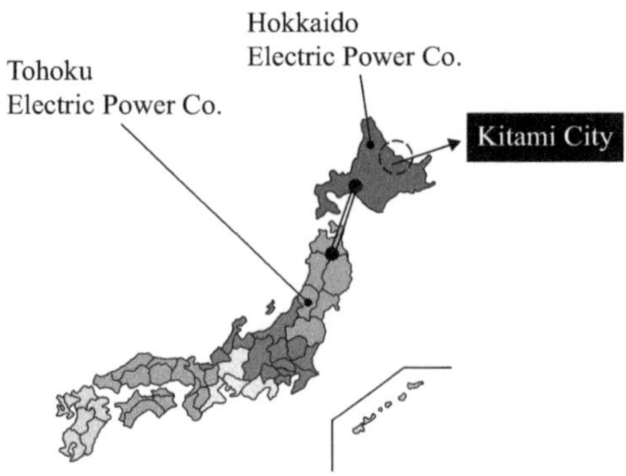

Figure 5.2   Kitami City location

5.4.2.2   Transmission network

The Japanese electricity industry consists of ten power companies that supply energy to specific and semi-independent regions. They are interconnected (except Okinawa Electric Power Company) through transmission lines with limited capacities. The Hokkaido Electric Power Company (HEPCO), shown in Figure 5.2, with a total installed capacity of 7,500 MW, supplies power to the Hokkaido Island, where Kitami City is located. HEPCO is connected to Tohoku Electric Power Company, located in Honshu by a High-Voltage Direct Current transmission system (indicated by a double line in Figure 5.2), with a capacity of 600 MW, approximately 8% of HEPCO's total installed capacity [46].

The Kitami City power system is connected to the local utility HEPCO, which currently provides the power for the entire city. The simplified scheme of the Kitami's power system considered for simulation, including the proposed location of renewable generators and storage systems is depicted in Figure 5.3. The names and rated capacities of the substations are shown in Table 5.1.

Selection of the location for the renewable generators was made on the basis of available resources in the area. The location of the NaS batteries was selected to be the Rubeshibe substation where the bulk power comes from the conventional power plants of HEPCO. This selection has no direct effect on the results of the transient stability analysis presented in this section because of the short distances involved. However, from an economical point of view, the most appropriate locations and sizes must be carefully considered.

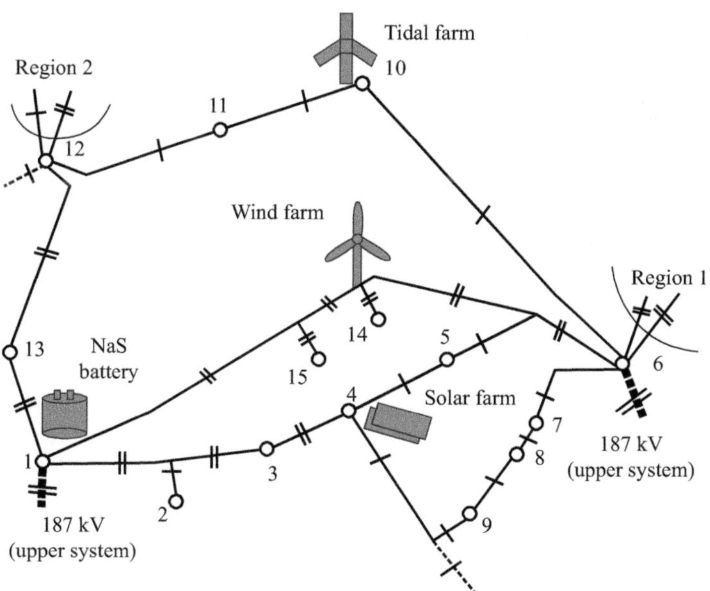

Figure 5.3   Power system of Kitami City

Table 5.1   Substation rated capacities

No.	Name	Capacity (MVA)
1	Rubeshibe	280
2	Kuneppu	6
3	Kitaminishi	20
4	Kitami	35
5	Tabata	22
6	Memanbetsu	200
7	Bihoro	20
8	Inami	10
9	Tsubetsu	12
10	Tokoro	12
11	Saroma	10
12	Engaru	18
13	Ikutahara	3
14	Kiyomi	30
15	Ainonai	12

As shown in Figure 5.3, there are mainly two points of connection to the local utility: The Rubeshibe and the Memanbetsu substations. Each of them is supplied by a double-circuit transmission line of 187 kV. Region 1 and Region 2 shown in this figure are two systems with no generating units. The total load for Region 1 is 124 MVA and that for Region 2 is 87 MVA. The two thin dashed lines represent the two weak connections to other systems which are not considered in this section. The small circles represent substations and single or double-circuit transmission lines are indicated by one or two transversal short lines over the lines connecting the substations. All lines are overhead with rated voltage of 66 kV and lengths of less than 40 km. Typical tower and conductor data for 66-kV transmission line is considered. Each substation is composed of 66 kV/6.6 kV step-down transformers with the rating indicated in Table 5.1. The reactive power of the loads is assumed to be compensated since it does not affect the present transient stability analysis.

5.4.2.3   Renewable generation

The three types of renewable generation considered in this section are wind, solar and tidal power.

Wind power

Currently, most WTs in the market are variable-speed types: doubly fed induction generator (DFIG) and full-scale converter types, which are capable of independently controlling active and reactive power injected to the system. In this section, the wind farm (WF) is simulated using an aggregated model of DFIG-based WTs, modelled by MATLAB/Simulink.

Solar power

PV type solar farms (SFs) are considered. They are connected to the transmission network via inverters which also may have the capability of independently controlling

Table 5.2   Aggregated renewable generator parameters

Generation	Rated capacity (MVA)	Point of connection (no.)
Wind farm	150	Kiyomi (14)
Solar farm	100	Kitami (4)
Tidal farm	1.5	Tokoro (10)

the amount of reactive power injected to the network for voltage regulation purposes. This can be exploited conveniently, especially for isolated systems. Since for frequency study purposes the faster dynamics are not considered, in this section the SF is modelled as a first-order system with a short-time constant of 10 μs.

Tidal power

Horizontal axis tidal turbines are considered [47]. They have similar structure and working principle as the WTs [48,49]. For short-term frequency studies the tidal turbines can be assumed to have constant power output. For long-term studies, the variability can be forecasted with a high degree of accuracy. The tidal farm (TF) is simulated using an adapted version of the DFIG-based WT. These assumptions are valid due to the similarity between the wind and tidal generation systems and the time scale considered for simulation.

Horizontal WTs and horizontal tidal turbines have essentially the same structure: A rotor composed of blades, connected to a generator. The rotor transforms the kinetic energy of the incoming fluid into rotating mechanical energy. The generator transforms the mechanical energy of the rotor into electricity. The differences in both types of turbines are mainly due to the interaction between the fluids with different densities, that is wind and water, with the rotor. For the same rated power, a tidal turbine will have a smaller rotor diameter than a WT [49].

The DFIG structure is the same in both types of turbines. The generator is connected to the power grid through the stator and through the rotor. The connection of the stator is made directly or via a transformer, and that of the rotor is made by a back-to-back converter, with or without a transformer. This arrangement allows control of the electrical torque by the rotor side converter to adjust the rotational speed of the rotor with wind speed variations in order to improve energy absorption efficiency. The detailed modelling of a tidal turbine including equations, similarities and differences with WTs can be found in [48,49].

The parameters of the renewable generators are shown in Table 5.2.

5.4.2.4   Storage systems

For long-term energy storage aiming at the seasonal and daily energy shifting, a storage system with slow dynamics but with high energy density should be utilised. For seasonal energy shifting, an organic chemical hydride type system can be employed [43].

For instantaneous and fast demand–supply imbalance compensation, NaS batteries are employed due to their fast charge–discharge capability. They have been satisfactorily applied in the levelling of power outputs fluctuations of WFs. They also can be used for load levelling and load peak shaving [50].

In this section, a simplified model of the NaS battery is considered. A first-order system with a small time constant of 10 µs, to represent the fast dynamics of this type of batteries, is considered.

5.4.3   Control strategy

For both normal and fault condition control strategies, a Model Predictive Control (MPC) approach is considered for the control of the NaS batteries. Due to the uncertainty and randomness involved in the study system, MPC is used as it can handle multiple inputs and multiple outputs, in contrast to conventional proportional-integral controllers [51,52]. Inertial control approach by WTs [53] provides primary frequency support to mitigate the reduced system inertia. The overall control strategy is shown in Figure 5.4.

5.4.3.1   Normal condition

During normal condition, the WTs are controlled for maximum power absorption and maximum power output, and to limit the power output during high wind speed conditions by applying pitch control. A similar control strategy is considered for the TF. The SF is operated for maximum power generation. The NaS batteries are controlled to match supply and demand in the system to keep frequency within permitted ranges.

5.4.3.2   Abnormal condition

In the case of abnormal or fault condition, the NaS batteries and the WF are operated with different strategies. WTs are allowed to release or absorb, temporarily the kinetic energy by their rotating parts in order to mitigate the reduced inertia of the system and to support the primary frequency control. At the same time, battery output is entirely devoted to damp the fast oscillations of the system by absorbing or injecting active power to the network.

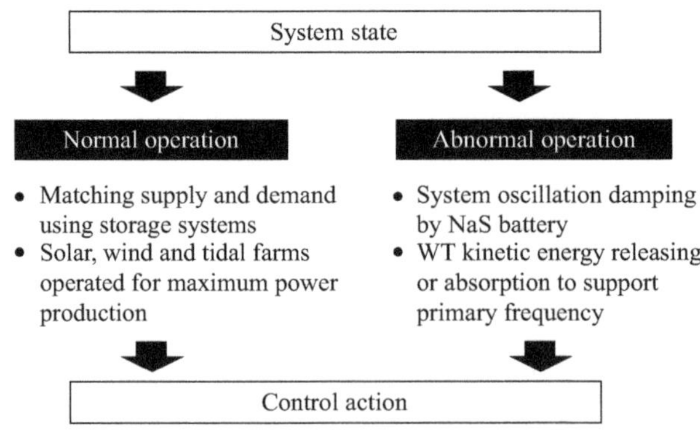

Figure 5.4   Overall control strategy

5.4.4   Scenario construction

Dynamic simulations are performed for three different scenarios to evaluate the effects of fast charging–discharging batteries in reducing the frequency oscillations and stabilising the system: Target system connected to the local utility (Scenario 1), target system isolated from the local utility (Scenario 2) and primary frequency support by WTs (Scenario 3).

Scenario 1 considers the target power grid connected to that of HEPCO at two points: Rubeshibe and Memanbetsu substations. At these points, two aggregated generator models simulate the synchronous generators of the rest of the system. These generators also perform voltage and frequency control tasks.

In Scenario 2, the target power system is considered completely isolated from that of HEPCO. Only small hydropower stations in the order of 80 MVA remained connected to the system. Since there is a considerable reduction of the inertia of the system, a more severe effect is expected during transient events.

Scenario 3 shows the effect of the support provided by WTs to the network frequency during disturbances, such as in the case of a loss of an important load in the system. This scenario evaluates this WT capability assuming that the battery's stored energy level is not enough to allow any additional charge–discharge operation. A loss of 10% of the system's load is considered, and the frequency profiles are analysed. The target system is assumed connected to the local utility HEPCO.

Two cases are considered for Scenarios 1 and 2: Loss of wind generation due to a storm (Case 1) and three-phase-to-ground-fault (Case 2).

In Case 1, a strong storm is assumed to affect the area where the WFs are located. A step increase in the wind speed at $t = 20$ s is followed by the disconnection of the WTs at $t = 30$ s caused by the wind speed exceeding the cut-out wind speed level, as shown in Figure 5.5.

In Case 2, a three-phase-to-ground fault, with a duration of five cycles, is considered in the grid at $t = 35$ s, causing strong oscillations in the synchronous units of the system.

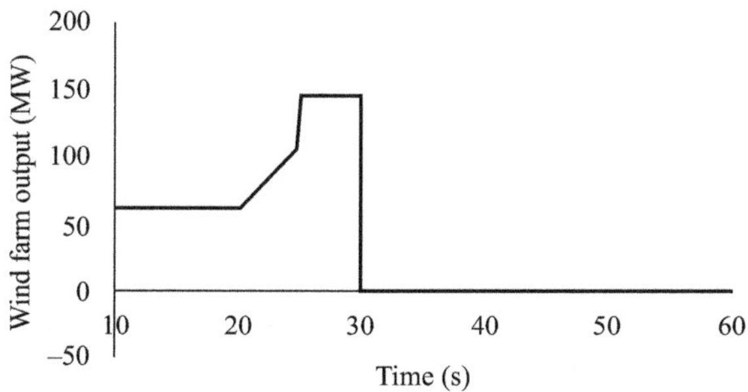

Figure 5.5   Wind farm power output

5.4.5   Simulation results

The overall scheme of the MATLAB/Simulink model utilised for simulation is shown in Figure 5.6. In this figure the WF, the SF, the TF, the NaS battery, the two aggregated generators (HEPCO-Rubeshibe and HEPCO-Memanbetsu), together with the MPC controller, are shown. The measured output is the FREQ with a reference setting of 50 Hz, and the measured disturbances are the WF, TF and SF outputs. The manipulated variable is the corresponding setting of the NaS battery active power. Reactive power setting of the NaS battery was considered equal to zero.

5.4.5.1   Scenario 1: Target system connected to the local utility

Case 1: Loss of wind generation due to a storm
In order to assess the stability level of the system, synchronous machine speed deviations are analysed with and without the support of the proposed control scheme. In per unit values, the speed deviations are equal to the frequency deviations. Figure 5.7 shows the variations in the system's frequency at Rubeshibe substation for this scenario. Without the support of the NaS batteries, the systems experiences instability. Using the fast acting batteries completely avoids the instability.

Case 2: Three-phase-to-ground fault
As shown in Figure 5.8, the utilisation of the NaS batteries allows faster damping of these oscillations, recovering to the normal frequency level, compared to the case without batteries.

5.4.5.2   Scenario 2: Target system isolated from the local utility

Case 1: Loss of wind generation due to a storm
Figure 5.9 shows the variations in the system frequency at Rubeshibe substation. As can be noticed, the system experiences instability if the NaS batteries are not in operation.

Case 2: Three-phase-to-ground fault
Results in Figure 5.10 show that with the NaS batteries in operation, the oscillations are damped completely, recovering to the normal frequency level, faster than in the case when the system is not provided with this support.

5.4.5.3   Scenario 3: Primary frequency support by WTs

Variable speed WTs are able to release or absorb kinetic energy through their rotating parts, temporarily and almost instantaneously after a disturbance in the system, emulating the dynamics of synchronous generators and contributing to the recovery of the system's frequency. This is achieved by additional control loops to regulate WT's power output. The support of WTs is particularly useful when the level of stored energy in the NaS batteries is not adequate for proper charging-discharging operation to support system's frequency.

Figure 5.11 shows the primary frequency profile of the system for this scenario, with and without WT support. It can be seen that the WTs reduce the impact of a load change in the system by absorbing kinetic energy, just after the disturbance, damping the frequency peak at this point.

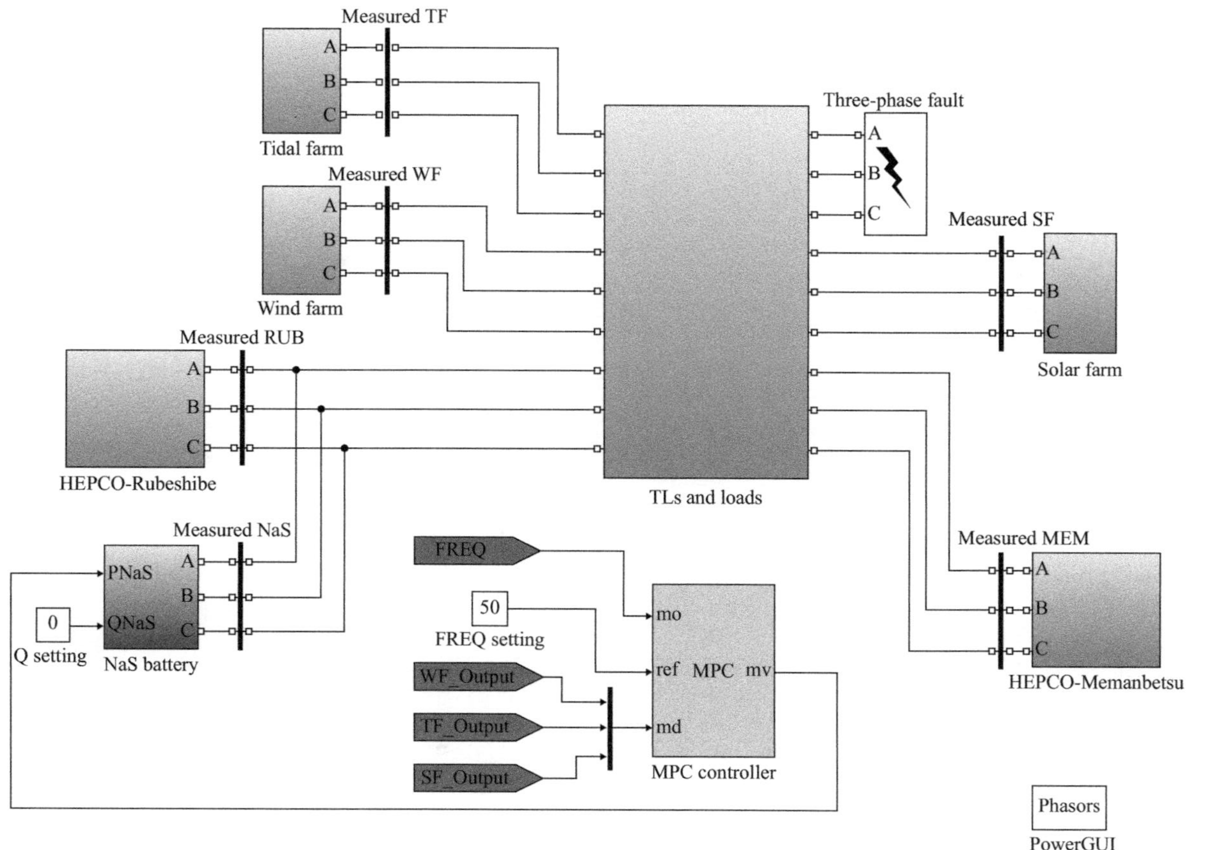

Figure 5.6   MATLAB/Simulink model diagram

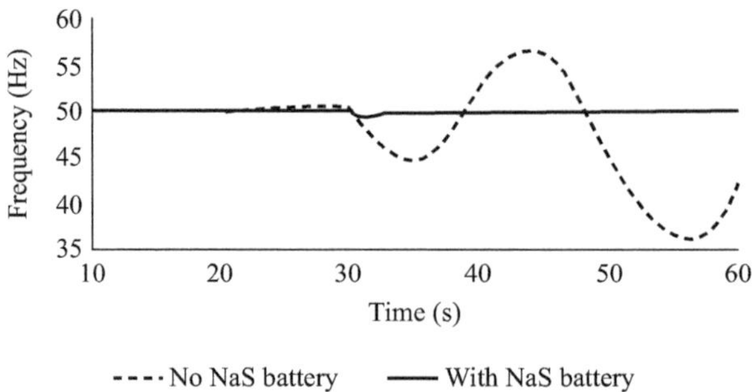

Figure 5.7   Frequency for loss of wind generation – Scenario 1

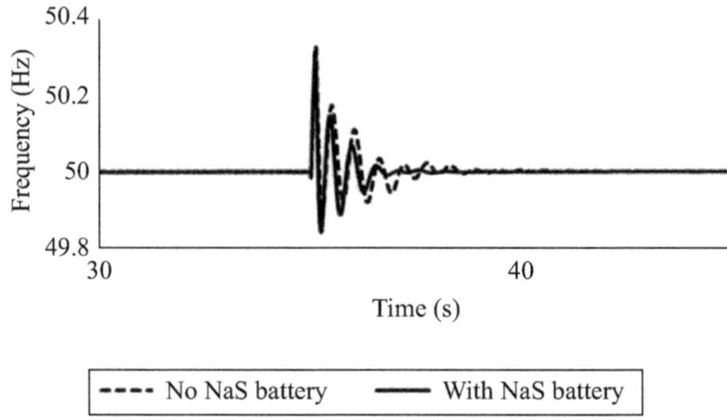

Figure 5.8   Frequency for fault condition – Scenario 1

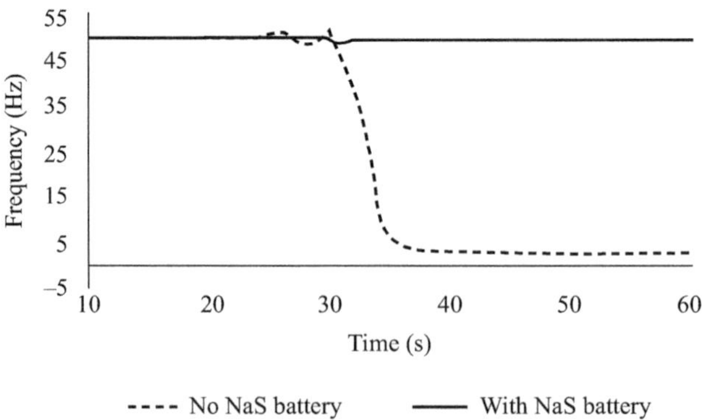

Figure 5.9   Frequency for loss of wind generation – Scenario 2

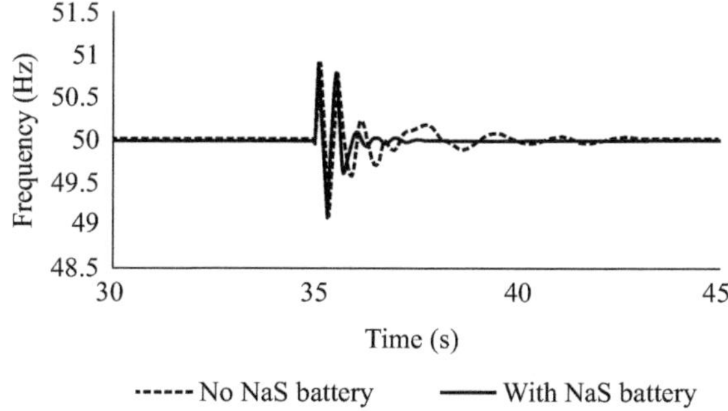

Figure 5.10   Frequency for fault condition – Scenario 2

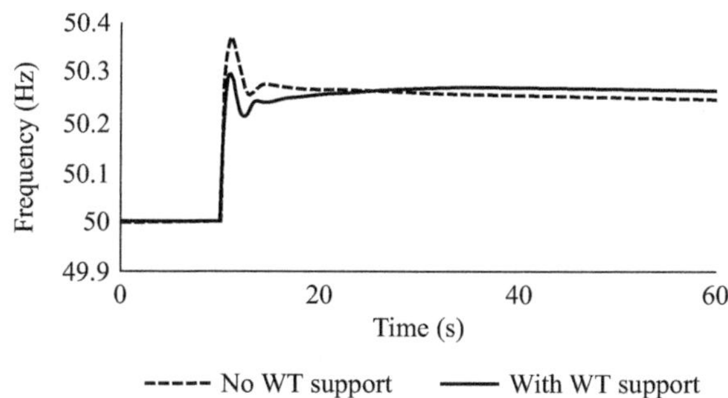

Figure 5.11   Frequency for loss of load – Scenario 3

5.4.6   Conclusion

This section introduced a strategy to improve the transient stability of a power system with a high-penetration of intermittent renewable sources. The stability degradation arises mainly due to the highly variable outputs of these types of generators, during normal and abnormal operations (e.g. disconnection of WTs due to a storm). Reduction in the system's total inertia because of the connection of renewable generators, connected through power converters that decouple the dynamics of the generators from that of the system, also degrades system stability.

Three scenarios were considered: First, the target power system was considered connected to the local public utility, and second, the target power system was considered isolated. Each of these two scenarios was studied with two types of disturbances in the system: disconnection of an important WF due to a storm and a three-phase-to-ground fault in the power network. Finally, a third scenario was presented to demonstrate that the WTs can damp oscillations in the frequency by absorbing electrical energy in their rotating parts for the case of a loss of an important load in the system.

Results show that the proposed scheme, designed to charge and discharge the NaS batteries according to the oscillations in the system, and to make WTs participate in primary frequency support, improved the stability level by considerably reducing the synchronous machine oscillations, keeping the system operating and stable for the various scenarios presented.

References

[1]   Ministry of Economy, Trade and Industry (METI), "FY2013 Annual Report on Energy (Energy White Paper 2014) Outline," [Online]. Available from: http://www.meti.go.jp/english/report/index_whitepaper.html [Accessed 9 December 2014], 2014.

[2]   Ministry of Economy, Trade and Industry (METI), "Cabinet Decision on the New Strategic Energy Plan," [Online]. Available from: http://www.meti.go.jp/english/press/2014/0411_02.html [Accessed 9 December 2014], 2014.

[3]   E. H. Fuchs and M. A. S. Masoum, *Power Quality in Power Systems and Electric Machines*, Burlington, MA: Elsevier Academic Press, 2008.

[4]   Smart Grid Consumer Collaborative, "Smart Grid and Power Quality," [Online]. Available from: http://www.whatissmartgrid.org/smart-grid-101/fact-sheets/smart-grid-and-power-quality [Accessed 14 September 2016].

[5]   US Department of Energy (DOE), "Technology Development: Smart Grid," [Online]. Available from: http://energy.gov/oe/services/technology-development/smart-grid [Accessed 14 September 2016].

[6]   P. Kundur, *Power Systems Stability and Control*, New York, NY: McGraw-Hill, 1994.

[7]   ENERGY.GOV, "How Microgrids Work," [Online]. Available from: http://energy.gov/articles/how-microgrids-work [Accessed 9 December 2014], 2014.

[8]   A. I. Sarwat and A. H. Moghadasi, "A Downsizing Strategy for Combinatorial PMSG Based Wind Turbine and Micro-SMES System Applied in Standalone DC Microgrid," *International Journal of Energy Science (IJES)*, vol. 4, no. 2 (April), pp. 50–59, 2014.

[9]   R. Kim, G.-H. Kim, S. Heo, M. Park, I.-K. Yu and K. Hak-Man, "SMES Application for Frequency Control During Islanded Microgrid Operation," *Physica C*, vol. 484, pp. 282–286, 2013.

[10] C. S. Ali Nandar, "Robust PI Control of Smart Controllable Load for Frequency Stabilization of Microgrid Power System," *Renewable Energy*, vol. 56, pp. 16–23, 2013.

[11] H. Bevrani, T. Ise and Y. Miura, "Virtual Synchronous Generators: A Survey and New Perspectives," *Electrical Power and Energy Systems*, vol. 54, pp. 244–254, 2014.

[12] Q.-C. Zhong and G. Weiss, "Synchronverters: Inverters that Mimic Synchronous Generators," *IEEE Transactions on Industrial Electronics*, vol. 58, no. 4, pp. 1259–1267, 2011.

[13] G. Delille, B. Francois and G. Malarange, "Dynamic Frequency Control Support by Energy Storage to Reduce the Impact of Wind and Solar Generation on Isolated Power System's Inertia," *IEEE Transactions on Sustainable Energy*, vol. 3, no. 4 (October), pp. 931–939, 2012.

[14] S. D'Arco and J. A. Suul, "Equivalence of Virtual Synchronous Machines and Frequency-Droops for Converter-Based MicroGrids," *IEEE Transactions on Smart Grid*, vol. 5, no. 1, pp. 394–395, 2014.

[15] U. Ganesan, K. Santhosh and N. B. Seshankar, "Improvement of Transient Stability of Microgrids for Smooth Mode Conversion Using Synchro-converters," *International Journal of Engineering and Technical Research (IJETR)*, vol. 2, no. 3, pp. 71–76, 2014.

[16] M. Torres and L. A. C. Lopes, "A Virtual Synchronous Machine to Support Dynamic Frequency Control in a Mini-Grid That Operates in Frequency Droop Mode," *Energy and Power Engineering*, vol. 5, pp. 259–265, 2013.

[17] M. Torres and L. A. C. Lopes, "Virtual Synchronous Generator: A Control Strategy to Improve Dynamic Frequency Control in Autonomous Power Systems," *Energy and Power Engineering*, vol. 5, pp. 32–38, 2013.

[18] Y. Du, M. J. Guerrero, J. Su, L. Chang and M. Mao, "Modelling, Analysis, and Design of a Frequency-Droop-Based Virtual Synchronous Generator for Microgrid Applications," Proceedings of the 2013 IEEE ECCE Asia DownUnder, pp. 643–649, 2013.

[19] D. Wu, F. Tang, T. Dragicevic, J. C. Vasquez and J. M. Guerrero, "Coordinated Primary and Secondary Control with Frequency-Bus-Signaling for Distributed Generation and Storage in Islanded Microgrids," Proceedings of the 39th Annual Conference of the IEEE Industrial Electronics Society, IECON 2013, pp. 7140–7145, 2013.

[20] GridWise Architecture Council (GWAC), "The Reliability Benefits of Interoperability," 2009. [Online]. Available from: http://www.gridwiseac.org/pdfs/reliability_interoperability.pdf [Accessed 14 September 2016].

[21] US Department of Energy, "Reliability Improvements from the Application of Distribution Automation Technologies – Initial Results," 2012. [Online] Available from: https://www.smartgrid.gov/files/Distribution_Reliability_Report_-_Final.pdf [Accessed 14 September 2016].

[22] National Institute of Standards and Technology, "Guidelines for Smart Grid Cybersecurity – Vol. 1," 2014. [Online]. Available: nvlpubs.nist.gov/nistpubs/ir/2014/NIST.IR.7628r1.pdf [Accessed 14 September 2016].

[23] Eaton, "Self-healing Grid," [Online]. Available from: http://www.cooper-industries.com/content/public/en/power_systems/solutions/self-healing.html [Accessed 14 September 2016].

[24] US Department of Energy (DOE), "Smart Grid: Demand Response," 2006. [Online]. Available from: http://energy.gov/oe/services/technology-development/smart-grid/demand [Accessed 14 September 2014].

[25] "Module 14: Demand-side Management," REEEP – UNIDO, [Online]. Available from: http://africa-toolkit.reeep.org/modules/Module14.pdf [Accessed 14 September 2016].

[26] International Electrotechnical Commission (IEC), "Smart Grid Standards," [Online]. Available from: http://www.iec.ch/smartgrid/ [Accessed 14 September 2016].

[27] The Institute of Electrical and Electronics Engineers (IEEE), "IEEE Smart Grid Standards," [Online]. Available from: http://smartgrid.ieee.org/resources/standards [Accessed 14 September 2016].

[28] National Institute of Standards and Technology, "Smart Grid," [Online]. Available from: https://www.nist.gov/engineering-laboratory/smart-grid [Accessed 14 September 2016].

[29] Ministry of Economy, Trade and Industry (METI), Annual Report on Energy, Outline of the FY2012 Annual Report on Energy (Energy White Paper 2013), 2012. [Online]. Available from: http://www.meti.go.jp/english/report/index_whitepaper.html#energy [Accessed 24 February 2014].

[30] US Department of Energy, Mission. [Online]. Available from: http://www.energy.gov/mission [Accessed 24 February 2014].

[31] European Commission, Energy Strategy for Europe. [Online]. Available from: http://ec.europa.eu/energy/index_en.htm [Accessed 24 February 2014].

[32] SmartGrid.gov, What is a Smart Grid? [Online]. Available from: https://www.smartgrid.gov/the_smart_grid#smart_grid [Accessed 24 February 2014].

[33] EnergyPLAN, Advanced Energy System Analysis Computer Model, Aalborg University. [Online]. Available from: http://www.energyplan.eu/about/ [Accessed 24 February 2014].

[34] H. Lund, A. N. Andersen, P. A. Østergaard, B. V. Mathiesen and D. Conolly, "From Electricity Smart Grids to Smart Energy Systems – A Market Operation Based Approach and Understanding," *Energy: The International Journal*, vol. 42, no. 1, pp. 96–102, 2012.

[35] D. Conolly, H. Lund, B. V. Mathiesen, *et al.*, "Heat Roadmap Europe: Combining District Heating with Heat Savings to Decarbonise the EU Energy System," *Energy Policy*, vol. 65, pp. 475–489, 2014.

[36] D. Conolly, H. Lund, B. V. Mathiesen and M. Leahy, "The First Step Towards a 100% Renewable Energy-system for Ireland," *Applied Energy*, vol. 88, no. 2, pp. 502–507, 2011.

[37] S. Obara, "Development of a Dynamic Operational Scheduling Algorithm for an Independent Micro-Grid with Renewable Energy," *Journal of Thermal Science and Technology*, vol. 3, no. 3, pp. 474–485, 2008.

[38] U.S. Department of Energy, Microgrids at Berkeley Laboratory, Nagoya 2007 Symposium on Microgrids, Overview of Micro-grid R&D in Japan, 2006. [Online]. Available from: https://building-microgrid.lbl.gov/sites/all/files/Morozumi_2006.pdf [Accessed 25 February 2014].

[39] M. Cellura, A. Di Gangi and A. Orioli, "Assessment of Energy and Economic Effectiveness of Photovoltaic Systems Operating in a Dense Urban Context," *Journal of Sustainable Development of Energy, Water and Environment Systems*, vol. 1, no. 2, pp. 109–121, 2013, DOI: http://dx.doi.org/10.13044/j.sdewes.2013.01.0008.

[40] S. Quoilin and M. Orosz, "Rural Electrification through Decentralised Concentrating Solar Power: Technological and Socio-Economic Aspects," *Journal of Sustainable Development of Energy, Water and Environment Systems*, vol. 1, no. 2, pp. 199–212, 2013, DOI: http://dx.doi.org/10.13044/j.sdewes.2013.01.0015.

[41] S. Obara, M. Kawai, O. Kawae and Y. Morizane, "Operational Planning of an Independent Microgrid Containing Tidal Power Generators, SOFCs, and Photovoltaics," *Applied Energy*, vol. 102, pp. 1343–1357, 2013.

[42] T. Ohtaka and S. Iwamoto, "A Method for Suppressing Line Overload Phenomena Using NaS Battery Systems," *Electrical Engineering in Japan*, vol. 151, no. 3, pp. 19–31, 2005.

[43] S. Obara, Y. Morizane and J. Morel, "Economic Efficiency of a Renewable Energy Independent Microgrid with Energy Storage by a Sodium–Sulfur Battery or Organic Chemical Hydride," *International Journal of Hydrogen Energy*, vol. 38, no. 21, pp. 8888–8902, 2013.

[44] S. Obara and J. Morel, "Microgrid Composed of Three or More SOF Combined Cycles without Accumulation of Electricity," *International Journal of Hydrogen Energy*, vol. 39, no. 5, pp. 2297–2312, 2014.

[45] Japan Meteorological Agency. [Online]. Available from: http://www.jma.go.jp/jma/indexe.html [Accessed 26 February 2014].

[46] Hokkaido Electric Power Company, Main Infrastructure. [Online]. Available from: http://www.hepco.co.jp/corporate/ele_power/ele_power.html [Accessed 26 February 2014] (In Japanese).

[47] Tocardo Tidal Turbines. [Online]. Available from: http://www.tocardo.com/ [Accessed 26 February 2014].

[48] H. H. Aly and M. E. El-Hawary, "Comparative Study of Stability Range of Proposed PI Controllers for Tidal Current Turbine Driving DFIG," *International Journal of Renewable and Sustainable Energy*, vol. 2, no. 2, pp. 51–62, 2013.

[49] S. E. Ben Elghali, M. E. H. Benbouzid and J. F. Charpentier, "Modeling and Control of a Marine Current Turbine Driven Doubly-Fed Induction Generator," *IET Renewable Power Generation*, vol. 4, no. 1, pp. 1–11, 2010.

[50] NGK Insulators Ltd. [Online]. Available from: http://www.ngk.co.jp/english/products/power/nas/ [Accessed 26 February 2014].

[51] A. Bemporad, M. Morarim and N. L. Ricker, *Model Predictive Control Toolbox User's Guide*, Natick, MA: The MathWorks, Inc., 1998.

[52] D. Ernst, M. Glavic, F. Capitanescu and L. Wehenkel, "Reinforcement Learning Versus Model Predictive Control: A Comparison on a Power System Problem," *IEEE Transactions on Systems, Man, and Cybernetics*, vol. 39, no. 2, pp. 517–529, 2009.

[53] J. M. Mauricio, A. Marano, A. Gomez-Exposito and J. L. Martinez Ramos, "Frequency Regulation Contribution Through Variable-Speed Wind Energy Conversion Systems," *IEEE Transactions on Power Systems*, vol. 24, no. 1, pp. 173–180, 2009.

Chapter 6

Clean generation in microgrids

Jorge Morel

6.1 Introduction

6.1.1 Overview

Clean generation in microgrids is one of the most important aspects to be considered in microgrids and smart communities in order to achieve a low-carbon society, not only for its capacity to reduce greenhouse gas (GHG) emissions, but also from the perspective of variability and uncertainty in the generation inputs, local-generation properties and economic aspects.

Sustainable generation of electricity depends on the development and adoption of renewable generation systems. According to Renewable Energy Policy Network for the 21st Century (REN 21), renewable energy sources contributed 23.7% of the total electricity generated in the world at the end of 2015. Figure 6.1 shows the share of each type of major renewable energy generation within this portion [1]. As can be seen, hydropower is still by far the largest contributor to clean generation in the world. In contrast, the proportion of geothermal, concentrated solar power (CSP) and ocean energy combined is still small. Wind energy is the second largest contributor followed by bioenergy and solar photovoltaics (PV).

In this chapter, a complete overview of the most important renewable and clean generation systems applied today in microgrids is presented.

6.1.2 Chapter's aim and scope

Even though the present chapter title includes the word 'clean', it should be understood in the sense of lower GHG emissions compared to the traditional fossil-fuel-based centralised generation systems.

The aim of this chapter is to present a broad overview of the main clean generation methods existing today in microgrids. Since a complete and detailed analysis of each of them would become impractical for a single chapter in a book, and also considering that there exist advanced technical literature addressing these types of generation, the clean generation systems are succinctly described, and the main characteristics, including current market status and some advantages and disadvantages, are presented.

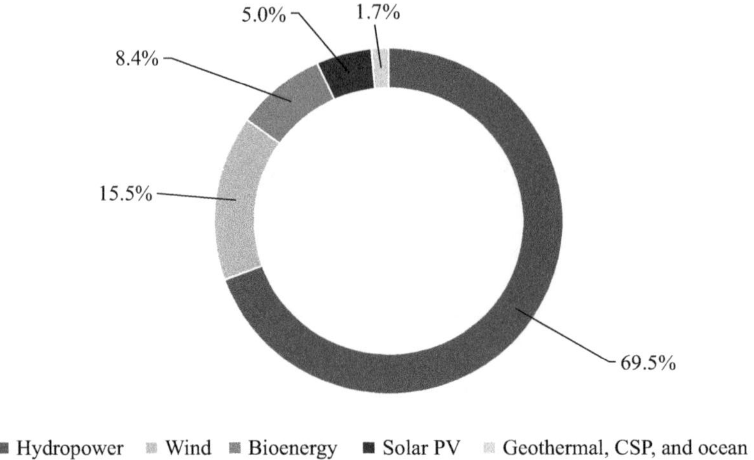

Figure 6.1   Share of renewables in the global electricity generation in 2015

Conventional centralised power generation, such as nuclear and fossil-fuel-based power centralised power plants, are not included in this chapter.

Following the sections presenting the different generation systems, the last section presents a case study of a microgrid covering a small city located in a cold region in Japan showing that in a given microgrid interconnected with a larger centralised grid, its annual total demand can be supplied with 100% of clean energy and renewable energy.

6.2   Conventional clean generation

6.2.1   Wind generation

Wind turbines (WTs) transform the kinetic energy of the wind into electrical energy by the rotation of the blades which turns an electrical generator. They can be broadly classified into vertical and horizontal axis WTs. For large WTs, in the MW range, the most common is the horizontal type. Vertical units are fewer and mainly for small units in the kW range, up to 100 kW. Most of the WTs today are connected to the grid using power converters.

The installation of WTs has been exponentially increasing around the world since the year 2000 reaching 433 GW in 2015, of which China represents one-third of this total value. Small WTs for microgrids represent only a small fraction of this total installation capacity. The number of cumulative installed small WTs at the end of 2014 was 945,000 units, 8.3% over the values of the previous year. The total installed capacity for small WTs at the end of 2014 was only 830 MW, for a total of 433 GW worldwide. Figure 6.2 depicts the distribution of this total installed capacity among the top three countries and the rest of the world [2]. China represents 41%, the United States 27% and the United Kingdom 16%

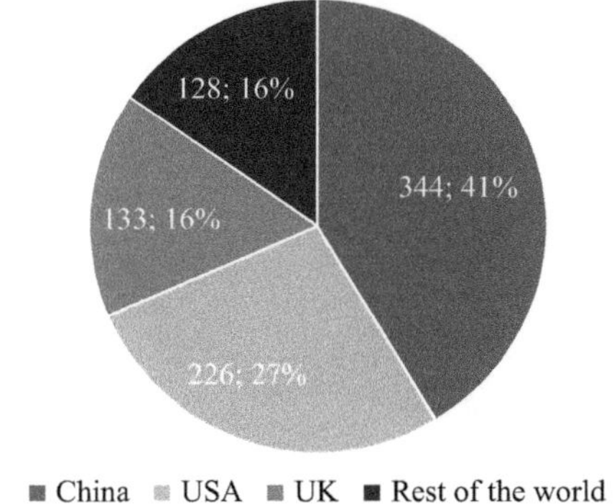

Figure 6.2   Cumulative installed capacity for small WTs by country, in MW

of the total capacity of small WTs in the world in 2014. China and the United States together have more than two-third of the total installed capacity of small WTs [1–3].

The main advantages of wind generation are the zero emission of CO_2 during operating and the distributed aspect of the resource. Cost is still a very important factor to consider compared to traditional centralised generation. Variability together with the unpredictability of the wind speed are other factors to be carefully considered. For large WTs, there may be issues associated with large space utilisation, bird strikes and aesthetics.

6.2.2   Solar generation

Solar generation is the transformation of the solar radiating energy to electricity, by photovoltaic panels (solar photovoltaic energy, or solar PV) or by concentrating the energy to produce heat to be utilised, usually to produce steam to move an electrical generator (concentrating solar power, or CSP).

The total installed capacity of solar PV in the world in 2015 was 227 GW, with Germany, China and Japan, representing more than half of the total installed capacity. The amount of new installed capacity in 2015 was 50 GW. China installed 15.2 GW, Japan 11 GW and Germany 11 GW. The introduction of solar thermal based electricity generation by using CSP is currently much less compared to the PV generation with only 4.8 GW in 2015, mainly in Spain, with 2.3 GW, and the United States with 1.7 GW. Figure 6.3 shows the proportion of PV and CSP in the total installed capacity in 2015 [1]. As can be seen, almost all the solar-based electricity is generated by PV systems in the world [1,4,5].

The main advantage of solar generation is zero CO_2 emission during operation and the distributed characteristic of the resources.

Costs still are important factor in the deployment of solar technology. Also, variability and uncertainty in the resource impose necessarily the utilisation of storage systems.

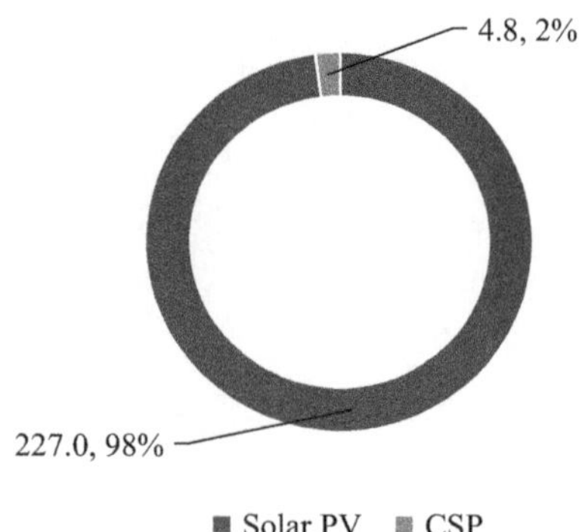

Figure 6.3   Share of electricity generation by solar energy in 2014, in GW

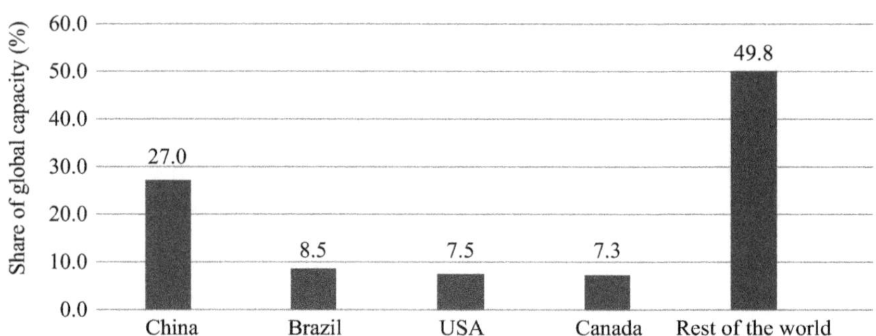

Figure 6.4   Share of the global installed hydropower capacity in the world

6.2.3   Hydro generation

Hydropower generation is the conversion of the kinetic energy of water to electricity by the rotation of a water turbine coaxially connected to an electric generator. Most of the hydroelectric power plants today are centralised, although small hydropower stations are also recently being installed. For microgrids, small hydro generation is utilised, which uses generators with capacities up to 10 MW.

They can be classified as run-of-river hydropower, storage hydropower, pumped-storage hydropower and ocean hydropower. The main roles of hydropower plants have been based load and energy storage in centralised power grids. The total global installed capacity was 1,064 GW in 2015. Figure 6.4 depicts the distribution of global installed capacity among the top four countries representing half of this total installed capacity: China, Brazil, the United States and Canada [1]. The global potential of small hydropower (up to 10 MW) is approximately 173 GW, with an installed capacity of 75 GW in 2012 [1,6–8].

One of the advantages of this type of generation is no CO_2 emissions and its sustainable and renewable properties. Unfortunately, dams can occupy thousands

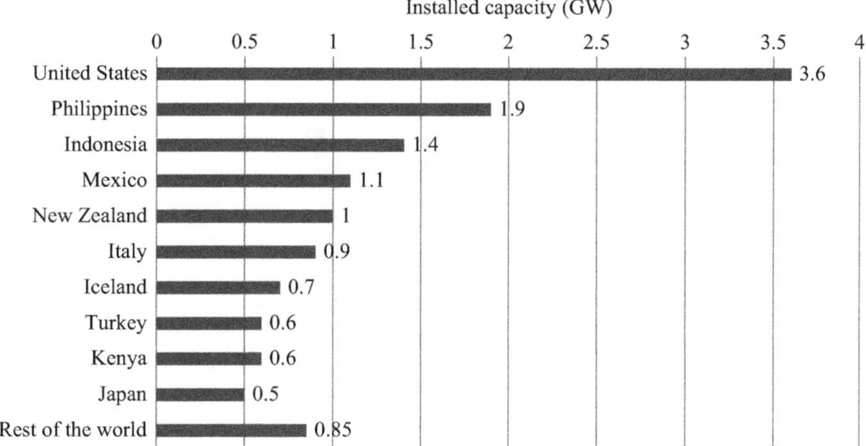

Figure 6.5   Share of geothermal power installed capacity in 2015

of kilometres affecting its surrounding. Detailed studies are necessary in hydrology, topography, geography, environment and social characteristics to more precisely assess the possible impacts.

6.2.4   Geothermal generation

Geothermal generation is the production of electricity by the transformation of the thermal energy coming from within the earth. It is mainly used for based load since its output is highly predictable. The technology to produce electricity from geothermal energy is mature. The heat can be utilised to produce electricity, directly as for geothermal heat pump.

The total installed capacity at the end of 2015 was 13.2 GW, a slight increase from 12.9 GW in 2014. The total production of energy was 75 TWh for heat and the same figure for electricity in 2015. Figure 6.5 shows the distribution of geothermal installed capacity by the top ten countries and the rest of the world in 2015 [1]. In this figure, the United States represents one quarter of the total installed capacity around the world.

There are no CO_2 emissions during generation. However, there is some CO_2 coming from underground fluids, but much less than fossil fuel plants. This CO_2 from the underground fluids is mainly for a high-temperature geothermal source. For low temperature, it is only a fraction of that corresponding to a high-temperature source. Also, it only can be applied in regions with geological conditions where fluids can practically transfer the heat from deep inside the earth to near surface [9,10].

6.2.5   Cogeneration

Cogeneration encompasses the simultaneous generation of heat and electricity with an increased efficiency of the conversion of fuel into usable energy in comparison

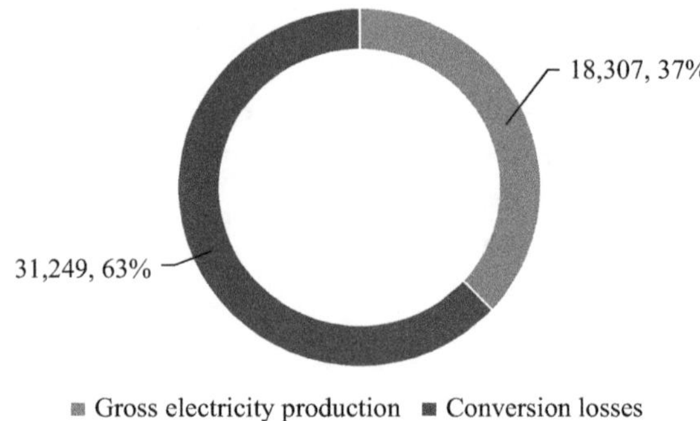

Figure 6.6   Proportion of energy converted to electricity in a conventional power plant, in TWh

with other conventional thermal generation systems. This increase in efficiency comes from the partial recovery of exhaust heat produced in the generation of electricity. A variety of fuels, such as coal, natural gas and biomass, can be utilised. Cogeneration is becoming more and more important in today's clean energy markets because renewables and cogeneration can be integrated together to complement each other. Biomass, geothermal and CSP can be utilised in the process. Also, for coal-based power plants, cogeneration is a more efficiency way to provide electricity and heat to customers. The efficiency of thermal power plants was 36% in 2011 and that of cogeneration generation systems was about 58%. State-of-the-art cogeneration units can achieve even 90% efficiency. Only 9% of electricity generation employs cogeneration technology. In contrast, in a conventional power plant, the losses may represent a high percentage of the primary fuel consumption, as shown in Figure 6.6 [11,12].

The main advantage of cogeneration is its higher efficiency, which converts it in a cleaner way of generating electricity and heat, and that it can be integrated with renewables.

6.3   Unconventional clean generation

6.3.1   Bioenergy

Bioenergy can be defined as [13]

> *Energy derived from biomass, which includes biological material such as plants and animals, wood, waste, (hydrogen) gas, and alcohol fuels. In essence bioenergy is the utilisation of solar energy that has been bound up in biomass during the process of photosynthesis. It is a renewable energy source.*

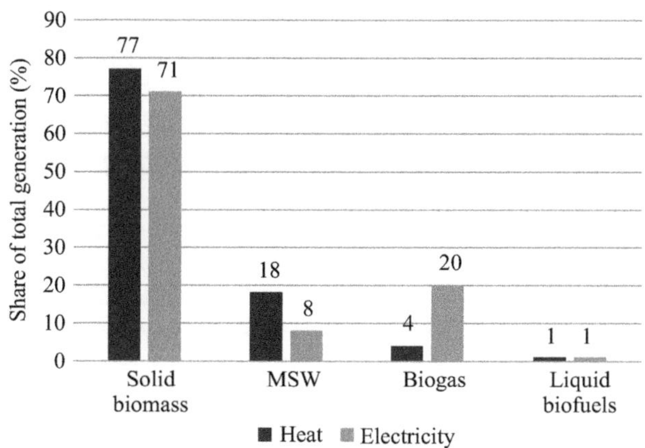

Figure 6.7   Share of heat and electricity generated by different biomass sources

Biofuels, or the fuels obtained from biomass, can be classified as [14] follows:

1. Liquid biofuels, where the main feedstocks are starch crops (maize, wheat, etc.), sugar crops (sugar cane, sweet sorghum, etc.), oil crops (soybean, animal fats, waste oils, etc.) and lignocellulosic biomass (bagasse, wood, algae, etc.).
    The first two are used to produce bioalcohols, such as ethanol and butanol, and the third one is to produce biodiesel. The fourth one is for second and third generation biofuels (biohydrogen, bio-methanol, etc.). Liquid biofuels are mainly applied to transport sector.
2. Solid biofuels, of which main feedstocks are forest, agricultural residues and wastes, are used to produce pallet, charcoal, biochar, etc., and are mainly applied to the generation of heat and power.
3. Gaseous biofuels, of which main feedstocks are solid and liquid biofuels, forest and agricultural residues and wastes, are used to produce biogas and syngas to be applied mainly in the generation of heat and power and transportation sector.

Biofuels are considered to be carbon neutral. If the generation of power is combined with carbon capture and storage technology, it can in fact contribute to the removal of GHGs from the atmosphere. Some issues still remain, such as its cost relative to fossil-fuel based generation and possible health impact to the communities close to the combustion plants [15].

Figure 6.7 shows the distribution of heat and electricity generated by different types of biomass sources, including municipal solid waste (MSW), in 2016 [1].

6.3.2   Ocean energy

Ocean or marine energy refers to the harvesting of the energy contained in the sea by technologies such as tidal energy, wave energy and ocean thermal energy conversion.

It is still in its initial stages of development, and more considerable contribution to CO_2 reduction is expected from 2020. The total installed capacity of ocean energy, mainly tidal generation, was 530 MW in 2014, with little capacity added in that year, mainly pilot projects. From this installed tidal capacity, a 254-MW tidal power plant is located in South Korea, and a 240-MW tidal power plant is located in France [1].

There are two main types of tidal generation technology: tidal range and tidal current. The first one uses a barrage (e.g. a dam) to exploit the height difference generated due to high and low tides. It has the most advanced technology and is currently the most common. The second one utilises the current produced during changes from high to low tide [10,16,17].

The main advantages are that there are no CO_2 emissions during operation and that it is highly predictable. The main drawbacks are that it is limited to areas where the sea is available, and the local conditions are proper to install the generation. Also, technology development and cost constitutes challenges still being faced by the industry.

6.3.3   *Waste-to-energy*

Waste-to-energy generation is the generation of electricity, heat and fuels for transportation, from several types of residual waste: semi-solid, liquid or gaseous. As main technologies, the following can be mentioned:

1. Combustion, or the burning of waste to produce electricity and heat
2. Pyrolysis and gasification, or the heating of fuel with low oxygen to produce synthesis gas or syngas. This gas can be used for generation or as input material to produce hydrogen, biofuels and methane, and
3. Anaerobic digestion, or the production of methane-rich biogas by using microorganisms from organic waste. Another method is to capture the methane produced in landfills.

An advantage of this type of generation is that the global warming potential of the methane in a landfill is higher than the CO_2 emitted during combustion of the waste, representing a reduction in the GHG emissions. Another important point is that it can be utilised as base load and that the fuel, the waste, has a negative price [10,18].

As disadvantages, the emission of pollutants, such as CO_2, N_2O and NOx, is still present during burning of the fuel.

6.3.4   *Fuel cell*

Fuel cells generate electricity by combining hydrogen and oxygen by an electrochemical process, producing heat and water as by-products. The process is more efficient and clean than a generation process where fuel is burnt. If the hydrogen, which acts as the energy carrier, is produced by a non-emitting process, it can be said that the generation is completely clean. The main types of fuel cells available today in the markets are as follows: molten carbonate fuel cell, solid oxide fuel cell,

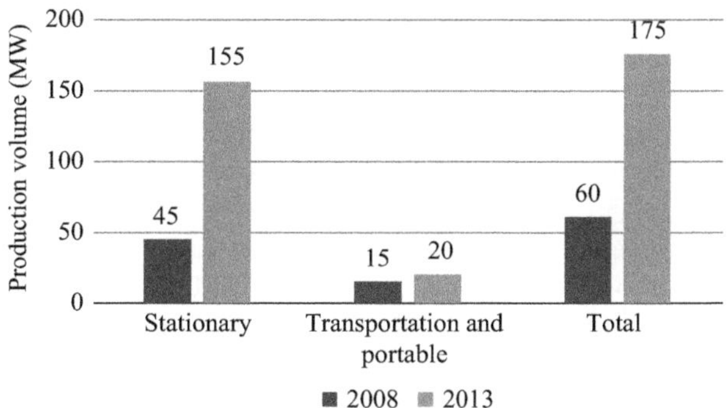

Figure 6.8   Production volumes of fuel cells for two representative years

phosphoric acid fuel cells and proton exchange membrane fuel cells. Each of the fuel cell technologies has its advantages and disadvantages. Fuel cells can be utilised for different applications, broadly in two groups: stationary (for power generation in a fixed location, such as back up or cogeneration) and movable (for transportation and for portable applications). The distributions of production volume for these two broad groups in the year 2008 and 2013 are shown in Figure 6.8 [19,20].

The main advantages of fuel cells are that they can be used not only for electricity generation in microgrids, but also for transportation, replacing oil-based fuels. Also, fuel cells can be used in cogeneration systems. Another important point is that hydrogen to be used with fuel cells can, with the proper technology, be stored for long periods and can be transported over long distance, which combined with low-carbon-based transport system, can represent a considerable global CO_2 emission reduction.

The disadvantages of fuel cells are that hydrogen requires, in most cases, special handling mechanism and that CO_2 may be emitted in the hydrogen production process.

6.4   Case study[*]

6.4.1   Introduction

Concerns about the adverse effects of global warming on the environment and the need of sustainable use of energy resources have been driving governments around the world in the definition of policies regarding the reduction of CO_2 emissions through the utilisation of renewable energy sources. The United States and the

[*]This section is based entirely on the paper 'Operation Strategy for a Power Grid Supplied by 100% Renewable Energy at a Cold Region in Japan' by Jorge Morel, Shin'ya Obara, Yuta Morizane, first published in the *Journal of Sustainable Development of Energy, Water and Environment Systems (JSDEWES)*, volume 2, issue 3, September 2014, pages 270–283.

European Union have been leading the development of clean and sustainable energy technologies through the implementation of specific policies and targets [21,22].

One of the technologies applied in power engineering that will allow the interconnection of large numbers of intermittent renewable sources, such as wind and solar, is the smart grid. Besides the possibility of interconnection of variable output renewable generators, smart grids offer other benefits such as active participation of customers in the control of the peak demand of the entire system as well as increased customer participation in the electricity market [23].

Japan, due to its strong and reliable power system, has been focusing mainly on nuclear power to achieve the CO_2 emission targets. The Fukushima Daiichi nuclear disaster revealed the real limitations of the system. Furthermore, forced by a strong public opinion resisting nuclear power generation, Japan now has the challenge to rebuild and adapt its power system based mainly on the deployment of intermittent and clean renewable energy generation, with a less centralised architecture, to increase its reliability in the case of natural disasters.

Tohoku Fukushi University's experimental microgrid in Sendai City demonstrated how important microgrids can be for a country frequently menaced by natural disasters. The microgrid was directly affected by the disaster, and it showed resilience [24]. The Japanese government has now established policies to achieve a more reliable and resilient power system [25]. Another important benefit of constructing a self-sufficient microgrid is the reduction of long-distance power transfer from centralised power plants, reducing losses and congestion in the transmission lines.

In a cold region, such as Kitami, where there is significant variability in seasonal demand with marked low consumption in summer and high demand in winter, there is a need to shift surplus renewable generation from summer to winter. Storage systems or a coordinated operation with the local power utility should be considered. It is also important to grasp the variability of the total supply considering the type and proportion of each intermittent generation. For example, solar photovoltaic power output changes much faster than a WT power output due to lack of inertia in photovoltaic systems. Also, a tidal farm power output is lower and much more predictable in the long and medium term, compared to that of a wind farm.

Development of energy systems, which considers a range of infrastructure, such as transport, heat and electricity, has been studied for an optimal design in a given region to create a sustainable energy supply system, reduce CO_2 emissions and utilise intermittent renewable generation. Among the leading research groups in this area is Aalborg University in Denmark that has developed an energy system analysis tool called EnergyPLAN and performed the design of energy systems for specific regions in Europe [26–29].

Japanese industry and academia, due to the reasons previously addressed (reliable power grid and focus on nuclear power), have shown relatively slower development on smart or microgrid technologies before the nuclear disaster in 2011. However, a major effort has been put in from the beginning by certain research groups [24,30,31]. After the nuclear disaster of 2011, research activities on independent microgrids for local generation and local consumption, containing sustainable renewable energy generation such as wind, solar and tidal power, have increased considerably. Operation of microgrids, aiming at the reduction of CO_2 emissions and the safety of energy supply, in cold, urban and remote areas has been studied [32–34].

The more rapid oscillations of renewable generator outputs in a microgrid make the utilisation of fast acting batteries necessary to match instantaneous imbalances, in contrast with a traditional power grid where the outputs of the generation units can be controlled and match the slower-changing aggregated demand. The utilisation of sodium–sulphur (NaS) batteries was mainly analysed in the past for suppressing instantaneous or fast changing impacts in the grid, including the possibility of independent active and reactive power control in this type of storage systems [35].

On a longer time frame, more availability of resources during certain seasons and low demand in the same seasons lead to surplus energy that may be stored for future use when needed. For this case, storage systems have been analysed for economic benefit and environmental impacts [36,37]. In [36], NaS batteries are compared to a storage system based on organic chemical hydride but the dynamic performance (fast charging/discharging capability) of the NaS battery was not analysed. In [37], a system with no storage system is analysed. Here, good CO_2 reductions are obtained despite the absence of battery storage. However, for a 100% renewable supply, batteries may be necessary to shift energy between seasons or to keep frequency balance, as well as to compensate for any transient faults in the system.

For a complete analysis of a power grid, consideration of dynamic properties of supply and demand is of vital importance. All the energy system designs mentioned above do not consider this aspect.

This work demonstrates that a medium size city located in a cold region, with particular annual load characteristics, can be supplied entirely by renewable energy, reducing completely its CO_2 emission without negatively affecting the quality of the electrical system, and completely breaking its dependence on nuclear power generation.

6.4.2 *System under study*

6.4.2.1 Location

Kitami City is located in a cold region, on the Hokkaido Island in the northern part of Japan, as shown in Figure 6.9. Kitami has an annual demand characteristic with a high heat-to-power demand ratio during winter. The temperature reaches a minimum of -20 °C in winter and a maximum of 35 °C in summer. Despite the low temperature in winter, the city gets little rainfall and snowfall. The city has rich natural resources such as wind, solar and tidal power that can be utilised for the generation of clean electrical energy. It is one of the richest areas in solar irradiation in Japan. Also, it has open areas with good average wind speeds which can also be exploited for the generation of electricity. Furthermore, the Saroma Lake tidal current speeds offer the possibility of exploitation of tidal generation [38].

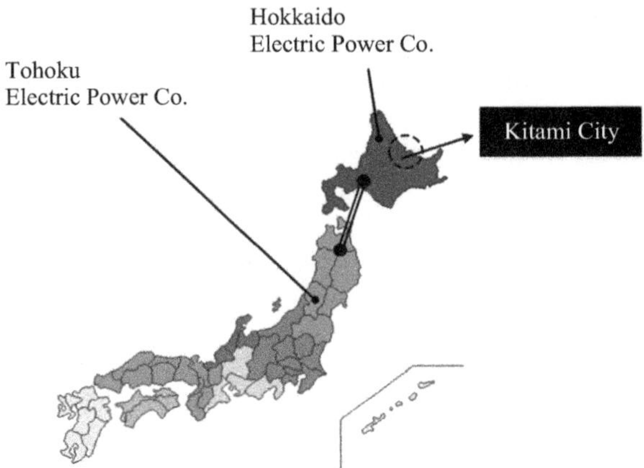

Figure 6.9   Kitami City location

6.4.2.2   Transmission network

The Japanese power system consists of ten power companies that supply energy to specific and semi-independent regions. They are interconnected (except Okinawa Electric Power Company) through transmission lines with limited capacities. The Hokkaido Electric Power Company (HEPCO), shown in Figure 6.9, with a total installed capacity of 7,500 MW, supplies power to the Hokkaido Island, where Kitami City is located. HEPCO is connected to Tohoku Electric Power Company, located in Honshu, the main island of Japan, by a high-voltage direct current (HVDC) transmission system (indicated by a double line Figure 6.9), with a capacity of 600 MW, approximately 8% of HEPCO's total installed capacity [39].

The power system of Kitami City is connected to the local utility HEPCO which currently provides the power for the entire city.

The simplified scheme of the Kitami's power system considered for simulation, including the proposed location of renewable generators and storage systems is depicted in Figure 6.10. The names and rated capacities of the substations are shown in Table 6.1.

Selection of the location of the renewable generators was made on the basis of the available resources in the area. Location of the NaS batteries was selected to be the Rubeshibe substation where the bulk power is coming from the conventional power plants of HEPCO. This selection has no direct effect on the results of frequency quality presented in this paper because of the short distances involved. However, from an economics point of view, the most appropriate locations and sizes must be carefully considered.

As shown in Figure 6.10, there are mainly two points of connection to the local utility: The Rubeshibe and the Memanbetsu substations. Each of them is supplied by a double-circuit transmission line of 187 kV. Regions 1 and 2 shown in this figure are two systems with no generating units. The total load for Region 1 is 124 MVA and that for Region 2 is 87 MVA. The two thin dashed lines represent the two

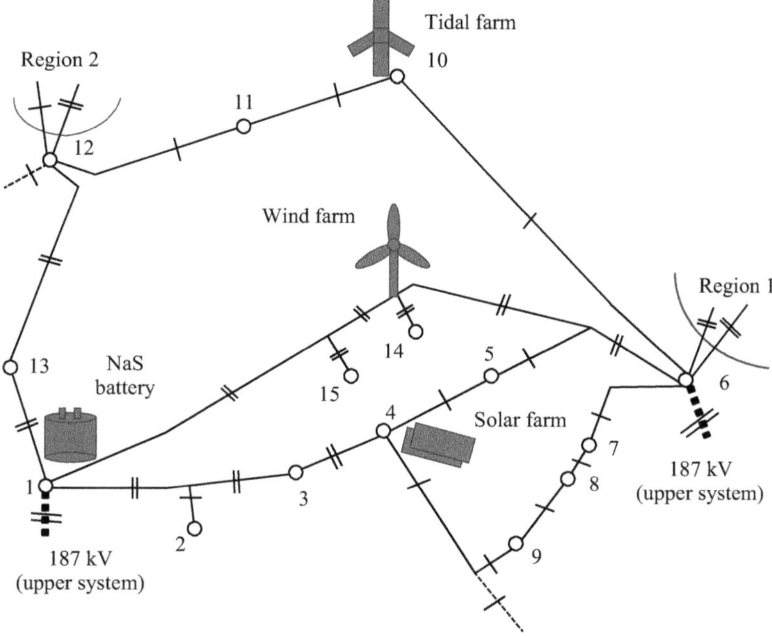

Figure 6.10   Power system of Kitami City

weak connections to other systems which are not considered in this work. The small circles represent substations and single or double-circuit transmission lines are indicated by one or two short transverse lines over the lines connecting the substations. All lines are overhead, with rated voltage of 66 kV, and short, with lengths of less than 40 km. Typical tower and conductor data for 66 kV transmission line is considered. Each substation is composed of 66 kV/6.6 kV step-down transformers with the rating indicated in Table 6.1. The reactive power parts of the loads are assumed to be compensated because they do not affect the frequency analysis.

6.4.2.3   Renewable generation

Wind power

Currently, most of WTs in the market are variable-speed types: doubly fed induction generator (DFIG) and full-scale converter types, which are capable of independently controlling active and reactive power injected to the system. In this work, the wind farm is simulated using an aggregated model of DFIG-based WTs, modelled by MATLAB®/Simulink®.

Solar power

Photovoltaic type solar farms are considered. They are connected to the transmission network via inverters which have the capability of independently controlling, the amount of reactive power injected to the network for voltage regulation purposes. For frequency study purposes, the faster dynamics are not considered;

Table 6.1   Substation rated capacities

No.	Name	Capacity (MVA)
1	Rubeshibe	280
2	Kuneppu	6
3	Kitaminishi	20
4	Kitami	35
5	Tabata	22
6	Memanbetsu	200
7	Bihoro	20
8	Inami	10
9	Tsubetsu	12
10	Tokoro	12
11	Saroma	10
12	Engaru	18
13	Ikutahara	3
14	Kiyomi	30
15	Ainonai	12

Table 6.2   Aggregated renewable generator parameters

Generation	Rated capacity (MVA)	Point of connection (no.)
Wind farm	150	Kiyomi (14)
Solar farm	100	Kitami (4)
Tidal farm	1.5	Tokoro (10)

the solar farm is modelled in this work as a first-order system with a short-time constant of 10 ms.

Tidal power

Horizontal axis tidal turbines are considered [40]. They have similar structure and working principle as the WTs. For short-term frequency studies, it can be assumed that the turbines have constant power output. For long-term studies, the variability can be forecast with a high degree of accuracy. The tidal farm is simulated using an adapted version of the DFIG-based WT. These assumptions are valid due to the similarity between the wind and tidal generation systems and the time scale considered for simulation.

In this paper, only the highly variable power output of the wind and solar farms are simulated as perturbations to the system due to their higher relative outputs and variability compared to those of the tidal farm.

The parameters of the renewable generators are shown in Table 6.2.

6.4.2.4   Storage systems

For long-term energy storage aiming at the seasonal and daily energy shift, a storage system with slow dynamics but with high energy density is utilised. For seasonal energy shifting, an organic chemical hydride type system is employed [36].

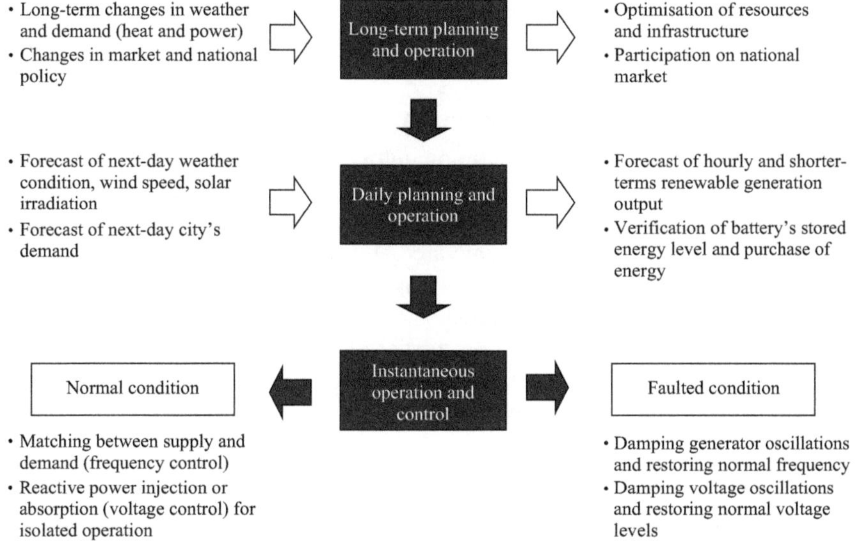

Figure 6.11   Overall operation scheme

For instantaneous and fast demand–supply imbalance compensation, NaS batteries are employed due to their fast charge and discharge capability. They have been satisfactorily applied in the levelling of power output fluctuations of wind farms. They can also be used for load levelling and load peak shaving [41].

In this work, a simplified model of the NaS battery is considered. A first-order system with a small time constant of 10 ms is used to represent the fast dynamics of this type of batteries.

6.4.3   Operation strategy

The operation strategy is divided into three parts, according to the time frames involved. The overall operation scheme is shown in Figure 6.11.

6.4.3.1   Long-term planning and operation

In order to determine the optimal utilisation of the available resources and infrastructure, together with the strategy for the participation of the proposed system in the national electricity market, careful assessment is necessary, focusing on key elements such as weather change, demand trends and change in the energy policy for the entire country.

From the point of view of available and exploitable resources in the city, hourly averaged balance evaluation between supply and demand, for 1-year operation, from April 2012 to March 2013, is presented.

The weather data affecting the output of the renewable generation units are taken from Japanese meteorological and marine agencies, and from the solar power facilities at Kitami Institute of Technology, Japan. The hourly annual energy demand is divided into demand for light and power, and demand for space heating. The power for heating is assumed to be provided by heat pumps.

6.4.3.2   Daily planning and operation

The demand for the city and the generation output of wind, solar and tidal farms are forecast in advance, for one-day operation. These predictions depend directly on the season, weather condition, tidal current and wind speeds and solar irradiation.

If the renewable energy to be generated and that stored in the batteries is not enough to supply the forecast demand at any instant in the day, the deficit is purchased from the local utility at previously agreed price and time of the day. On the contrary, if there is a surplus of energy, this is sold to the local utility at previously agreed conditions. This daily operation is made in order to set the net amount of energy interchanged with the local power utility equal to zero.

6.4.3.3   Instantaneous operation and control

The balance between demand and supply is performed instantaneously and automatically by a controller with optimal characteristics since a certain degree of randomness is involved. If the daily forecast of demand and supply is properly established, the last stage in balancing supply and demand is the instantaneous balance of the outputs of generators and energy storage, and the corresponding demand. The instantaneous balancing controls the frequency of the system. Since the renewable generators are highly uncontrollable, with fast changes that may not be compensated by the conventional generators in the system, the control task is performed by the NaS battery through charge and discharge operations.

Furthermore, during faulted conditions, the NaS battery must be capable of dealing with the disturbance by injecting and absorbing active and reactive power to recover the normal operating condition. In the case of isolated operation, WTs can support frequency by emulating conventional synchronous generators and by injecting reactive power to the system [42,43]. In this section, only the normal condition case shown in Figure 6.11 is considered.

Another important parameter of electricity quality, the voltage level, is assumed to be controlled by the local utility. This assumption is valid since the distances involved are short and the relative size of the target system is small.

For the instantaneous control, during normal operating conditions a model predictive control (MPC) approach is considered due the degree of uncertainty and randomness involved, and due to its capability to deal with multiple inputs and multiple outputs, in contrast to conventional proportional-integral controllers [44].

In Figure 6.12, the actual frequency of the system f_sys is compared with the reference frequency f_ref of 50 Hz, and a signal is sent to the battery for charging and discharging operation (active power, or P control).

The dashed arrows indicate possible inputs and outputs in the case of operation of isolated systems where voltage must be entirely controlled by renewable generator (reactive power, or Q control), and the WTs can participate in the frequency regulation of the system during abnormal conditions.

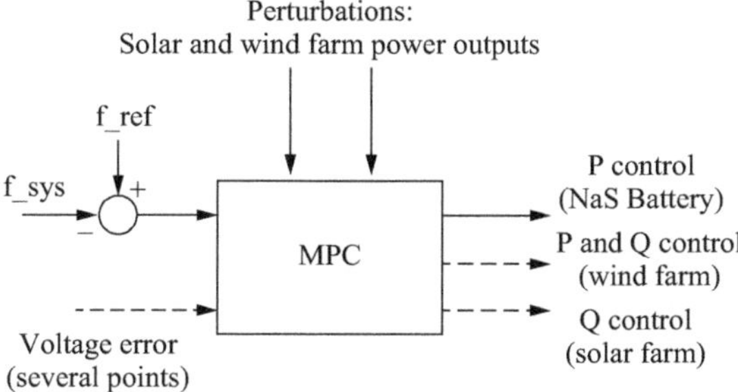

Figure 6.12   MPC control scheme

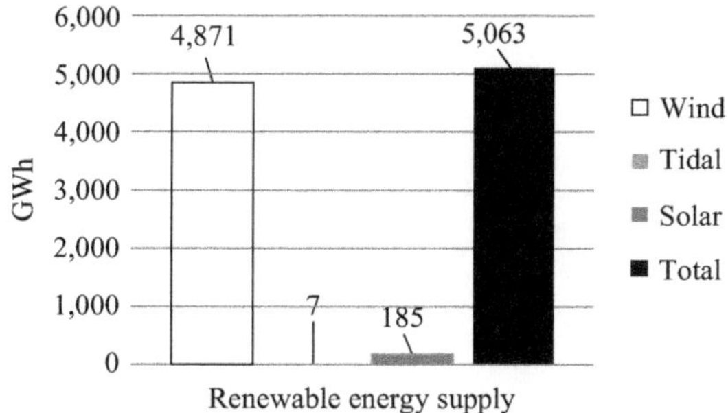

Figure 6.13   Exploitable renewable energy in Kitami

6.4.4   Simulation results

First, static simulation is used to assess the total annual renewable generation and the total annual demand in the city, based on actual past data. Second, dynamic simulations are performed to evaluate the impact of the intermittent renewable power outputs on the frequency of the system and to demonstrate the favourable effects of a fast charging–discharging battery in reducing the frequency oscillations and keeping them within permitted ranges.

6.4.4.1   Annual supply and demand evaluation

The exploitable renewable power generation together with the annual demand of the city are shown in Figures 6.13 and 6.14, respectively. According to results, the total annual demand can be supplied completely by renewable generation. However, shorter term balances, such as monthly, daily and hourly balances, must be carefully considered in order to define the characteristics of the storage systems needed.

6.4.4.2   Instantaneous energy balance

Three cases are considered. Step change in wind speed, loss of an important load in the system and random variation in solar irradiation.

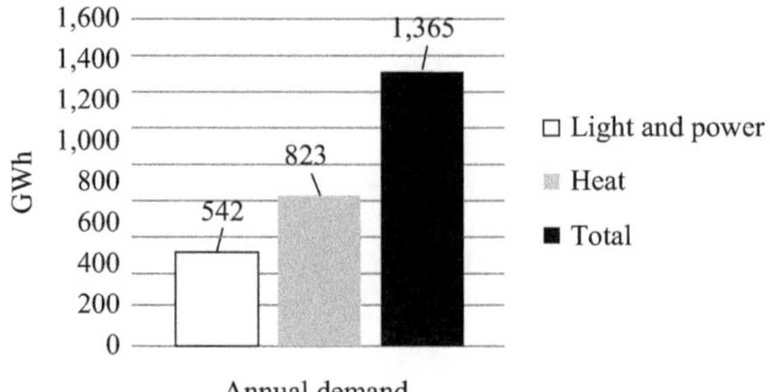

Figure 6.14   Annual demand of Kitami

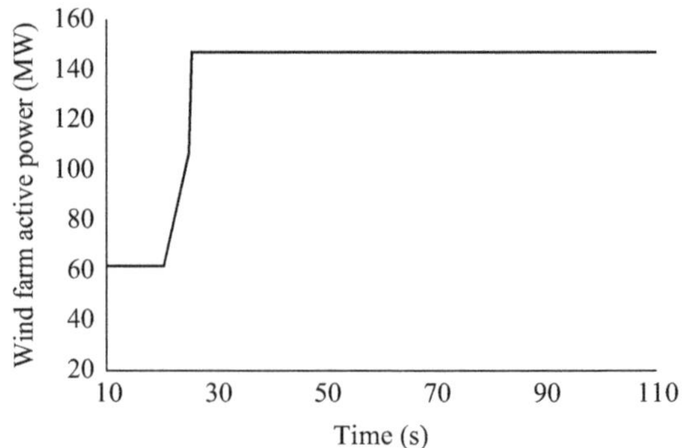

Figure 6.15   Wind farm power output

In this paper, a proportional-integral controller showed similar results to those of the MPC controller because of the single-input–single-output characteristic of the cases analysed.

Step change in wind speed

In order to capture the impact of the output of the wind farms, a simple step change in wind speed is selected, keeping other outputs constant. Figure 6.15 shows the power output of the wind farm due to an increase in wind speed from 10 to 15 m/s, at $t = 10$ s.

It can be noticed that despite the step change in the wind speed, the WT output follows a ramp profile. This shows one of the main differences between the characteristics of a wind and a solar farm outputs. Solar farm output follows a much closer profile to the shape in the input due to its lack of inertia.

Figure 6.16 shows the variations in frequency for the case of wind speed step increase, with and without the action of the controller applied to the NaS battery. Without the support of the fast charging and discharging capability of the battery, the load frequency control (LFC) of the conventional generators is not capable of keeping the frequency within the permitted range of ± 0.3 Hz.

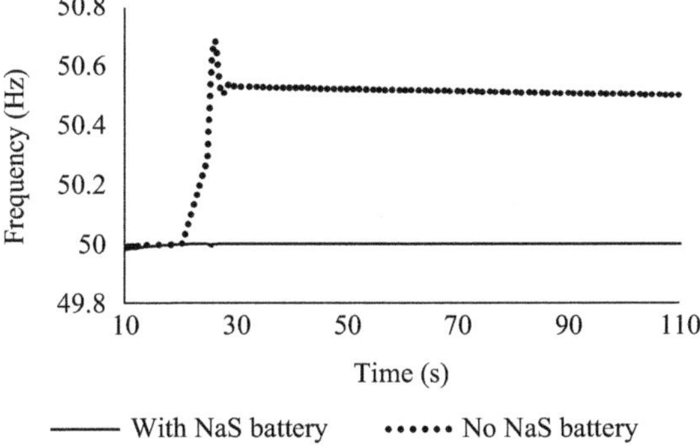

Figure 6.16   *Frequency variation for a step increase in wind speed*

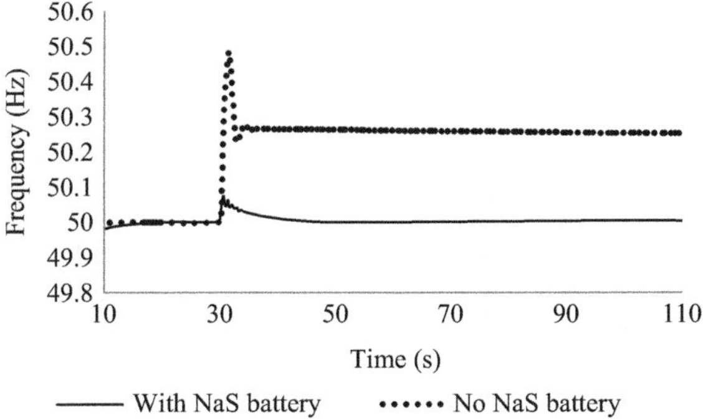

Figure 6.17   *Frequency variation for a loss of load in the system*

Loss of a large load in the system

The loss of a load of 40 MVA at $t = 30$ s is simulated in order to analyse a large instantaneous impact on the system. The loss of a generator or a load behaves as step changes in power that affects the total balance in the system.

Figure 6.17 shows the variations in frequency for the case of a step change in the system's load, with and without the action of the controller applied to the NaS battery.

Here, as in the case of wind step change, the LFC of conventional generators is not capable of keeping the frequency variation within the range of ± 0.3 Hz. On the other hand, the fast dynamics of the NaS battery allows the variation to remain bounded in this range.

Figures 6.10 and 6.11 show the active power variation of the NaS battery for the two scenarios simulated above. It can be noticed how the NaS battery absorbs satisfactorily the two types of variations in the system, that of wind farm output and that of loss of load.

Figures 6.18 and 6.19 show the active power variation of the NaS battery for the two scenarios simulated above. It can be noticed how the NaS battery absorbs

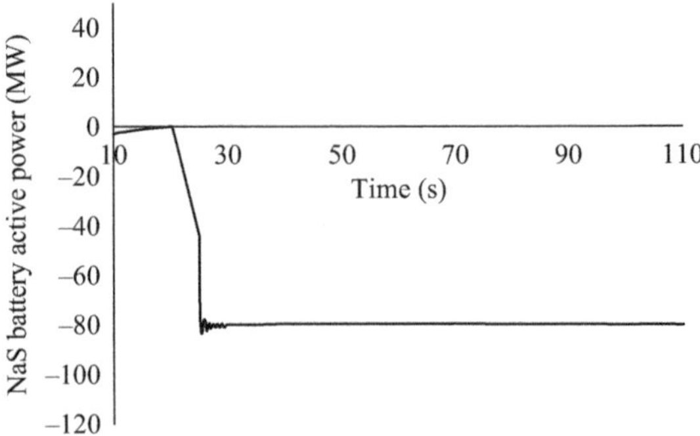

Figure 6.18   NaS battery active power for a step increase in wind speed

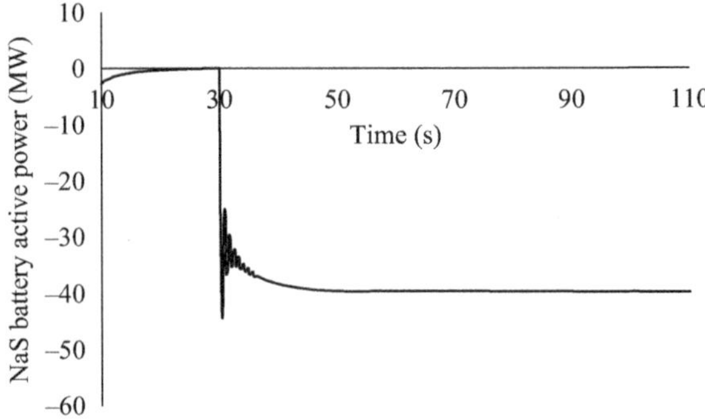

Figure 6.19   NaS battery active power for a loss of load in the system

satisfactorily the two types of variations in the system, that of wind farm output and that of loss of load.

Random solar irradiation

A random variation in solar irradiation is considered as the third scenario. The simulations are intended to show how the controller, despite the rapid and random changes, can keep frequency variations within the permitted range of ± 0.3 Hz.

Figure 6.20 shows the power output of the solar farm for random irradiation changes. Due to the fast dynamics of the solar cell, the output follows the random irradiation pattern.

Figure 6.21 shows the variation in frequency for the random variation in solar irradiation with and without the action of the controller applied to the NaS battery. It is clear that the LFC of the conventional generators, working without the support of the fast acting NaS battery, is not capable of properly regulating the frequency, due the fast and random changes. The use of a NaS battery allows the frequency variations to remain within the permitted range.

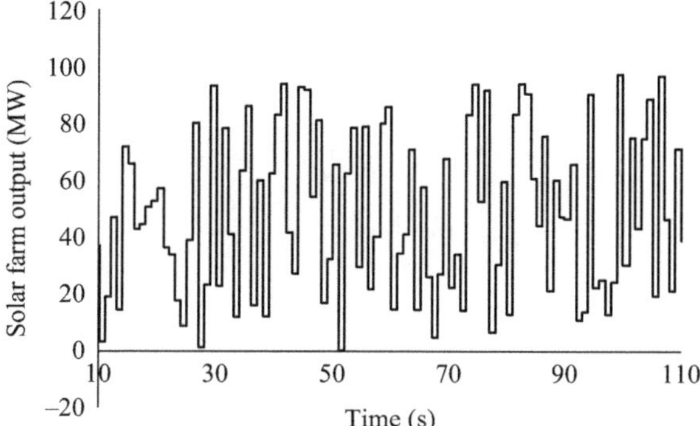

Figure 6.20   Solar farm power output for random solar irradiation

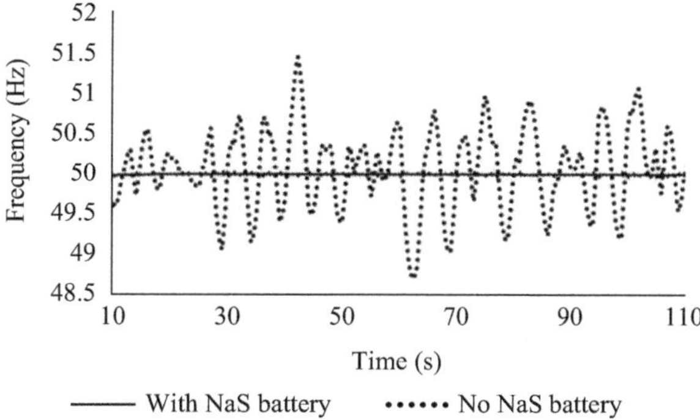

Figure 6.21   Frequency variation for random solar irradiation

Figure 6.22 shows the active power absorption or injection for the case of random variation in solar irradiation. It can be seen how fast the NaS battery changes the output set point to follow the changes in solar farm outputs.

6.4.5   Conclusion

In this section, a strategy for the operation of renewable energy generators and energy storage devices for the development of a clean smart city, with zero CO_2 emissions, independent from nuclear generation, located at a cold region in Japan, is presented. According to simulations based on actual hourly weather and demand data for 1 year, the city's annual demand can be supplied entirely by solar, wind and tidal generation.

Instantaneous, daily and long-term operational strategies are proposed. The instantaneous strategy that uses a fast dynamic NaS battery and is based on a MPC approach is detailed and dynamically simulated for changes in selected wind speed and solar irradiation, and for a loss of an important load in the system. According to

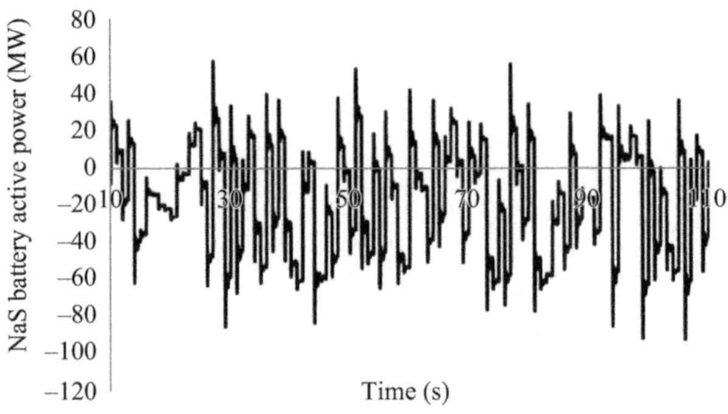

Figure 6.22   NaS battery active power for random solar irradiation

simulations, despite the highly variable outputs of the intermittent sources, the frequency variations can be kept within the permitted range of ±0.3 Hz by the proposed control system. An MPC-base control system is proposed, as it is capable of dealing with discrepancies between the parameters of real systems and that of models utilised for the controller design, as well as randomness, uncertainties and multiple-input–multiple-output control in the system.

The area where Kitami city is located is particularly rich in wind, solar and tidal resources. Although the results presented are not directly applicable to an area with less favourable conditions, the results regarding the instantaneous strategy are still applicable as they are more related to short-term power balance.

Since the studied system is considered connected to the local utility's power grid, the voltage control is assumed performed by this stronger system. For an eventual isolated operation scenario, the voltage control must be performed by renewable generators which have converters that are capable of independently controlling active and reactive power.

Nomenclature

DFIG   doubly fed induction generator
f_ref   frequency reference
f_sys   system frequency
HEPCO   Hokkaido Electric Power Company
LFC   load frequency control
MPC   model predictive control
P   active power
NaS   sodium–sulphur
Q   reactive power
WT   wind turbine

References

[1] Renewable Energy Policy Network for the 21st Century (REN21), "Renewables 2016 global status report," 2016, [Online]. Available from: http://www.ren21.net/wp-content/uploads/2016/05/GSR_2016_Full_Report_lowres.pdf [Accessed 10 September 2016].

[2] World Wind Energy Association (WWEA), "2016 Small wind world report," 2016. [Online]. Available from: http://distributedwind.org/wp-content/uploads/2016/03/2016-Small-Wind-World-Report.pdf [Accessed 10 September 2016].

[3] Global Wind Energy Council (GWEC), "Global wind report – Annual market update 2015," 2016. [Online]. Available from: http://www.gwec.net/wp-content/uploads/vip/GWEC-Global-Wind-2015-Report_April-2016_19_04.pdf [Accessed 10 September 2016].

[4] International Energy Agency (IEA), "Technology roadmap – Solar photovoltaic energy 2014," 2014. [Online]. Available from: https://www.iea.org/publications/freepublications/publication/TechnologyRoadmapSolarPhotovoltaicEnergy_2014edition.pdf [Accessed 10 September 2016].

[5] International Energy Agency (IEA), "Technology roadmap – Solar thermal electricity 2014," 2016. [Online]. Available from: https://www.iea.org/publications/freepublications/publication/technologyroadmapsolarthermalelectricity_2014edition.pdf [Accessed 10 September 2016].

[6] International Energy Agency (IEA), "Key world energy statistics 2015," 2016. [Online]. Available from: https://www.iea.org/publications/freepublications/.../KeyWorld_Statistics_2015.pdf [Accessed 10 September 2016].

[7] World Energy Council, "World energy resources – Charting the upsurge in hydropower development," 2015. [Online]. Available from: https://www.worldenergy.org/wp-content/uploads/2015/05/World-Energy-Resources_Charting-the-Upsurge-in-Hydropower-Development_2015_Report2.pdf [Accessed 10 September 2016].

[8] Small Hydropower World, 2013. [Online]. Available from: http://www.smallhydroworld.org/fileadmin/user_upload/pdf/WSHPDR_2013_Final_Report-updated_version.pdf [Accessed 10 September 2016].

[9] Intergovernmental Panel on Climate Change (IPCC), "Geothermal energy – IPCC special report on renewable energy sources and climate change mitigation," 2012 [Online]. Available from: https://www.ipcc.ch/pdf/special-reports/srren/Chapter%204%20Geothermal%20Energy.pdf [Accessed 10 September 2016].

[10] World Energy Council, "World energy resources – Survey 2013," 2013. [Online]. Available from: https://www.worldenergy.org/wp-content/uploads/2013/09/Complete_WER_2013_Survey.pdf [Accessed 10 September 2016].

[11] International Energy Agency (IEA), "Co-generation and renewables," 2011. [Online]. Available from: https://www.iea.org/publications/free-publications/publication/CoGeneration_RenewablesSolutionsforaLowCarbonEnergyFuture.pdf [Accessed 10 September 2016].

[12] International Energy Agency (IEA), "Linking heat and electricity systems," 2014. [Online]. Available from: https://www.iea.org/publications/free publications/publication/LinkingHeatandElectricitySystems.pdf   [Accessed 10 September 2016].

[13] Sustainable Energy Authority of Ireland, "Bioenergy," [Online]. Available from: http://www.seai.ie/Renewables/Bioenergy/ [Accessed 17 September 2016].

[14] Food and Agriculture Organization (FAO), "Biofuels and the sustainability challenge," 2013. [Online]. Available from: www.fao.org/docrep/017/i3126e/i3126e.pdf [Accessed 18 September 2016].

[15] Congressional Research Service, "Is biopower energy neutral?," 2016. [Online]. Available from: https://fas.org/sgp/crs/misc/R41603.pdf [Accessed 17 September 2016].

[16] Intergovernmental Panel on Climate Change (IPCC), "Ocean energy – IPCC Special Report on Renewable Energy Sources and Climate Change Mitigation," 2011. [Online]. Available from: https://www.ipcc.ch/pdf/special-reports/srren/Chapter%206%20Ocean%20Energy.pdf [Accessed 10 September 2016].

[17] International Renewable Energy Agency (IRENA), "Tidal energy – Technology brief," 2014. [Online]. Available from: http://www.irena.org/documentdownloads/publications/tidal_energy_v4_web.pdf [Accessed 10 September 2016].

[18] Renewable Energy Association (REA), "Energy from waste: A guide for decision makers," 2011. [Online]. Available from: http://www.r-e-a.net/pdf/energy-from-waste-guide-for-decision-makers.pdf [Accessed 10 September 2016].

[19] International Energy Agency (IEA), "Technology road map – Hydrogen and fuel cells," 2015. [Online]. Available from: https://www.iea.org/publications/freepublications/publication/TechnologyRoadmapHydrogenandFuelCells.pdf [Accessed 10 September 2016].

[20] U.S. Department of Energy, "Fuel cell technologies – Market report 2014," 2015. [Online]. Available: energy.gov/sites/prod/files/2015/10/f27/fcto_2014_market_report.pdf [Accessed 10 September 2016].

[21] US Department of Energy, Mission. Available from: http://www.energy.gov/mission [Accessed 24 February 2014].

[22] European Commission, Energy Strategy for Europe. Available from: http://ec.europa.eu/energy/index_en.htm [Accessed 24 February 2014].

[23] SmartGrid.gov, What is a Smart Grid? Available from: https://www.smartgrid.gov/the_smart_grid#smart_grid [Accessed 24 February 2014].

[24] Japan Smart Community Alliance, "The Operational Experience of Sendai Microgrid in the Aftermath of the Devastating Earthquake: A Case Study," 2013. Available from: https://www.smart-japan.org/english/reference/13/Vcms3_00000020.html [Accessed 24 February 2014].

[25] Ministry of Economy, Trade and Industry (METI), "Annual Report on Energy, Outline of the FY2012 Annual Report on Energy (Energy White

Paper 2013)," 2012. Available from: http://www.meti.go.jp/english/report/ index_whitepaper.html#energy [Accessed 24 February 2014].

[26] EnergyPLAN, Advanced Energy System Analysis Computer Model, Aalborg University. Available from: http://www.energyplan.eu/about/ [Accessed 24 February 2014].

[27] H. Lund, A. N. Andersen, P. A. Østergaard, B. V. Mathiesen and D. Conolly, "From Electricity Smart Grids to Smart Energy Systems – A Market Operation Based Approach and Understanding, Energy," *The International Journal*, Vol. 42, No. 1, pp. 96–102, 2012.

[28] D. Conolly, H. Lund, B. V. Mathiesen, *et al.*, "Heat Roadmap Europe: Combining District Heating with Heat Savings to Decarbonise the EU Energy System," *Energy Policy*, Vol. 65, pp. 475–489, 2014.

[29] D. Conolly, H. Lund, B. V. Mathiesen and M. Leahy, "The First Step Towards a 100% Renewable Energy-system for Ireland," *Applied Energy*, Vol. 88, No. 2, pp. 502–507, 2011.

[30] S. Obara, "Development of a Dynamic Operational Scheduling Algorithm for an Independent Micro-Grid with Renewable Energy," *Journal of Thermal Science and Technology, The Japan Society of Mechanical Engineers (JSME)*, Vol. 3, No. 3, pp. 474–485, 2008.

[31] US Department of Energy, Microgrids at Berkeley Laboratory, "Nagoya 2007 Symposium on Microgrids, Overview of Micro-grid R&D in Japan," 2006. Available from: https://building-microgrid.lbl.gov/sites/all/files/ Morozumi_2006.pdf [Accessed 25 February 2014].

[32] S. Obara, M. Kawai, O. Kawae and Y. Morizane, "Operational Planning of an Independent Microgrid Containing Tidal Power Generators, SOFCs, and Photovoltaics," *Applied Energy*, Vol. 102, pp. 1343–1357, 2013.

[33] M. Cellura, A. Di Gangi and A. Orioli, "Assessment of Energy and Economic Effectiveness of Photovoltaic Systems Operating in a Dense Urban Context", Journal of Sustainable Development of Energy, Water and Environment Systems, Vol 1, No. 2, pp. 109–121, 2013, DOI: http://dx.doi. org/10.13044/j.sdewes.2013.01.0008.

[34] S. Quoilin and M. Orosz, "Rural Electrification through Decentralised Concentrating Solar Power: Technological and Socio-Economic Aspects", *Journal of Sustainable Development of Energy, Water and Environment Systems*, Vol 1, No. 2, pp. 199–212, 2013, DOI: http://dx.doi.org/10.13044/j. sdewes.2013.01.0015.

[35] T. Ohtaka and S. Iwamoto, "A Method for Suppressing Line Overload Phenomena Using NaS Battery Systems," *Electrical Engineering in Japan*, Vol. 151, No. 3, pp. 19–31, 2005.

[36] S. Obara, Y. Morizane and J. Morel, "Economic Efficiency of a Renewable Energy Independent Microgrid with Energy Storage by a Sodium–Sulfur Battery or Organic Chemical Hydride," *International Journal of Hydrogen Energy*, Vol. 38, No. 21, pp. 8888–8902, 2013.

[37]   S. Obara and J. Morel, "Microgrid Composed of Three or More SOF Combined Cycles without Accumulation of Electricity," *International Journal of Hydrogen Energy*, Vol. 39, No. 5, pp. 2297–2312, 2014.

[38]   Japan Meteorological Agency. Available from: http://www.jma.go.jp/jma/indexe.html [Accessed 26 February 2014].

[39]   Hokkaido Electric Power Company, Main Infrastructure. Available from: http://www.hepco.co.jp/corporate/ele_power/ele_power.html [Accessed 26 February 2014] (in Japanese).

[40]   Tocardo Tidal Turbines. Available from: http://www.tocardo.com/ [Accessed 26 February 2014].

[41]   NGK Insulators Ltd. Available from: http://www.ngk.co.jp/english/products/power/nas/ [Accessed 26 February 2014].

[42]   J. M. Mauricio, A. Marano, A. Gomez-Exposito and J. L. Martinez Ramos, "Frequency Regulation Contribution through Variable-Speed Wind Energy Conversion Systems," *IEEE Transactions on Power Systems*, Vol. 24, No. 1, pp. 173–180, 2009.

[43]   J. Morel, H. Bevrani, T. Ishii and T. Hiyama, "A Robust Control Approach for Primary Frequency Regulation through Variable Speed Wind Turbines," *IEEJ Transactions on Power and Energy*, Vol. 130, No. 11, pp. 1002–1009, 2010.

[44]   A. Bemporad, M. Morarim and N. L. Ricker, *Model Predictive Control Toolbox User's Guide*, Natick, MA: The MathWorks, Inc., 1998.

Chapter 7

Microgrids in Japan

Jorge Morel

7.1 Introduction

7.1.1 Overview

The Japanese energy sector has experienced a large and abrupt change following the 2011 Tohoku Earthquake and Tsunami which unchained the series of catastrophic events, including the Fukushima Daiichi nuclear disaster of 11 March 2011.

The sudden need to rely on 'dirty' coal generation and clean renewable energy, to replace nuclear energy at least temporarily, is currently calling for an extraordinary effort to face the challenge of reducing carbon dioxide (CO_2) emission without the help of the 'clean' nuclear generation which is currently facing a strong public opposition.

Nuclear generation was viewed, previous to the accident, as the 'ideal' electricity generation system for Japan, since it does not emit CO_2 and the energy density is much higher, that is lower space needed for same amount of power generated, than the emerging and costly clean wind and solar generation. To date, more than 5 years after the shutdown of all nuclear reactors, the restarting of nuclear generation still faces strong public opposition which preludes a much slower restarting of the reactors than the originally predicted and planned by the current Japanese energy policy.

Therefore, there is a strong need for the country to develop its clean energy technology, not only to cut CO_2 emissions, but also to guarantee the supply of energy in areas with high probability of natural disasters; in the event of an earthquake, a tsunami or other natural disasters.

Although development of clean energy technology in Japan has relatively lagged in time compared to Europe and the United States due to the reasons above, there is no doubt that the Japanese, with their capacity to quickly innovate and adapt, will make its energy sector in general and clean energy sector in particular, perfectly concordant with the high standards needed to face its large responsibility in reducing global warming to foster a sustainable planet.

Nowadays, the development of smart communities, smart grids and microgrids (MGs) in Japan is accelerating and starting to keep pace with technologies of other developed countries in the world.

7.1.2   Chapter's aim and scope

This chapter aims to present to the reader an overview of the current status of the Japanese clean energy technology, in perspective with the current Japanese Energy Policy, putting emphasis on MGs in the country and its interrelation with, and its role within the whole energy sector in Japan.

The main current trends, reflected in the new revised Japanese energy policy following the Fukushima Daiichi nuclear accident, including the pilot projects in the country or carried out by Japanese companies and institutions abroad, are addressed in this chapter. The roles and characteristics of MGs can be better appreciated from the perspective of a country with strict requirements for energy reliability due to its risky location regarding natural disasters and demanding geography.

It is expected that from this chapter the reader can appreciate how largely Japanese clean energy sector has changed and is now evolving in order to establish a future energy mix to achieve the targets of CO_2 emission reduction, enhanced energy security and a sustainable energy generation.

7.1.3   Status before the Fukushima nuclear accident

Japan, as the host of the Kyoto Protocol in 1997, and one of the largest emitters in the world, carries a large responsibility in the reduction of greenhouse gas emissions. However, from the perspective of its power grid, the Japanese electric system was considered reliable even under the continuous risk of disruptions in the energy generation and supply system, due to earthquakes, floods and tsunamis.

The policies and activities regarding the massive introduction of smart grids and large renewable generation plants, such as large wind and solar farms, in the country are still few compared to Europe and the United States.

Part of the reason for this was the Japanese geography, which was then considered to have 'no space' to deploy large wind and solar farms, and also due to the fear of the possible disruption from large variable power sources in the particularly fragmented power system with limited connection between regional utilities. Due to this, nuclear generation was viewed as a good alternative to solve the problem of CO_2 emissions and less emphasis was put on renewable generation.

7.1.4   Impact of Fukushima nuclear accident on microgrid development

The accident called for an urgent revision of the policies of the energy sector in general and especially of the clean energy sector. The potentially catastrophic environmental impact of a nuclear accident became clear after the events of March 2011. Deployment of wind farms on the sea, large solar farms (or 'mega solar') and the consideration of MGs for a reliable operation during outages started to show more impetus in the country.

Japan acknowledged the vulnerability of its energy sector and called for a complete review of its energy policy and the reduction of the dependence on nuclear power. The official Japanese Energy Policy states in its Part III: The Direction of the

Energy Policy Review in Japan, of the 'Outline of the 2011 Annual Report on Energy' [1].

> *'1- The Great East Japan Earthquake and the accident at the TEPCO Fukushima Nuclear Stations significantly damaged public trust in the safety of nuclear power. The disruption of energy supplies, including electricity, oil and gas, revealed the vulnerability of the Japan's energy system.*
> *2- Reflect on the current energy policies and review without any exceptions. The Basic Energy Plan needs to be reviewed with zero based thinking methodology. In the medium to long term, a reduction of nuclear power dependency will be targeted as much as possible. At the same time, it is essential to thoroughly promote energy saving awareness and the development and popularization of renewable energy'*

7.1.5   Author's personal experience

The author has lived in Japan since 2005. In the time of the Fukushima nuclear accident on March of 2011, he was living in the island of Kyushu located in the southern part of the Japanese archipelago. Kyushu Island was one of the least impacted by the catastrophe, including the radiation levels due to the long distance from the earthquake epicentre and the Fukushima nuclear plant.

His experience is more related to the enormous shift in the paradigm in the country, especially the public opinion on the utilisation of clean renewable generation, such as wind and solar, which occupy a lot of space compared to the more compact nuclear and coal generation. The public and the media considered in those days that 'large wind and solar farms' were not possible in Japan due to the lack of space and cost.

Since 11 March 2011, the media started to change its view following the accident. The news, for the first time, started to put strong emphasis on the importance of wind and solar generation for the country and on how lowly the penetration of wind in Japan was compared to other countries such as Denmark and Germany.

Common people also started to realise how important, despite the additional cost associated with the deployment of large wind and solar farms in a country with reduced space, was the urgent massive inclusion of wind and solar to replace nuclear generation. And then also, efforts started to intensify on the development of MGs, smart grids and smart communities. The author's experience in Japan, first as graduate student and then as a researcher, allowed him to be a witness of a transition that can be called a 'before and after Fukushima' of the energy sector in Japan.

7.2   Current Japan energy policy

7.2.1   The Japanese energy sector

The Japanese energy supply is highly dependent on imported resources, such as coal and crude oil, thus making it very dependent on external conditions that put its energy security in a very vulnerable position.

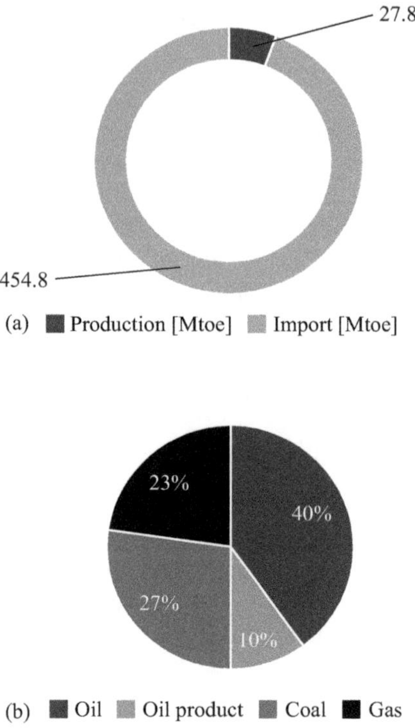

*Figure 7.1   Production and import of energy sources for Japan in 2013:
(a) distribution of primary energy sources and (b) percentage of
imported primary energy sources*

Figure 7.1 depicts the distribution of primary energy for Japan in 2013 published by the International Energy Agency [2], together with the distribution of the imported portion.

1.   Distribution of primary energy sources.
2.   Percentage of imported primary energy sources.

According also to [3], Japan is the fourth world net importer of crude oil, the first net importer of natural gas, and the third net importer of coal in the world. As can be seen in Figure 7.2, Japan has one of the largest dependence of primary energy resources, according to the Federation of Electric Power Companies of Japan [4], with the total imported resources forming about 94% of the total primary energy consumption in 2013. Japan, even though it has the third largest net installed capacity of nuclear generation, it represents only a small part of the total energy generated in the country. The high dependence on external resources calls for an urgent need to enhance the country energy security by, for instance, local production for local consumption concept.

Looking at the amount of CO_2 emitted, Japan occupies the fifth position, after China, the United States, India and the Russian Federation, as shown in Figure 7.3.

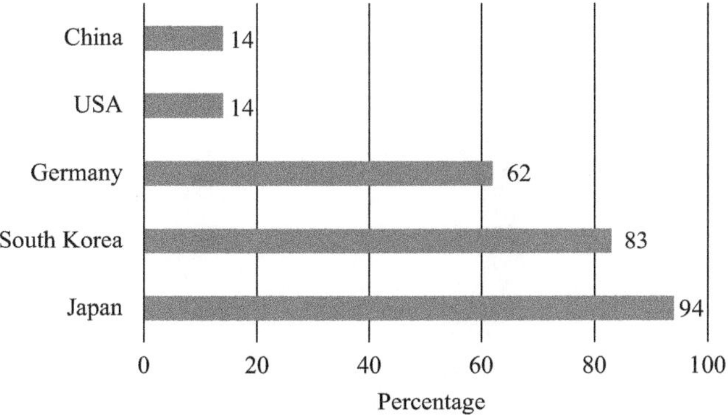

Figure 7.2   Dependence on imported energy sources for Japan and four representative countries in 2013

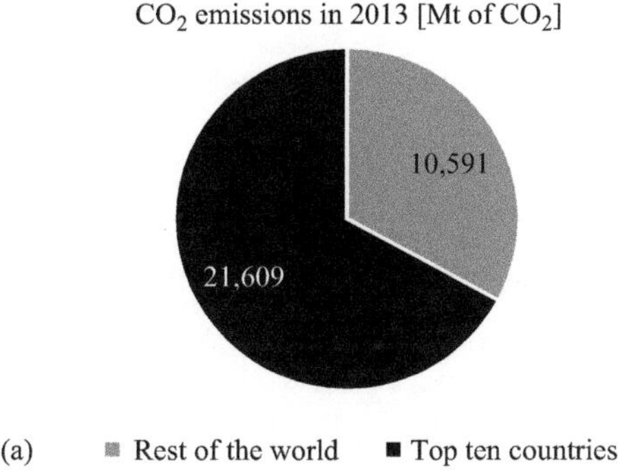

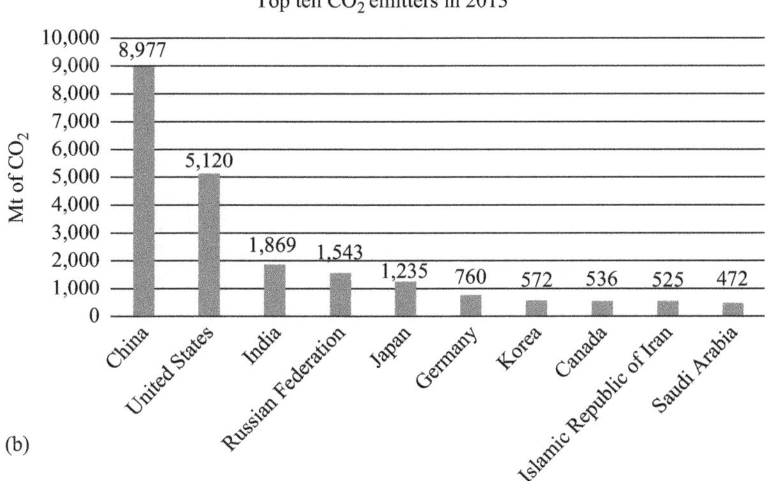

Figure 7.3 CO_2 emissions in 2013 by country and region: (a) CO_2 emission distribution by region and (b) CO_2 emissions of the top ten emitters

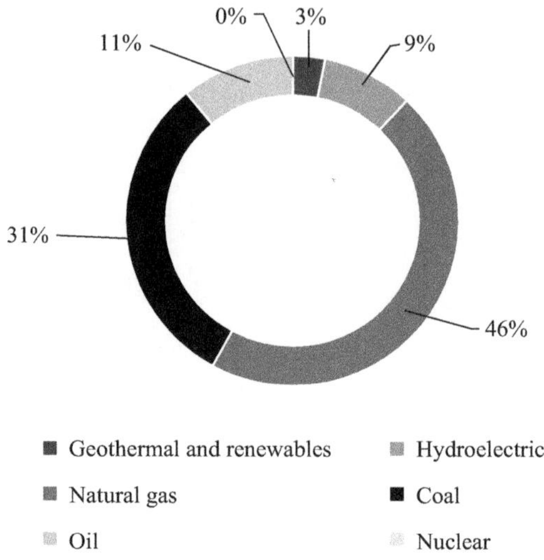

Figure 7.4   Distribution of sources for electricity generation in 2014

Thus, there is an urgent need for the country to reduce CO_2 emissions. The top ten countries are responsible for more than two-third of the world's total emissions in 2013 [5].

7.2.2   The Japanese electricity sector

The Japanese electricity sector has experienced important and sudden changes in the last 5 years, mainly due to the Fukushima Daiichi nuclear accident. There has been a sharp reduction in nuclear generation that was replaced mostly by the coal generation, which put a further burden in the target of reduction CO_2 emissions.

The sector is represented by ten companies whose power grids are inter-connected (except the one in Okinawa Island) by transmission lines and converter stations with limited capacities, which imposes distinctive and limited electric dynamic characteristics to the Japanese electricity grid.

Figure 7.4 depicts the distribution of generated electricity in 2014 according to the Federation of Electric Power Companies of Japan [6]. The total amount of generated electricity was 910 TW h with the energy resources distribution indicated in the figure. The amount of renewable generation is still minimum, with a total of 3% (together with geothermal generation), and in 2014, the nuclear generation was zero.

7.2.3   New policy following the Fukushima nuclear accident

The current Japanese energy policy is described in its 'Strategic Energy Plan' (April 2014). The most relevant aspects of the plan are described below [7,8].

Japan has three energy challenges:

1. Self-sufficiency rate (in its highest level due to coal dependence)
2. Electricity cost (sharp increase in price, high compared to other major economies, and affecting Japan's international competitiveness)
3. CO_2 emissions (growing due to more reliance on fossil fuel).

The targets for the year 2030 are as follows:

1. To increase self-sufficiency rate to 25% (over 2011 levels, of about 20%): By using renewable and nuclear generation.
2. To lower electricity cost: By utilising nuclear and coal-based generation
3. To set CO_2 reduction targets comparable to those of the EU and the United States by utilising nuclear and renewable generation, increasing the efficiency of coal-based thermal generation and using liquefied natural gas based generation.

   The three aspects above: Energy security, energy efficiency and environment, together with safety form what is called the 3E+S concept.

The challenges are believed to be possible to solve by identifying and acting on two pillars: Energy Conservation Promotion and Balanced Energy Supply

1. First pillar: For the household sector, by high-efficiency appliances and to improve energy conservation in buildings home energy management system (HEMS). For the industrial and business sector, by high-efficiency equipment and introduction of innovative technologies [factory energy management system (FEMS) and building energy management system (BEMS)]. Also Cool-Biz and Warm-Biz campaigns, and energy saving diagnosis.
2. Second pillar: The main goal is to ensure economically efficient and environmentally sustainable energy supply (by promoting renewable energy introduction and more efficient thermal power generation) to reduce dependency (but not to zero) on nuclear generation (still 20%–22%).

The projected energy mix for 2030 is shown in Figure 7.5 with the expected renewable distribution.

The expected impact of the two pillars of Japanese energy policy:

1. Self-sufficiency rate increase from 6.1% (2013) to 24.3% by 2030.
2. Electricity cost reduction by 2%–5% from current levels.
3. Reduction in the CO_2 emissions by 21.9% by 2030 compared to 2013.

The increase in renewable generation in the year 2030, compared to year 2013, is expected to be as follows: Four times increase in geothermal power generation, three times increase in biomass generation, 1.1 times increase in hydroelectricity, four times increase in wind generation and seven times increase in solar PV generation.

In the Japanese Strategic Energy Plan, the government presents its view on different types of generations focusing on each technology's characteristics and

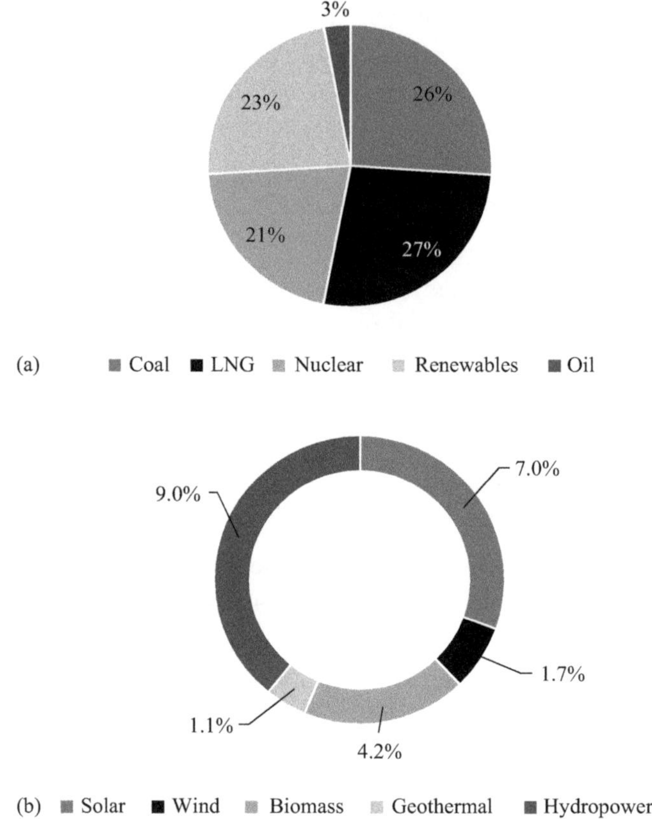

*Figure 7.5.   Projected Japanese energy mix in 2030 and expected renewable
energy distribution: (a) projected Japanese energy mix in 2030
and (b) expected Japanese renewable energy distribution
in 2030*

possible roles in the Japanese energy mix, and also, presents the policy direction for each of them.

The roles of each of the generation system are divided in base load (nuclear, coal and large hydropower), intermediate (natural gas) and peaking power sources (pumped storage hydropower and oil based generation). Detailed position and policy direction for each of the power sources are described in much more detail in the document [7].

It is worth mentioning that nuclear power will still play a very important role in the Japanese energy mix 'on the major premise of ensuring its safety' due to:

1.   Low carbon emission (environmental factor).
2.   Quasi-domestic energy source (supply stability factor).
3.   Low and stable operational cost (cost factor).

A complete section of the Japanese Strategic Energy Plan is devoted to nuclear generation taking into account safety factors in the light of the Fukushima nuclear disaster.

The Japanese Strategic Energy Plan also sets the foundation for the promotion of the use of cogeneration system, which generates electricity and heat at the same time, for a more efficient generation system, and also for the utilisation of heat generated from renewable energy, such as solar heat, underground heat and sewage heat.

Promotion of the use of renewable energy has been undertaken in Japan by the feed-in tariff system since July 2012 showing an increase, for instance, of solar generation of 34% in less than 1.5 years. Feed-in tariff system has the objective of promoting investment in renewable energy by reducing uncertainty in the investment through establishing a long term electricity price scheme. Despite the promising results, studies are needed to assess the impact of massive introduction of renewables in the power grid and also the cost structure in order to reduce the burden on the final customer [9].

The Japanese energy policy also puts strong emphasis on the reduction of market barriers to improve efficiency in the distribution of the electricity, gas and heat resources. As stated in the plan, this can be achieved by institutional reforms, technological innovations and more efficiency management methods. Electricity Market Reform and Gas Market Reform are currently advancing in the country [10].

Japan also focuses on strategic international energy cooperation in order to mitigate the impact of international geopolitical events in the supply–demand structure of energy. Also, it puts strong emphasis on the promotion of energy-related strategic technology development in order to assure a supply–demand structure in the long term.

7.3   Pilot projects in Japan and abroad

7.3.1   Earliest microgrid projects

7.3.1.1   The Hachinoe Project

The Hachinohe Project constitutes the first stage of Microgrid projects in Japan, run from 2003 to 2007 by NEDO [11–14]. It is located in Hachinohe City, Aomori Prefecture, in the northern part of Japan. The project aimed to implement an energy management system (EMS) for balancing supply and demand within 3% tolerance, with a 6-min moving average (Figure 7.6).

Figure 7.7 depicts the configuration of the MG. The electricity demand is 610 kW, distributed between the Hachinohe City Hall, four schools and an office building. The heat demand is 10 Gcal/day for a sewage plant.

The supply side is formed by tree biogas engines of 170 kW each, PV panels with total rated output of 130 kW, wind turbines with total rated power output of 20 kW. The system also includes a lead–acid battery with a maximum output/input of ±100 kW. The heat was supplied by a 1 t/h wood boiler and a 4 t/h gas boiler.

A 5.4-km private line connected the different components of the MG, and the system was interconnected to the public utility grid at a single point.

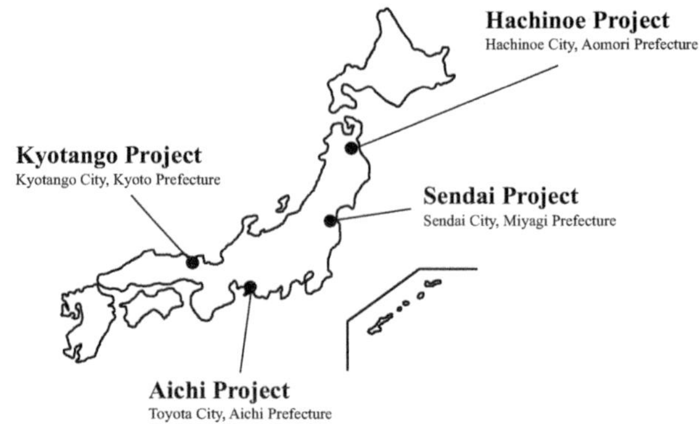

Figure 7.6   Location of the earliest microgrid projects in Japan

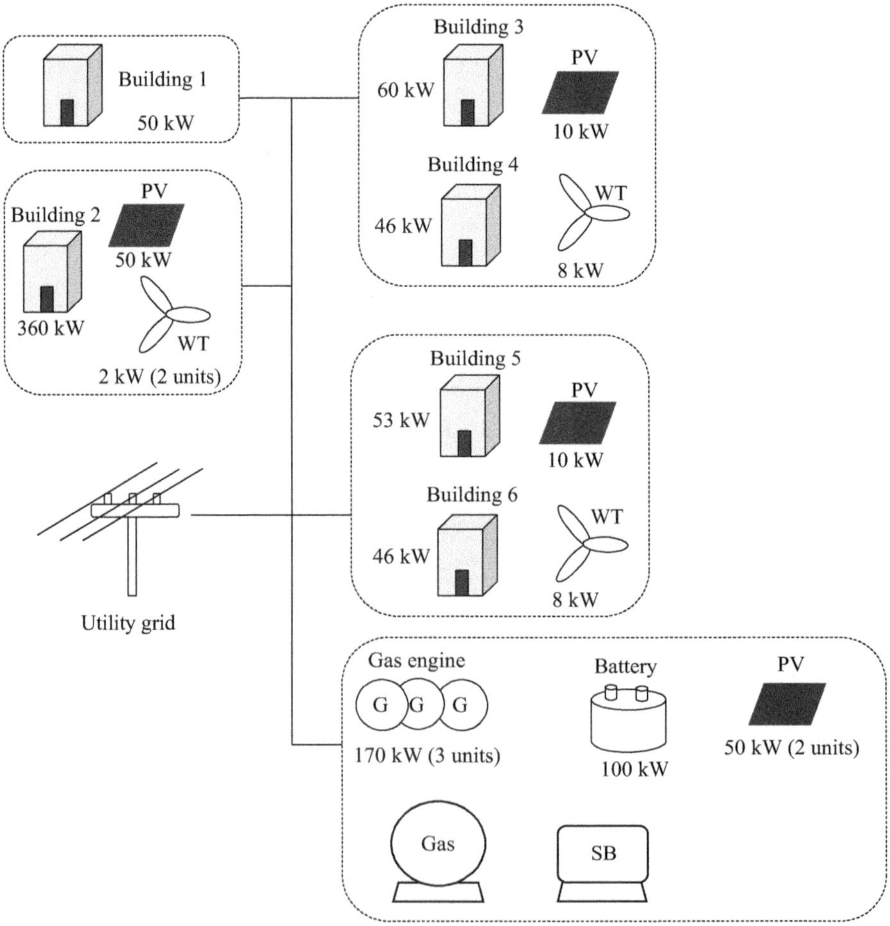

Figure 7.7.   Schematic diagram of the Hachinoe Project

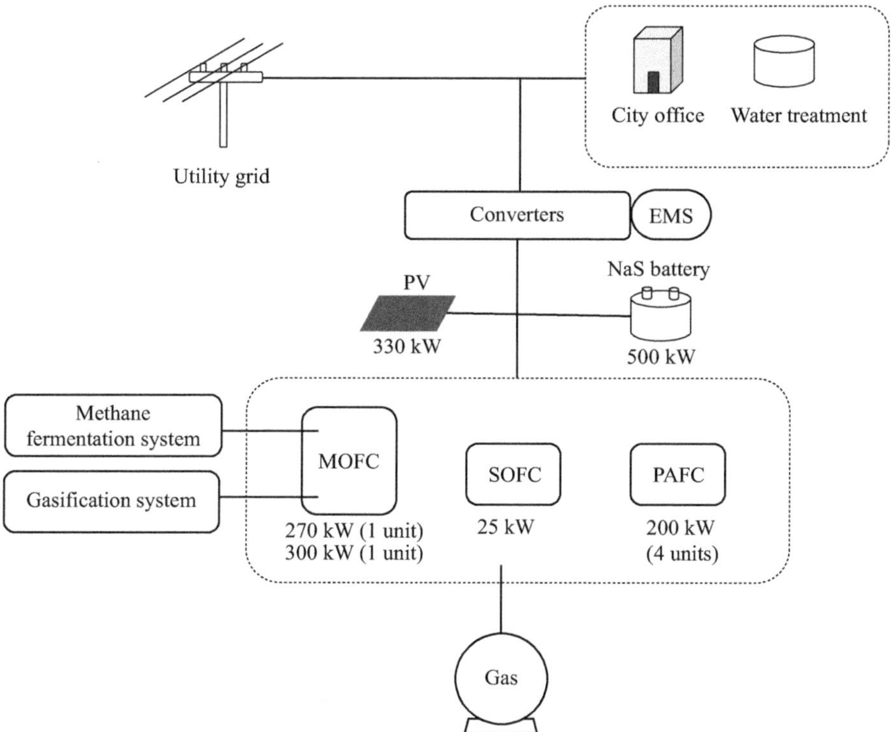

Figure 7.8   Schematic diagram of the Aichi Project

7.3.1.2   The Aichi Project

The Aichi Microgrid is also part of the first Japanese microgrid developments, run from 2003 to 2007, by NEDO [11–14]. It was constructed for demonstration at the Aichi Expo 2005. It also includes Tokoname City in the Aichi Prefecture. The targets of the projects were to study the optimum use of city gas and to mitigate the effect of variable generation in distribution grid. Also, it aimed at the autonomous control in the MG under islanding operation.

Figure 7.8 illustrates the Aichi Microgrid. All generators are connected to the power utility through power converters. The generators are PV systems, a molten carbonate fuel cell (MCFC), a solid oxide fuel cell and four units phosphoric acid fuel cells (PAFCs), with capacities shown in the figure. The fuel cells are supplied by city gas. The capacities of each unit are shown in the figure.

The generators supply electricity to an office and to a water treatment facility. The system can be connected to the power utility grid. The methane fermentation system has a capacity of 4.8 t/day and the gasification system 20 kg/h. The storage system implemented is a sodium–sulphur battery with a capacity of ±500 kW.

7.3.1.3   The Kyotango Project

The Kyotango Project is also part of NEDO's projects, run from 2003 to 2007 at the initial stage of Japanese microgrid development [11–14]. It is located in Kyotango City, Kyoto Prefecture. The project implements an EMS using public communication (Internet base control system) to balance supply and demand for time frame of 50 min.

Figure 7.9 shows the simplified scheme of the Kyotango Microgrid. The supply side is composed of five biogas engine generators of 80 kW, a MCFC of 100 kW, a wind turbine of 50 kW, two PV systems of 30 kW and 20 kW. The storage system is formed by a lead–acid battery with a capacity of ±100 kW. The demand side is made up of four buildings with the capacities shown in the figure and also 450 kW in the biogas plant.

7.3.1.4   The Sendai Project

The Sendai Microgrid is the last of the four initial Japanese projects for the implementation of MGs under the frame 'Experimental Study of Multi Power quality Supply systems' conducted by NEDO from 2003 to 2007, but perhaps the most important and iconic one [11,15,16]. The MG was constructed on the campus of Tohoku Fukushi University, Sendai City, capital of Miyagi prefecture, one of the most affected areas of the 2011 Tohoku Earthquake and Tsunami.

Figure 7.10 depicts the simplified scheme of the Sendai Microgrid in 2011 which differs slightly from the original system at the initial stage of the project. The system has remained in operation after the end of the project in 2007.

The supply side is composed of two gas engines of 350 kW, one PV system of 50 kW and one PAFC, which was originally a MCFC at the beginning of the project.

The demand side is formed by loads which are divided into five types, according to the level of power quality. Basically, the 'standard load' is not compensated at all, the B3 load is compensated for voltage dips in less than 15 ms, and the C load is compensated for outage (only when engines are under operation) in less than 15 ms. The B1 load is compensated for voltage dips and outage (provided engines are in operation), also in less than 15 ms. Finally, the C loads and the B3 loads, which require a high reliability, are compensated for all parameters (no interruption, voltage dips, outage, voltage fluctuations, voltage harmonics and for voltage unbalance and frequency deviation (for A class only)).

A comprehensive description of the entire system and the operation in the aftermaths of the 2011 Tohoku Earthquake and Tsunami can be found in a NEDO's report presented in [16].

7.3.2   Smart community projects

7.3.2.1   The Toyota Project

The Toyota Project, or the Toyota Low-Carbon Society Project (also Toyota City Low-Carbon Verification Project) is a smart community project subsidised by NEDO and located in Aichi Prefecture [17–20]. The project ran from 2010 to 2014

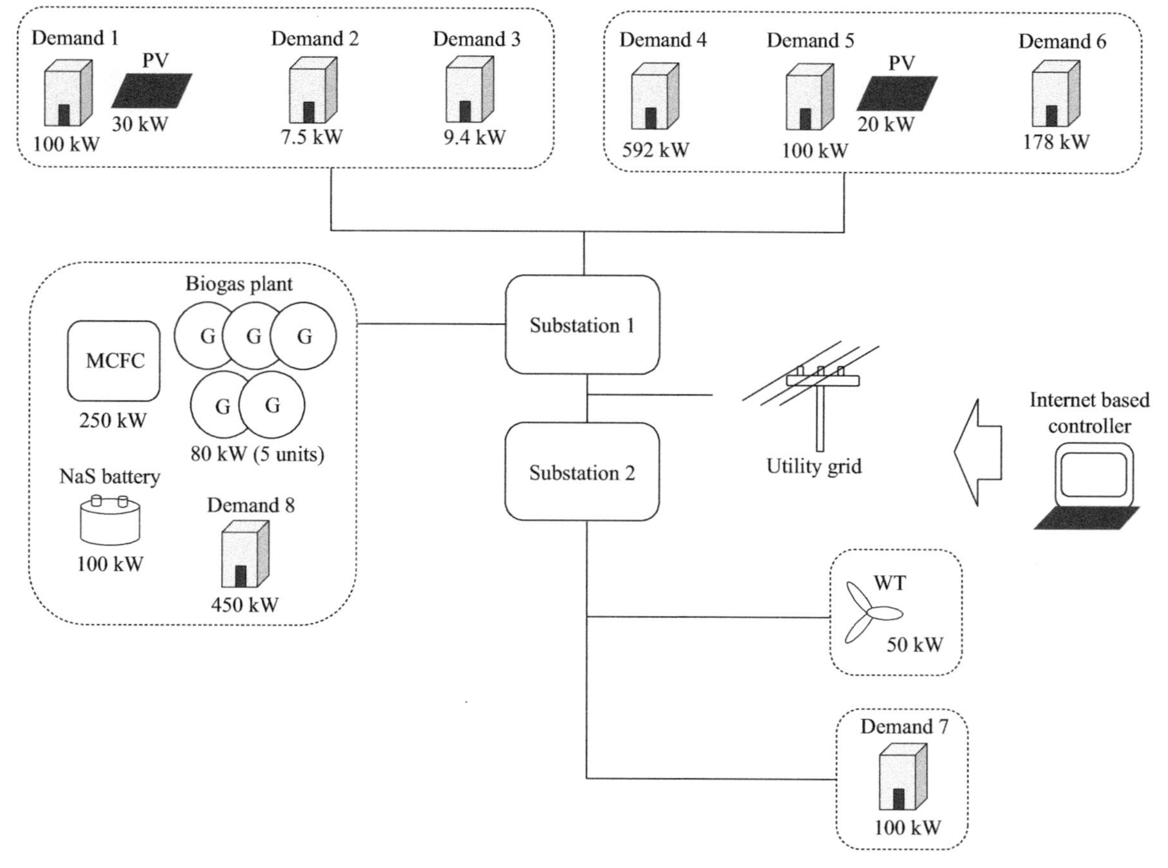

Figure 7.9   Schematic diagram of the Kyotango Project

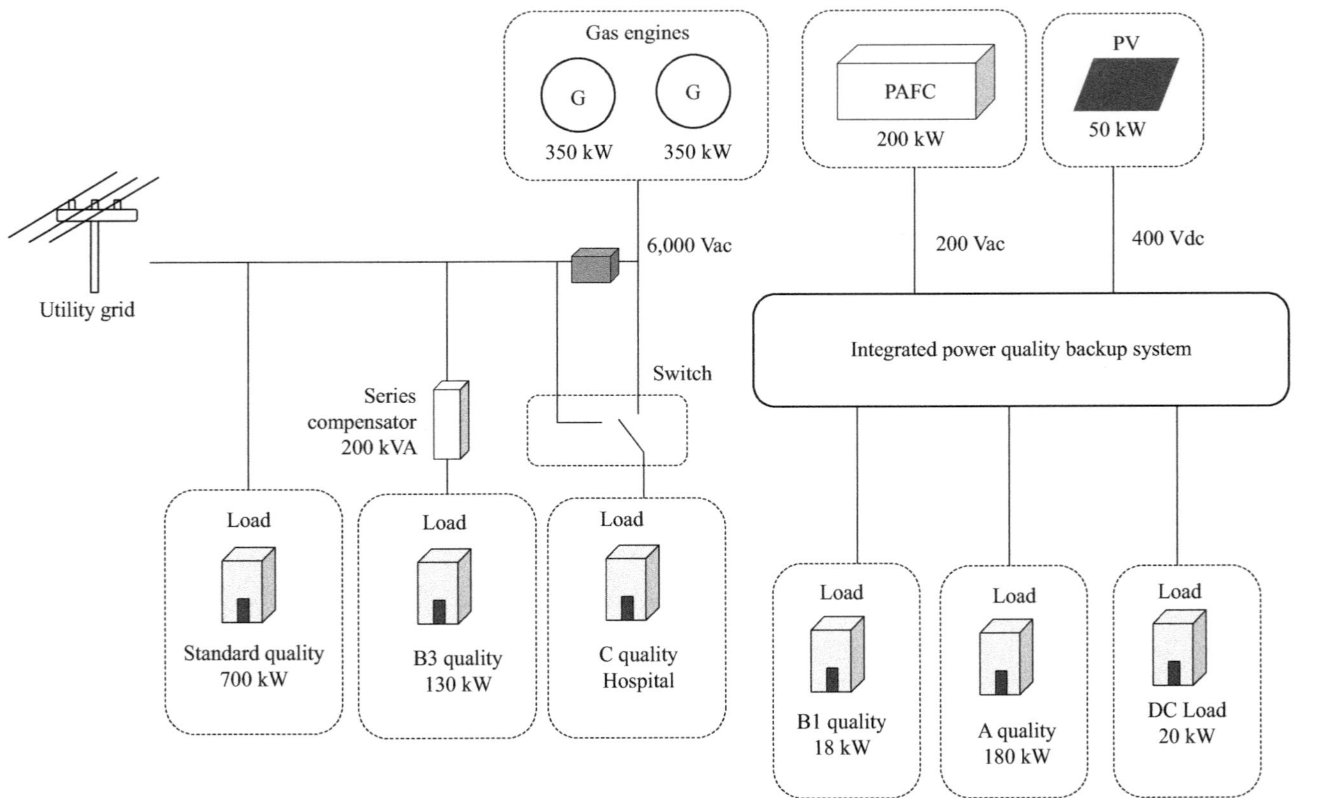

Figure 7.10   Schematic diagram of the Sendai Project

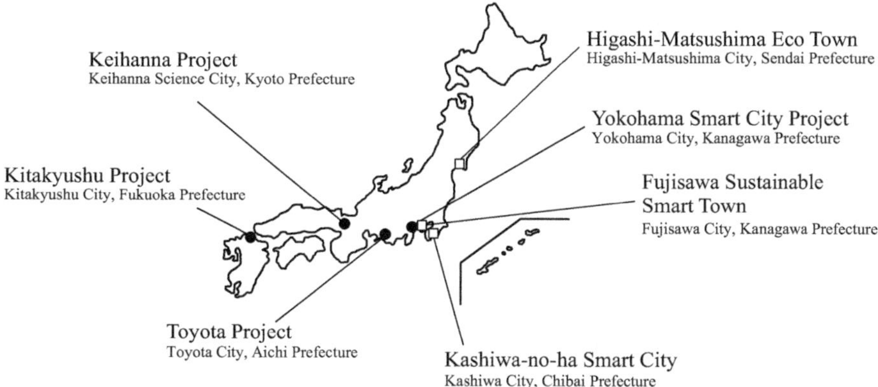

Figure 7.11   Location of the main smart community projects in Japan

and targeted 227 households, with a budget of 22.7 billion JPY (Approximately 206 million USD). The Toyota City Low-Carbon Society Verification Promotion Council, comprising 26 entities, including Toyota City and private companies, was in charge of the project (Figure 7.11).

The objectives of the project include 61.2% of houses provided with renewable energy generation, about 230 smart homes, and to promote the optimal use of energy in the living spaces in the community.

Also, the use of unused energy is considered. Demand response in more than 70 homes and 3,100 electric vehicles, vehicle to home (V to H) and vehicle to grid (V to G).

The targets of the projects are a reduction of CO_2 emissions by 20% for households and 40% for transportation. Also, the use of 3,100 electric vehicles was expected.

7.3.2.2   The Yokohama Project

The Yokohama Project, or the Yokohama Smart City Project, is a smart community demonstration project selected by the Ministry of Economy, Trade and Industry (METI) in April 2010, as Next Generation Energy Infrastructure and Social System Demonstration Area [19–23]. The project initially ran from 2010 to 2014 with a cost of 74 billion JPY (Approximately 670 million USD).

The Yokohama City and the private sector are working together in the various projects forming part of the larger project: renewable energy introduction, EMSs and next generation transportation system.

Three areas with distinctive characteristics (residential, urban and industrial) are part of the project, the Kohoku Newtown area (residential), the Minato Mirai 21 area (urban) and the Yokohama Green Valley area (industrial). The total area of the three areas is approximately 60 km^2, including approximately 170,000 households.

Five different technologies are considered as key initiatives: community energy management system (CEMS), HEMS, BEMS, FEMS and electric vehicle (EV).

The project aims at the reduction of CO_2 emissions by 30% (from the 2005 levels) by applying wind-area EMS. Also, the introduction of more than 27 MW of PV system, 4,000 HEMS in houses and 2,000 EVs is considered.

7.3.2.3　The Kyoto Keihanna Project

The Keihanna Project, or the Keihanna Eco City, is another project forming the four smart community projects subsidised by METI, under the program 'Test Projects for the Next Generation Energy and Social Systems' [19,20,24]. It is located in Kyoto Prefecture. The project involves 24 companies, local governments and universities. Mitsubishi Heavy Industries, Ltd. (MHI) leads the project. The target population of the project is around 100,000 people. It ran from 2010 to 2014, with a budget of 13.5 billion JPY (Approximately 120 million USD).

The served population is 170,000, with 60,000 households in an area of 154 km^2. The number of vehicles estimated is 80,000.

The objective of the project is the introduction of a CEMS to supervise the sectors involved (transportation, residential and commercial) and to optimise the utilisation of energy in the community.

The targets of the projects are as follows: A reduction of CO_2 emissions by 20% in residences by 2030 (from the 2005 level), and by 40% in transportation, and achieve the installation of PV systems in 1,000 houses.

7.3.2.4　The Kitakyushu project

The Kitakyushu Smart Community project is one of the four smart community projects subsidised by METI, part of the 'Test Projects for the Next Generation Energy and social Systems' [19,20,25]. It is located in Kitakyushu City, Fukuoka Prefecture, in the southern island of Kyushu. The targeted population is 225 households. The Kitakyushu Smart Community Council, composed by 77 groups and companies is in charge of the project. It ran from 2010 to 2014 and included 26 projects with a budget of 12 billion Japanese Yen (Approximately 110 million USD).

The objectives of the project are as follows: First, to transform current energy consumers into 'prosumers' (producer-consumer) by installing PV arrays and other generation systems; second, to implement a demand-side management system in order to allow the prosumers to manage energy together with the current energy providers; and finally, to implement dynamic pricing and other incentive programs.

The project covers part of the Higashida District which has an area of 120 ha and a population of approximately 1,000.

The targets of the project are as follows: reduction of CO_2 emissions by 50% in residential and transport demand by 2030, 10% increase in production of clean energy, and the installation of smart meters for 70 firms and 200 households.

The generation units are composed of a cogeneration plant of 33 MW rated capacity, fuelled by natural gas, PV generation with a total capacity of 5 MW, wind generation with a capacity of 30 kW, fuel cell of 300 W and 400 kW geothermal power.

7.3.2.5   The Kashiwa-no-ha Project

The Kashiwa-no-ha Project is located in Kashiwa City, Chiba Prefecture [26,27]. The project is being leaded by Mitsui Fudosan Co. Ltd. The project has three groups of stakeholders: public sector (Chiba Prefectural Government, Kashiwa City, and nonprofit organisations), private sector (business and citizens) and the academia (The University of Tokyo and Chiba University).

The broad objective of the project is to tackle the problems humankind may face in the future such as environment, energy and food, by considering three key concepts: Creation of an Environmental-Symbiotic City (i.e. creation of solutions for environmental and energy problems), a City of New Industry Creation (i.e. creation of solutions for vitalising the economy) and a City of Health and Longevity (i.e. creation of solutions for an aging society). Figure 7.12 shows the basic configuration of the project.

7.3.2.6   The Fujisawa Sustainable Smart Town Project

The Fujisawa Smart Town Project is located in the Fujisawa City, Kanagawa Prefecture, around 45 km from the capital Tokyo [28,29]. It is a long-term project which has as objectives, besides the need to construct a sustainable and disaster proof eco-town development, to promote the use of clean natural resources and the local production for local consumption of energy. Also, by using electricity and information networks, the project aims to build a safe and peaceful life for residents. Panasonic and other eleven partner companies, together with Fujisawa City, are in charge of the project.

The project will have 600 houses and 400 apartments for a total of 3,000 people. Each house has its own solar panel, a Panasonic ECO-CUTE heat-pump-driven hot water system, and a domestic size ENE-FARM household fuel cell.

The project possesses a long-term vision of 100 years. After completion of the project, the town is expected to develop for 30 years, to mature for another 30 years and then further develop for 30 years. The total time for three generations to leads prosperous lives.

The targets of the project are as follows: Environmental targets of 70% CO_2 reduction compared to 1990 level and 30% reduction in water consumption taking as reference water used in 2006. The energy target is the use of more than 30% renewable energy. Finally, for safety and security target (Lifeline Maintenance) the target is 3 days.

7.3.2.7   The Higashi Matsushima Eco Town Project

Higashi-Matsushima City is located in Miyagi Prefecture, in the Tohoku Area, and is one of the areas severely hit by the 2011 Tohoku Earthquake and Tsunami [30]. The 'Higashi-Matsushima Smart Disaster-Resilient Eco-Town Project' was officially opened on 12 June 2016. The objective of this Eco-Town is to offer the community a resilient MG in the presence of natural disasters, while preventing global warming by promoting efficient use of energy and reducing CO_2 emissions.

The basic configuration of the eco-town is shown in Figure 7.13. The supply of power is performed, internally, by PV systems, a 500-kW bio-diesel emergency

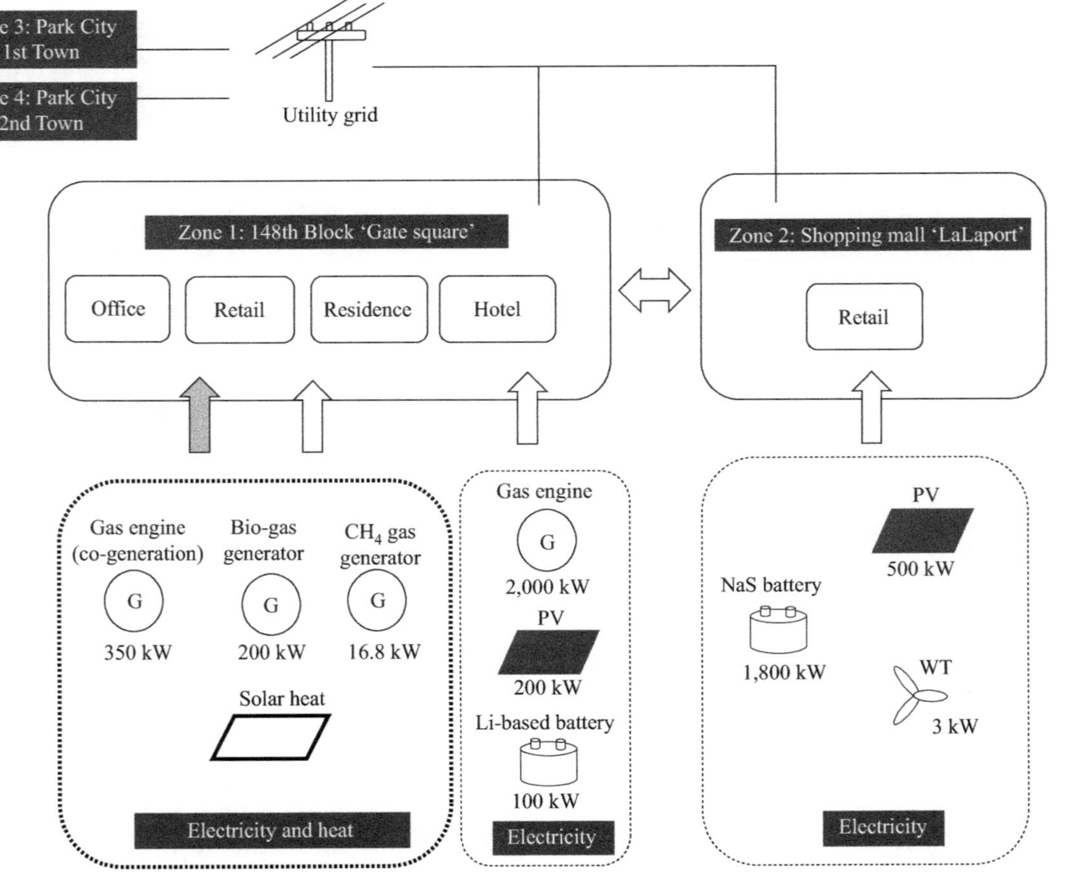

Figure 7.12   Schematic diagram of the Kashiwa-no-ha Project

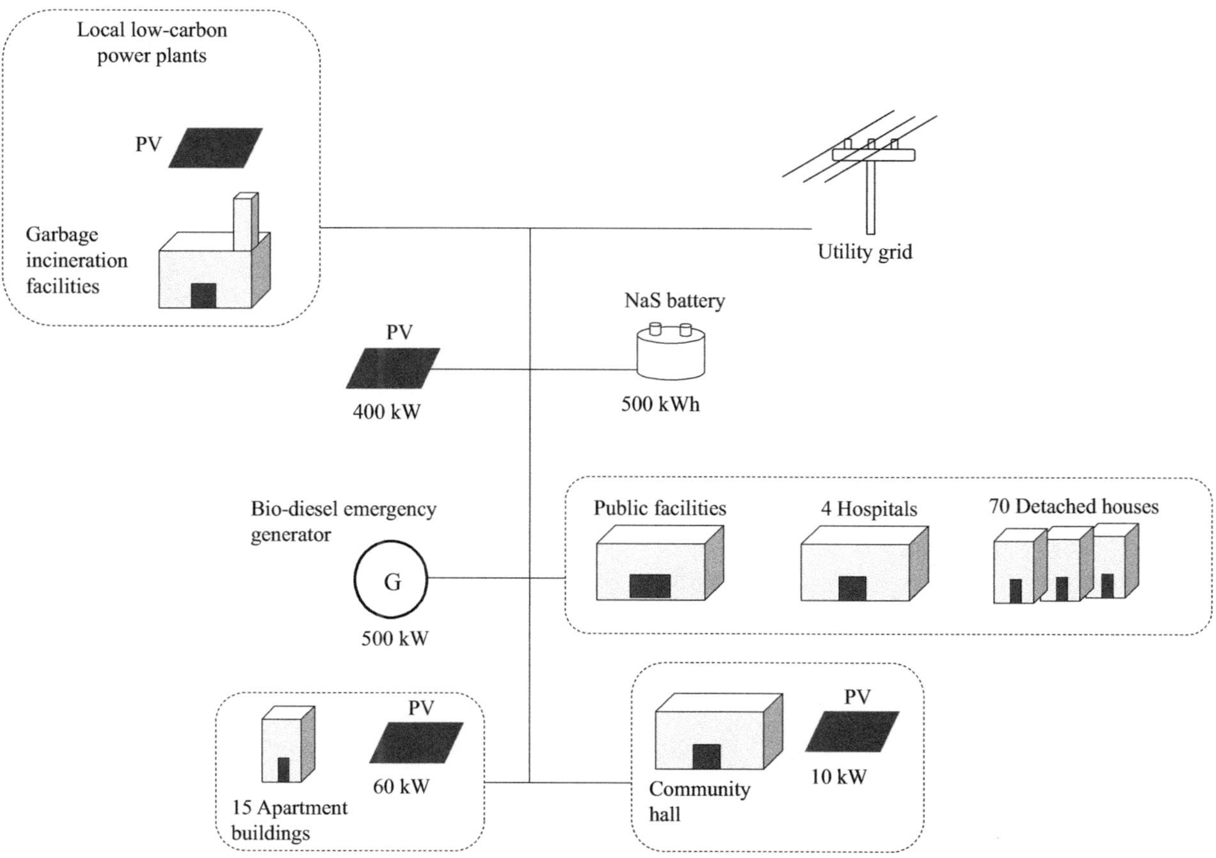

Figure 7.13 Schematic diagram of the Higashi-Matsushima Eco Town Project

power generation and a 500-kW h storage system. The 470-kW PV systems are installed independently (400 kW), on the apartment buildings (60 kW) and on the community hall (10 kW). Externally, power is supplied when needed, by low-carbon power plants (garbage incineration facilities, etc., and solar farms). The externally generated power is used to balance supply and demand during peaks in energy demands or other types of imbalances. The demand side is formed by 70 detached houses, 15 apartment buildings, the community hall, four hospitals and other public facilities.

The houses are equipped with HEMS. The energy is supplied by implementing a CEMS administered by a power producer and supplier (PPS) who also is the owner of the PV system.

The low-carbon power plants are located in Higashi-Matsushima City and in this way the whole system can be regarded as fitting in the concept of local-production for local-consumption of energy.

In case the supply of energy by the main power line is interrupted by a natural event, the PV systems, the storage systems and the bio-diesel power generator can ensure a normal supply of electricity for three days.

7.3.3   *International projects*

The New Energy and Industrial Technology Development Organization (NEDO) has been implementing several international demonstration projects around the world. The following is a summary of the most important projects, which is based entirely on the information available at the NEDO's homepage as indicated in the table.

No	Place	Period	Description
1	California, USA	From 2014	A demonstration project for Electric Vehicle Driving Behaviour is being conducted in the Northern California metropolitan areas with the aim of expanding electric vehicle (EV) travel distances A demonstration project for validation of Redox Flow Battery Performance is being carried out in San Diego to promote the use of storage batteries [31]
2	New Mexico, USA	2009–2014	It was comprised of two sites in New Mexico (Los Alamos and Albuquerque). The objective of the project was to resolve issues associated with integrating large volumes of renewable energy into a power distribution systems [32]
3	Maui, Hawaii, USA	2011–2015	The objectives of the project were: Realisation of a smart grid including EV by an energy management system for an efficient use of renewable generation, and the mitigation of the effect of variation in renewable energy outputs by the implementation of direct load control [33]

No	Place	Period	Description
4	Oshawa, Canada	2015–2017	Installation of hybrid inverter systems for unified control of solar panels and storage systems in 30 houses in the city of Oshawa. The project aims to demonstrate the system's use as an emergency power source during power outages while verifying its ability to stabilise the power grid [34]
5	Manchester, UK	2014–2016	The objectives are to aggregate residential negawatt (saved electricity by the demand side) for making a shift of the principal source of energy from gas to electricity as well as in for establishing a low-carbon society in England [35]
6	Malaga, Spain	2011–2015	The project utilises quick chargers and 200 units of Japanese-made EVs to evaluate highly advanced smart community technology for the large-scale introduction and dissemination of EVs [36]
7	Slovenia	From 2014	A preliminary investigation for a smart community demonstration project started in Slovenia in 2014 [37]
8	Speyer, Germany	2015–2017	The project aims to introduce and demonstrate technologies to establish a 'self-consumption model' where PV generated power in the community is consumed locally [38]
9	Lyon, France	From 2014	Project on positive energy building which produces energy in excess of the amount consumed made possible through the introduction of PV power generation, storage batteries and heat storage materials that are controlled by an energy management system [39]
10	Lisbon, Portugal	2016–2018	The project aims to demonstrate an automated demand response system which can adjust power demand and supply as needed in public buildings and households in Lisbon. http://www.nedo.go.jp/english/news/AA5en_100005.html
11	Poland	From 2015	The project aims to perform experiments for interconnecting Polish domestic wind power generation with the grid, by using a stabilisation/control systems and storage batteries, and fully exploiting the transmission capacity of existing equipment, without enhancing the power grid [40]
12	Panipat, India	From 2014	This smart grid demonstration project aims to make Indian distribution network smarter and to the dissemination of Japanese technologies in India [41]
13	Suryacipta City of Industry, Indonesia	2013–2016	This project is the NEDO's first smart community demonstration project in Asia [42]
14	Putrajaya, Malaysia	2015–2017	The project objective is to develop a smart community based on EV (electric vehicle) buses, utilising Japanese energy storage and charging technologies [43]

References

[1]   Ministry of Economy, Trade and Industry (METI), "Outline of the 2011 annual report on energy (Energy white paper 2011)," 2011. [Online]. Available: http://www.meti.go.jp/english/report/downloadfiles/2011_outline. pdf [Accessed 4 September 2016].

[2]   International Energy Agency (IEA), "Energy balance flow," [Online]. Available: http://www.iea.org/sankey/#?c=Japan&s=Balance [Accessed 4 September 2016].

[3]   International Energy Agency (IEA), "2015 key world statistics," 2015. [Online]. Available: https://www.iea.org/publications/freepublications/ publication/KeyWorld_Statistics_2015.pdf [Accessed 6 September 2016].

[4]   The Federation of Electric Power Companies of Japan (FEPC), "Japan's energy supply situation and basic policy," [Online]. Available: http://www. fepc.or.jp/english/energy_electricity/supply_situation/index.html [Accessed 4 September 2016].

[5]   International Energy Agency (IEA), "CO_2 emissions from fuel combustion," 2015. [Online]. Available: https://www.iea.org/publications/freepublications/ publication/CO2EmissionsFromFuelCombustionHighlights2015.pdf [Accessed 6 September 2016].

[6]   The Federation of Electric Power Companies of Japan (FEPC), "History of Japan's Electric Power Industry," [Online]. Available: http://www.fepc.or. jp/english/energy_electricity/history/index.html [Accessed 4 September 2016].

[7]   Ministry of Economy, Trade and Industry (METI), Agency for Natural Resources and Energy (ANRE), "Strategic energy plan," 2014. [Online]. Available: http://www.enecho.meti.go.jp/en/category/others/basic_plan/pdf/ 4th_strategic_energy_plan.pdf [Accessed 4 September 2016].

[8]   Ministry of Economy, Trade and Industry (METI), Agency for Natural Resources and Energy (ANRE), "Japan's energy plan," 2015. [Online]. Available: http://www.enecho.meti.go.jp/en/category/brochures/pdf/energy_ plan_2015.pdf [Accessed 4 September 2016].

[9]   Ministry of Economy, Trade and Industry (METI), Agency for Natural Resources and Energy (ANRE), "Feed-in tariff scheme in Japan," 2011. [Online]. Available: http://www.meti.go.jp/english/policy/energy_environment/ renewable/pdf/summary201207.pdf [Accessed 4 September 2016].

[10]   Ministry of Economy, Trade and Industry (METI), Agency for Natural Resources and Energy (ANRE), [Online]. Available: http://www.enecho. meti.go.jp/en/category/electricity_and_gas/electric/electricity_liberalization/ [Accessed 4 September 2016].

[11]   Y. Shimizu, "Micro-grid related activities in Japan," New Energy and Industrial Technology Development Organization (NEDO), 11 September 2013. [Online]. Available: http://www.ct-si.org/events/APCE2013/program/ pdf/YasuhiroShimizu.pdf [Accessed 4 September 2016].

[12] S. Morozumi, "Overview of micro-grid R&D in Japan," Micro-grid Symposium in Nagoya, 2007. [Online]. Available: http://microgrid-sympo siums.org/wp-content/uploads/2014/12/nagoya_morozumi.pdf [Accessed 4 September 2016].

[13] O. Onodera, "Micro-grid activities related to Japan," Microgrid World Forum, 14 March 2013. [Online]. Available: http://microgridworldforum. com/pdf/osamu-onodera.pdf [Accessed 4 September 2016].

[14] R. Fujimori, "Status of microgrids in Japan," The 4th Korea Smart Grid Week, 17 October 2013. [Online]. Available: https://www.google.com. py/?gws_rd=ssl#q=microgrids+in+japan+korea+smart+week [Accessed 4 September 2016].

[15] Berkeley Lab, "The Sendai microgrid," [Online]. Available: https://building-microgrid.lbl.gov/sendai-microgrid [Accessed 4 September 2016].

[16] New Energy and Industrial Technology Development Organization (NEDO), "The Sendai microgrid operational experience in the aftermath of the Tohoku earthquake: A case study," 2013. [Online]. Available: http:// www.nedo.go.jp/content/100516763.pdf [Accessed 4 September 2016].

[17] Toyota Motor Co., Ltd., "Toyota City low-carbon project model homes completed," 2011. [Online]. Available: http://www2.toyota.co.jp/en/news/ 11/06/0630.html [Accessed 4 September 2016].

[18] T. Ohta, "Verification project for the establishment of a low-carbon society system in Toyota City, Aichi Prefecture," Toyota City Government, 2012. [Online]. Available: http://www.uncrd.or.jp/content/documents/140SMT% 20-%20P12_2_Case%20Study_Toyota.pdf [Accessed 4 September 2016].

[19] EU-Japan Centre for Industrial Cooperation, "Smart cities in Japan: An assessment on the potential for EU-Japan cooperation and business development," October 2014. [Online]. Available: http://www.eu-japan.eu/sites/ default/files/publications/docs/smartcityjapan.pdf [Accessed 4 September 2016].

[20] Netherlands Enterprise Agency, "Japan's four major cities," 2014. [Online]. Available: http://www.rvo.nl/sites/default/files/Smart%20Cities%20Japan. pdf [Accessed 4 September 2016].

[21] City of Yokohama, "Yokohama smart city project (YSCP)," 2011. [Online]. Available: http://esci-ksp.org/wp/wp-content/uploads/2012/05/Yokohama-Smart-City-Project-YSCP.pdf [Accessed 4 September 2016].

[22] City of Yokohama, "Yokohama smart city project – YSCP," [Online]. Available: http://www.city.yokohama.lg.jp/ondan/english/yscp/ [Accessed 4 September 2016].

[23] Toshiba Co., Ltd., "Yokohama smart city project (YSCP)," [Online]. Available: https://www.toshiba.co.jp/csr/en/highlight/2012/smart02.htm [Accessed 4 September 2016].

[24] Mitsubishi Heavy Industries Technical Review, "Smart community demonstration projects – initiative in Keihana and Malaga for EV management," 2013.

[25] T. Sasakura, "Results of the Kitakyushu smart community creation project," Fuji Electric Co., Ltd., 18 June 2015. [Online]. Available: http://www.nedo. go.jp/content/100750432.pdf [Accessed 4 September 2016].

[26] Nikken Sekkei Ltd., "Kashiwa-no-ha Smart City," 2015. [Online]. Available: http://unctad.org/meetings/en/Presentation/CSTD_2015_ppt08_Shinji %20Yamamura_en.pdf [Accessed 4 September 2016].

[27] Kashiwa-no-ha Smart City, [Online]. Available: http://www.kashiwanoha-smartcity.com/en/ [Accessed 4 September 2016].

[28] Fujisawa Sustainable Smart Town, "Fujisawa SST," [Online]. Available: http://fujisawasst.com/EN/ [Accessed 4 September 2016].

[29] Panasonic, "Fujisawa Sustainable Smart Town," [Online]. Available: http:// panasonic.net/es/solution-works/fujisawa/ [Accessed 4 September 2016].

[30] Sekisui House, "Promoting Net-Zero-Energy Housing," 2015. [Online]. Available: https://www.sekisuihouse.co.jp/english/sr/datail/__icsFiles/afield file/2015/08/19/P21-P52.pdf [Accessed 4 September 2016].

[31] New Energy and Industrial Technology Development Organization (NEDO), "News release – implementation of smart community demonstration projects in California," 2015. [Online]. Available: http://www.nedo.go. jp/english/news/AA5en_100015.html [Accessed 18 September 2016].

[32] New Energy and Industrial Technology Development Organization (NEDO), "News release – smart grid demonstration project completed in New Mexico," 2015. [Online]. Available: http://www.nedo.go.jp/english/ news/AA5en_100001.html [Accessed 18 September 2016].

[33] New Energy and Industrial Technology Development Organization (NEDO), "News release – starting of joint Japan-US smart grid demonstration project in Hawaii (in Japanese)," 2013. [Online]. Available: http:// www.nedo.go.jp/news/press/AA5_100240.html [Accessed 18 September 2016].

[34] New Energy and Industrial Technology Development Organization (NEDO), "News release – NEDO launches its first smart community demonstration project in Canada," 2015. [Online]. Available: http://www. nedo.go.jp/english/news/AA5en_100006.html [Accessed 18 September 2016].

[35] New Energy and Industrial Technology Development Organization (NEDO), "News list 2014 – NEDO and UK Government concludes MOU for smart community project in Manchester," 2014. [Online]. Available: http://www.nedo.go.jp/english/whatsnew_20140318.html [Accessed 18 September 2016].

[36] New Energy and Industrial Technology Development Organization (NEDO), "Press release – smart community demonstration project in Málaga City," 2013. [Online]. Available: http://www.nedo.go.jp/content/ 100523327.pdf [Accessed 18 September 2016].

[37] New Energy and Industrial Technology Development Organization (NEDO), "News list 2014 – launch ceremony for preliminary investigation of smart community demonstration project in Slovenia," 2014. [Online].

Available: http://www.nedo.go.jp/english/whatsnew_20141210.html [Accessed 18 September 2016].

[38] New Energy and Industrial Technology Development Organization (NEDO), "News release – local energy production and consumption model smart community demonstration project launches in Speyer, Germany," 2015. [Online]. Available: http://www.nedo.go.jp/english/news/AA5en_100004.html [Accessed 18 September 2016].

[39] New Energy and Industrial Technology Development Organization (NEDO), "News release – demonstration of positive energy building begins in Lyon, France," 2015. [Online]. Available: http://www.nedo.go.jp/english/news/AA5en_100016.html [Accessed 18 September 2016].

[40] New Energy and Industrial Technology Development Organization (NEDO), "News list 2014 – agreed to start preliminary survey for smart grid demonstration project in the Republic of Poland," 2014. [Online]. Available: http://www.nedo.go.jp/english/whatsnew_20140530.html [Accessed 18 September 2016].

[41] New Energy and Industrial Technology Development Organization (NEDO), "News release – NEDO launches smart grid demonstration project in Haryana, India," 2015. [Online]. Available: http://www.nedo.go.jp/english/news/AA5en_100026.html [Accessed 18 September 2016].

[42] New Energy and Industrial Technology Development Organization (NEDO), "News list 2013 – MOU signing with Indonesia MEMR for smart community project," 2013. [Online]. Available: http://www.nedo.go.jp/english/whatsnew_20130718.html [Accessed 18 September 2016].

[43] New Energy and Industrial Technology Development Organization (NEDO), "News release – smart community project launches in Putrajaya, Malaysia," 2015. [Online]. Available: http://www.nedo.go.jp/english/news/AA5en_100007.html [Accessed 18 September 2016].

Chapter 8

Microgrids in Europe

Sergio Rivera and Tomas Valencia

8.1 Introduction: the European electrical power system

8.1.1 General description

Europe's power system is one of the largest power systems in the world. It consists of five interconnected areas [1]: Continental Europe, the Nordic Countries, the Baltic Countries, Great Britain and Ireland. There exist high voltage direct current links and meaningful energy exchanges between these five regions. The energy consumption of the whole system was 3,210 TW h in 2014, with a total installed capacity of 1,024 GW. In 2014, the peak load reached 522 GW, and the historical maximum peak load is 557 GW that happened in 2012 during an unusually harsh winter [1].

Since the end of the twentieth century, European authorities and researchers have been aware of the importance of fighting climate change. That is why today Europe has one of the highest grid penetration rates of renewable energy sources (RESs) in the world [2]. In 2014, the proportions of prime energy sources used in electric generators across Europe looked as follows: 40.5% fossil fuels, 26.3% nuclear, 18.5% hydraulic and 14.4% RES. There was a 6.6% increase of RES in the last years and a 7% decrease for fossil fuels [1]. Overall, Europe's electrical power system is currently the most environment-friendly in the world.

However, this composition of prime energy sources in electric generation is by no means homogenous across the continent. In fact, numbers vary significantly for individual countries, ranging from over 95% hydraulic in Norway to over 75% nuclear in France, over 88% fossil in Poland and over 42% wind and solar in Denmark [1].

That same diversity can be found on the other end of the power line: consumers. Economic development and activities are very different in the various regions of Europe. In addition, demand in Europe has a seasonal nature. Therefore, final usage of electricity varies greatly with location and throughout the year [3]. Adding up the whole continent, the final electricity consumption by sector in 2010 was 1.8% agriculture, 29.7% households, 29.7% services, 36.5% industry and 2.4% transport.

Significant efforts have been made to make European electrical consumers more efficient. This has resulted in a steady decrease of the overall consumption in

Europe since 2010. It is worth noting that because of this initiative to improve energy efficiency, along with industry relocation to other continents, which has been happening in Europe for some years now, and will continue in the years to come, the proportion of industrial consumers in the total electricity demand is expected to shrink in the near future [4]. On the other hand, with more and more trains, undergrounds, tramways and electric automobiles, proportion of electric demand from transportation systems, especially public transportation, is expected to grow [5].

8.1.2   *The transformation of the European power system*

Introduction of microgrids in Europe is happening as part of the large-scale energy system change of the Old Continent. This change affects the whole grid and is mostly caused by the necessity of replacing two of the current principal energy sources: fossil fuels and nuclear reactors. We will next explore the main reasons for this replacement.

Europe's second most important energy source, nuclear power, has always had enemies, especially amongst ecological activists [6]. Environmental concerns regarding the handling of radioactive waste and lack of confidence on the safety of nuclear plants operation have been the main arguments against this technology [7]. Since the Fukushima nuclear accident in 2011, pressure has increased on European countries that use nuclear power to begin phase-out programmes. Two of the major nuclear users, France and Germany, have recently decided to reduce or phase out completely their nuclear generation in the next decade [8,9]. Other countries, like Switzerland and Spain, are currently hosting social debates on the matter and might make similar choices in the next few years [6].

Motivation for replacing fossil fuels relies on two main reasons: decreasing dependence on fossil fuel suppliers and decreasing emissions of greenhouse gases [10,11]. In 2013, Europe imported over 50% of its gross energy consumption. For fossil fuels, the dependency rates (a measure of how much the continent relies on imports to satisfy its demand, and how many different suppliers it has at hand to make the said imports) were very high: 88% for crude oil and over 65% for natural gas. Under certain political circumstances, this could lead to a very undesirable situation, considering that the main suppliers are often politically unstable and unreliable countries (Russia, Saudi Arabia, etc.). Achieving a certain level of energy independence has thus become a European priority [12].

The main reason for Europeans to stop burning fossil fuels, however, seems to be reducing their carbon footprint. Many collective initiatives have been undertaken by Europeans to reduce emission of greenhouse gases. We mention here three of the initiatives that show how committed Europe is with reaching this goal.

In December 2015, at the UN Climate Change Conference COP21, in Paris, 175 countries, among them the whole European Union (EU) and virtually all European countries, signed an agreement in which all governments committed to undertake measures necessary to maintain the global temperature rise in the years to come under 2 °C. This is drive by awareness of the threat that a more pronounced global warming will pose to the whole planet, especially insular states [13]. This means all countries are now obliged to reduce their CO_2 emissions (among other greenhouse gases) [14].

On an EU level, the Strategic Energy Technology Plan (SET-Plan) provides the framework to promote 'research and innovation efforts across Europe by supporting technologies with the greatest impact on the EU's transformation to a low-carbon energy system' [15].

Another EU initiative is the 20 20 20 Directive, which sets the goal, enforced by a legislative package, of reducing carbon emissions by 20%, increasing RES to 20% Europe-wide and improving efficiency in consumption by 20% by the year 2020 [16].

These directives, plans and agreements are general initiatives to promote and coordinate projects rather than actual projects. This means each of these directives englobes a large number of projects that aim to make the transformation of the European energy system possible.

This social and political situation has forced European countries to find a substitute for two technologies that currently make up over 65% of the energy sources used for electrical generation. Two options emerge as substitutes: RESs and power efficiency.

Microgrids are one of the new grid concepts that have been developed as an answer to these new requirements, as they permit both increasing penetration of RES and improving energy efficiency. After this review of the context that has led to the development of microgrids in Europe, in the next section we take a closer look at research on this subject.

8.2 Microgrids research in Europe

As mentioned in the previous section, the primary focus of European research in power-grid-related areas is replacing nuclear power and fossil fuel consumption through two strategies: increasing penetration of RES and improving energy efficiency. Microgrids are one way of reinventing the traditional electricity grid to achieve both objectives. In this section, we review how microgrids research is being oriented in Europe. The first significant observation to make is that microgrids do not occupy a place amongst power grid research subjects as important as they do in North America (see Chapter 9).

In fact, alongside microgrids, other concepts that are meant to transform power grids, such as smart grids, distributed generation and virtual power plants, are also subject of research in European universities and Research & Development (R&D) centres. A major advantage of a microgrid over other such alternatives is that it allows its users to disconnect from the main power grid when it is economically attractive to do so or when there is a fault on the electricity grid side. The latter case means consumers can still have access to electrical energy, even when power from the main electrical utility is unavailable. Thus, microgrids can increase source of energy reliability for small communities. This advantage has made microgrid development very attractive for sensitive power users in the United States, such as jails, health centres and military facilities (see Chapter 9).

In Europe, however, blackouts are not as big a concern as they are in the United States (see Chapter 9). On the one hand, the European power grid, especially in mainland Europe, is a highly interconnected one, which reduces blackout risks. On the other hand, natural disasters do not pose a threat as considerable as the one represented in the United States, where hurricanes and tornados are very frequent [17]. Hence, improving reliability, although not a neglected issue, is not one of the primary motivations for R&D in Power Grid related subjects in Europe. This is why other alternative power system concepts, such as smart grids, which also help RES penetration and energy efficiency, are much more favoured in Europe than microgrids.

Indeed, looking at large-scale European initiatives in this area of study, it is easily possible to find more smart grids-related projects than microgrids-related. Among the most important smart grids projects worth mentioning [18–23] are as follows:

- Grid4EU
  Grid4EU is the largest smart grids project to be funded by the EU to date. The project aimed to test RES integration, energy integration, electric vehicle development, power network automation, energy storage, energy efficiency and load reduction. The Grid4EU project consisted of six demonstrators, located in six European countries (Sweden, Germany, France, Austria, Italy and Spain). All six sites were tested and analysed for four years. The technologies and systems included advanced meter management, medium voltage grid automation through heuristic methods, and automatic grid recovery [19,20].
- Smart cities (previously Concerto)
  This project makes information about smart grids projects that have been or are being carried out in over 50 cities all around Europe more readily available for investors, local authorities and other people interested. It focuses on projects related to improving energy efficiency in buildings and communities. This project hopes to make this sort of smart grids projects easier to develop, more attractive and more accessible, and thus aim to promote the development of such smart grids all across Europe [21,22].
- Numerous other projects can be consulted in [23].

Nevertheless, microgrids are still an important subject of research, on which there have been large-scale projects worth mentioning here:

- The *Microgrids* project, and its follow-up, the *More Microgrids* project.
  This project was the pioneer and most important large-scale project on microgrids that has been developed in Europe to date. The first stage of the project dates from 2002 and consisted in designing and studying the dynamic behaviour of a single microgrid. The second stage started in 2005 and consisted in designing and implementing a series of microgrids in locations all

around Europe for demonstrative purposes. Many of these pilot microgrids are described in detail in the following section. Although some of them were decommissioned after the end of the project, most are still operational and have inspired other projects [24,25].

- DISPOWER [26,27]
  Distributed Generation with High Penetration of Renewable Energy Sources. Although not exclusively a microgrids project, some of the work developed in this project included microgrids indeed. The research areas of this project are
  – Grid stability and control
  – Power quality and safety
  – Socio-economic issues
  – Planning, training and operation tools
  – Information, communication and electricity trading
  – Test facilities for grid stability, control and power quality
  – Implementation of distributed generation technology in distributed power generators in low voltage grids
  – Overall assessment of distributed generation in local power supply systems.

Besides the projects above, which focus on smart grids and microgrids only, other wider European programmes often host microgrids and smart grids-related projects. Among these are, for example, the LIFE13 programme, which is the *EU's funding instrument for the environment and climate action* [28], and the Framework Programmes, in which projects can be presented in order to get funded.

Focus in this section has been in Europe-wide initiatives, which are often related to EU agencies. It is important, however, to note that some European countries also have programmes and initiatives of their own. Germany, for instance, has set itself the goal of achieving 80% RES by 2050. In the United Kingdom, there has been discussion for a few years about the regulatory aspects that will be necessary to make nation-wide microgrid development possible [29].

8.3 Microgrid patents and companies in Europe

8.3.1 Microgrids patents in Europe

According to our search in the WIPO (World Intellectual Property Organization) [30] around 282 patents related with any branch of microgrid research were processed in the European patent offices (European Patent Office – EPO); Spain Patent Office – OEPM (Oficina Española de Patentes y Marcas); German Patent and Trade Mark Office – DPMA (Deutsches Patent- und Markenamt); the Intellectual Property Office of United Kingdom – IPO; Portugal Patent Office – INPI (Instituto Nacional da Propriedade Industrial); the Eurasian Patent Organization – EAPO; Russian Federation Patent Office – ROSPATENT (Federal Service for Intellectual Property).

The percentages of these patents proposing new techniques and knowledge in the different microgrid areas are control of microgrids, 84%; microgrids with controllable loads, 48%; microgrids with storage, 36%; Direct Current microgrids, 20%; communication within microgrids, 21%; cooperation within microgrids, 8%; microgrid optimization, 41%; microgrid modelling, 45%; and clean energy microgrids, 13%.

The number of patents published per year in the European patents offices in the last 10 years are in 2006, 12 patents; in 2007, 12 patents; in 2008, 10 patents; in 2009, 10 patents; in 2010, 23 patents; in 2011, 26 patents; in 2012, 33 patents; in 2013, 28 patents; in 2014, 29 patents; in 2015, 29 patents; and half of 2016, 22 patents.

Similarly, in The International Patent System WIPO/PCT (World Intellectual Property Organization/Patent Cooperation Treaty) around 499 patents related with any branch of microgrid research have been processed. The percentages of these patents proposing new techniques and knowledge in the different microgrid areas are control of microgrids, 89%; microgrids with controllable loads, 63%; microgrids with storage, 55%; Direct Current microgrids, 33%; communication within microgrids, 41%; cooperation within microgrids, 18%; microgrid optimization, 61%; microgrid modelling, 53%; and clean energy microgrids 16%.

The number of patents published per year in the WIPO/PCT patents system in the last 10 years are in 2006, 26 patents; in 2007, 26 patents; in 2008, 33 patents; in 2009, 31 patents; in 2010, 25 patents; in 2011, 31 patents; in 2012, 36 patents; in 2013, 44 patents; in 2014, 48 patents; in 2015, 48 patents; and half of 2016, 34 patents. The 10% of applicants of microgrids patents in WIPO/PCT are from universities.

8.3.2   *European companies with services on microgrids*

In this section, some of the leading European companies with services on microgrids are presented. ABB is a multinational corporation headquartered in Zurich, Switzerland. Central business focuses of ABB are power generation and industrial automation. The microgrid services offered by ABB are automation and intelligent control solutions. They enable very high levels of renewable energy penetration in isolated diesel-powered grids. They provide a tool that computes the optimal power configuration in order to guarantee a proper balance of energy generation and demand that maximizes renewable energy integration. The primary applications are related to microgrid communities and microgrids for industry applications (more information in [31]).

Eaton Corporation Plc is a multinational company focused on power management; its corporate headquarters is in Dublin, Ireland. The microgrid services offered by this company are called Microgrid Energy Systems. These services help to guarantee electrical energy security independent of the utility grid availability. In addition, it is possible to provide demand management through, a combination of multiple renewable generation sources integrated into a grid structure in an area with the necessary loads (more information in [32]).

Schneider Electric is a European company operating worldwide. Central business focuses of Schneider are heavy and electric industries. The Schneider microgrid solutions are in the branches of metering and power reliability. These solutions offer grid independence without losing benefits of the main electricity grid. They provide Microgrid Intelligence through the following: Power Monitoring Solutions; Energy Performance Services; Invoice & Data Management; Power Reliability, Availability & Quality; Energy Management & Sustainability and Sustainability Reporting (more information in [33]).

Siemens AG is a German multinational company operating in the industrial, energy, health and infrastructure sectors. Siemens considers a broad view of microgrid infrastructure and includes Power Generation and Power Consumption Solution not only for electricity, but also for water and gas. They have a Microgrid Management System able to integrate and optimize decentralized power generation and controllable loads. This advanced software solution can optimally manage the grid assets and ensure that economic goals are reached, while meeting energy demand (more information in [34]).

Alstom is a French corporation focused on the business of electricity generation and manufacturing trains. The Alstom Grid and General Electric alliance offer an optimization solution for permanently islanded or grid connected microgrids called Grid IQ. It is a central supervisory controller of a microgrid control system able to maximize the use of renewable resources. It gives guidelines to dispatchable generation resources to provide power to the load in the most economical way (more information in [35]).

DNV-GL was created in 2013 as a result of a merger between two leading organizations DNV (Det Norske Veritas from Norway) and GL (Germanischer Lloyd from Germany). The organization is a reference one on renewable, alternative and conventional energy. They provide advice on how to integrate distributed energy resources into the main power grid and how to maximize the value of microgrid assets (more information in [36]).

8.4   Current microgrids projects in Europe

A feature often deemed necessary for a small-scale power system to be considered a microgrid is ability to operate both isolated and connected to the power grid. As was explained in the previous section, being able to isolate from the power grid is not a priority in Europe. Besides, the primary motivation in Europe for microgrids development is reducing the carbon footprint of power systems. Many projects have been carried out in isolated areas, chiefly islands, where virtually all electrical power generation has traditionally relied on fossil fuels. For obvious reasons, connecting to the main electricity grid is not an option there.

Thus, abiding by the strictest definition, all these projects, which are considered microgrids in Europe, because of their situation and requirements, would be left out. This is why in our review of current projects, we also consider some microgrids that cannot operate both isolated and connected to the power grid.

However, other microgrids features are present in all these projects, such as having distributed generation, that is, power sources close to the power demand and some sort of energy storage. Table 8.1 presents a list of significant microgrids

Table 8.1   Current microgrids projects in Europe

Project	First year of operation	Conventional sources	Renewable sources	Storage	Type	Total capacity (kW)
Isle of Eigg (Scotland)	2008	H	W, PV	Batt	CC	166
Island of Rum (Scotland)	2008	H, D	–	Batt	CC	45
Isle of Muck (Scotland)	2013	D	W, PV	Batt	CC	30
Isle of Foula (Scotland)	2008	D, H	PV	Batt	CC	35
Horse Island (Scotland)	2009	D	W	Batt	CC	30
Flat Holm (Bristol Channel – GB)	2006	D	W, PV	Batt	CC	10
Centre for Alternative Technology (Wales)	2009	H	PV, W	Batt	CC	30
UK-Microgrid (Uni-Manchester) Decommissioned (United Kingdom)	2005	–	–	Fly	UC	20
University of Strathclyde's Distribution Network and Protection Laboratory (United Kingdom)	–	D	W, PV	Batt	UC	80
Cork Inst of Tech (Ireland)	2012	CHP	W	Batt	UC	50
Life Factory Microgrid (Spain)	2016	–	W, PV	Batt, V2G	In	160
Atenea (Spain)	2013	D	W, PV	Batt	In	50
Smart City Málaga (Spain)	2008	Cog	W, PV	Batt, V2G	CC	6
Labein Experimental Centre (Spain)	<2007	D	W, PV	Batt, Fly, Cap	In	60
EDP – Swimming Pool (Portugal)	<2011	CHP, D, M	–	–	P	180
Açores – Island of Graciosa (Portugal)	2014	D	W, PV	Batt	CC	5,500
Açores – Island of Flores (Portugal)	2013	D, H	W	Batt, Fly	I	2,500
Açores – Island of Faial (Portugal)	2014	D	W	–	I	17,000
Instituto Politécnico de Bragança (Portugal)	–	BG, H	W, PV	V2G, Batt	UC	5
CESI Ricerca DER Test Facility (Italy)	2005	CHP, D	PV, W	Batt, Fly	P	350
University of Genoa – Savona (Italy)	2014	CHP	PV	Batt	UC	200
Kythnos (Greece)	2005	D	PV	Batt	CC	15
ICCS-NTUA (Greece)	2014	D	PV	Batt	UC	2
Mannheim – Wallstadt (Germany)	2006	CHP	PV, FC	Fly	CC	30
Am Steinweg – Stutensee (Germany)	2005	CHP	PV	Batt	CC	35
Pellworm (first smart microgrid in Germany)	2013	BG	PV, W	Batt	U	1,000
EUREF Campus – Berlin (Germany)	2011	CHP	PV, W	V2G, Batt	P	–

Project	Year	Conventional	Renewable	Storage	Type	Size
AEG Warstein-Belecke (Germany)	2013	CHP	PV	Batt	–	300
IRENE/iren2 (Wildpoldsried im Allgäu) (Germany)	2012	BG, CHP	PV, W	Batt	–	–
ISET (Kassel) (Germany)	2005	D	PV, W	Batt	UC	20
Feldheim (independent from utility, exports wind energy) (Germany)	2013	–	PV, W, Biomass	Batt	CC	74,400
EKZ (Energie Kanton Zürich) (Switzerland)	2005	D	PV	Batt	CC	1,000
i-BATS (Valais, Suisse) (more smart-grid than μgrid) (Switzerland)	2013	–	PV	Batt	In	15
PowerMatching City (phases 1 and 2) (The Netherlands)	2009 (2011)	CHP	PV, W	Batt	CC	2,020
Continuon (The Netherlands)	2009	–	PV	–	CC	315
ENECO (NOT a microgrid rather a virtual power plant) (The Netherlands)	2015 (still ongoing project)	–	–	–	–	100,000
Armines (France)	2005	D	PV, FC	Batt	UC	7.46
Nice Grid (France)	2012	–	PV	Batt	CC	2,000
Murmansk, Murmansk Oblast, Chavanga, Kola Peninsula (Russia)	2015	–	W, PV	Batt	RC	195
Bornholm Danish Island (Denmark)	2007	CHP, D, BG,	W	–	I	127,000
Avedøre Power Plant Unit 1 (Denmark)	2012	CHP	–	–	CM	250,000
Utsira Island (Norway)	2004–2008	–	W, FC	Fly, Batt	I	2,000
Helen Ltd. (Helsingin Energia) Helsinki (Finland)	–	CHP, TH	W	Batt	CM, CC	630,000
Agria (Macedonia)	2010	BG	–	–	CC	55
Kozut (Macedonia)	2007	BG	PV	–	RC	5
SGEM (Finland)	2012	D	W	–	U	2,000
UTC Copenhagen (Denmark)	–	–	FC, PV	–	UC	10
Samso Island (Denmark)	2005	CHP	W, Solar	–	I, CC	21,000
Lolland Hydrogen Community (Denmark)	2007	CHP	FC	Hydrogen tank	I	8.5

Conventional sources: D, diesel; CHP, combined heat and power; S, stream; G, gas; H, hydraulic; M, motor driven gen; BG, biogas; Cog, cogeneration.

Renewable sources: W, wind; PV, photovoltaic; FC, fuel cell; Biomass.

Storage: Batt, battery; Fly, flywheel; Cap, capacitor; TH, thermal; V2G, vehicle to grid; CSP, concentrated solar power; hydrogen tank.

Microgrid type: RC, remote community; I, island; UC, university campus; CC, city/community; P, public institution; MI, military; CM, commercial; U, utility; In, interconnected.

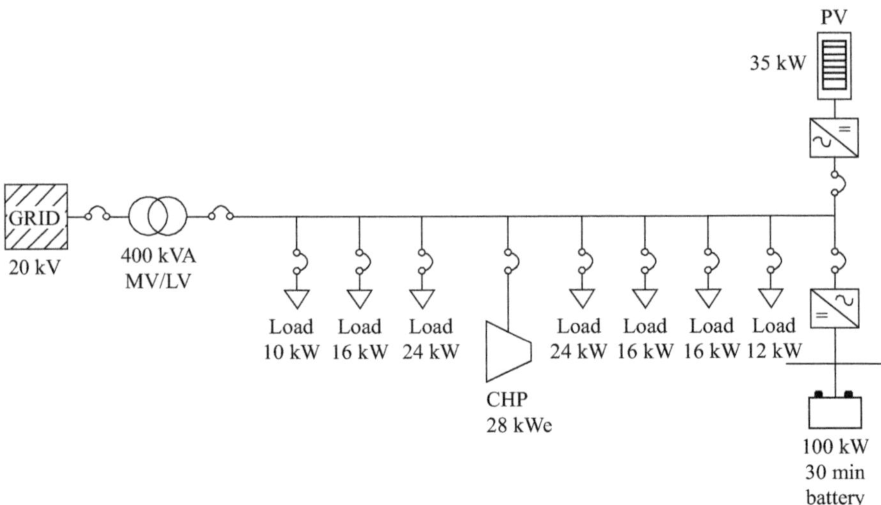

Figure 8.1   Am Steinweg microgrid in Stutensee, Germany

projects in Europe. This list was built by the authors based on the data found in
[37–48].

In the next section, we describe some of the most important microgrids that
have been developed in Europe to date with a little more detail.

8.4.1   Am Steinweg (Germany)

The *Am Steinweg* microgrid in Stutensee, close to Karlsruhe, in Germany, was one
of the first microgrids projects in small communities or neighbourhoods being
developed in Europe. It was part of the DISPOWER project and began operating in
2005. It consists of 101 apartments fed by a 28-kW CHP generator and 35 kWp of
PV installations. The system also has both thermal and electrical energy storage. It
is connected to the medium voltage utility grid through a 400-kVA transformer.
The electric storage system consists of a 1,000 kW, 880 Ah lead-free battery bank.
The microgrid is capable of working as an island at certain hours of the day, when
local generation matches demand (Figure 8.1).

Aside from implementing distributed generation and energy storage, this
microgrid project also included some features of energy management. Namely,
consumers were informed when PV generation was at its peak, so certain home
appliances were turned on (for instance washing machines). Power consumption
from the utility grid can thus be minimised [49].

8.4.2   Mannheim-Wallstadt (Germany)

The Mannheim-Wallstadt microgrid near Mannheim, Germany was one of the
More Microgrids projects. It started in 2006 and was developed by the utility

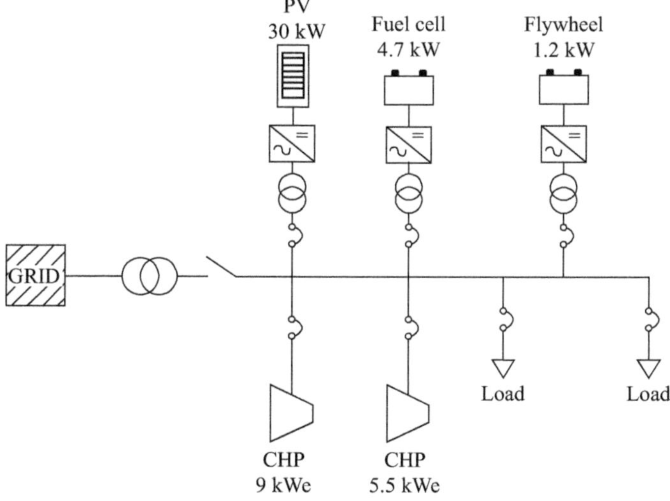

Figure 8.2   Mannheim-Wallstadt microgrid in Germany

operator, MVV Energie. The project aimed to implement a microgrid capable of islanding in this residential and commercial area, which has around 1,200 inhabitants. This microgrid features a 4.7-kW fuel cell, 30 kW of PV generation and two CHP units of 9 and 5.5 kW of electrical capacity. Amounts of 1.2- kW electric power storage are also available by means of a flywheel system (Figure 8.2).

The project was developed as a pilot to research the frequency behaviour of an islanded microgrid and to find out about the social issues that should be taken into account when developing such projects. Results were satisfying [42].

8.4.3   AEG microgrid in Warstein-Belecke (Germany)

Allgemeine Elektricitäts-Gesellschaft Power Solutions (AEG PS) or General Electricity Company in German is a German company specialized in the design and construction of solutions to power supply needs. AEG PS has been doing research together with the University of Paderborn on industrial microgrids for SMEs (Small- and Medium-sized Enterprises). A microgrid in AEG's factory in Belecke, Germany is being constructed for demonstrative purposes (Figure 8.3).

It features a CHP plant, 330 kWp of PV generation and 2,500 Ah of battery storage. The microgrid is controlled by a smart energy management system that minimizes costs of operation. Other SMEs-oriented features are also being planned to be added to the microgrid, such as power-to-gas energy storage [38].

8.4.4   Wildpoldsried im Allgäu (Germany)

Wildpoldsried is a small town in the region of Allgäu in southern Germany with a population of 2,500. Several wind turbines, PV generators and biogas units have

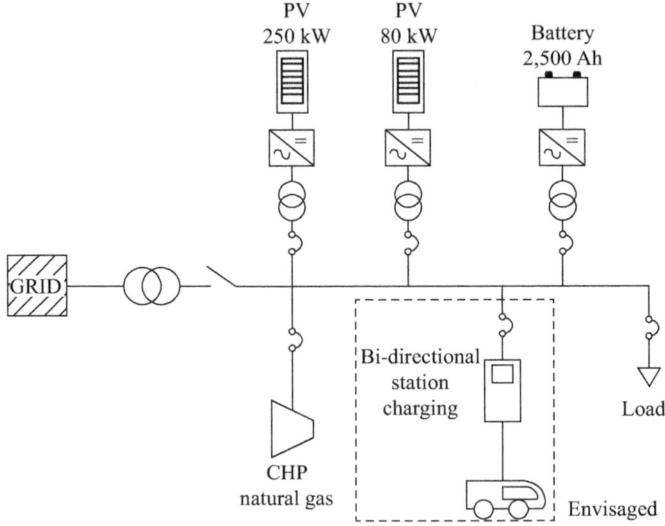

Figure 8.3   AEG microgrid for SMEs (Germany)

been installed in the area since 2010. For some years now, Wildpoldsried has produced about three times more energy from its RES than it consumes. Since 2012, as part of the German IRENE (Integration of Regenerative Energy and Electric Mobility) project, and its follow-up project IREN2, a smart microgrid is being developed to pursue three main objectives. The first one is to manage the energy consumption to balance generation and demand in order to maintain grid stability. The second one is to allow the community grid to disconnect from the utility grid and thus become a true microgrid (which was not possible before, because excess generation destabilized the grid). The third one is to research how the microgrid can provide conventional system services (reactive power control, voltage control, etc.) [50].

8.4.5   ISET, Kassel (Germany)

The microgrid developed at the Institute of Solar Energy Technology (ISET) in Kassel, Germany, was one of the pilot microgrids developed during the *Microgrids* project. The microgrid had the following features: a 16-kW diesel generator set, 2-kW PV and 11-kW wind generation, two 13-kW battery banks, loads with different priority levels, possibility of breaking the microgrid into three islands, connection to medium voltage utility grid through a 100-kVA transformer and automatic supervision and operation control (Figure 8.4).

The project aimed to test many different situations to which a real microgrid could face: island to grid transition, grid to island transition, black start, reconnection to utility grid after fault, different supervision and control schemes and others. The results obtained were positive [51].

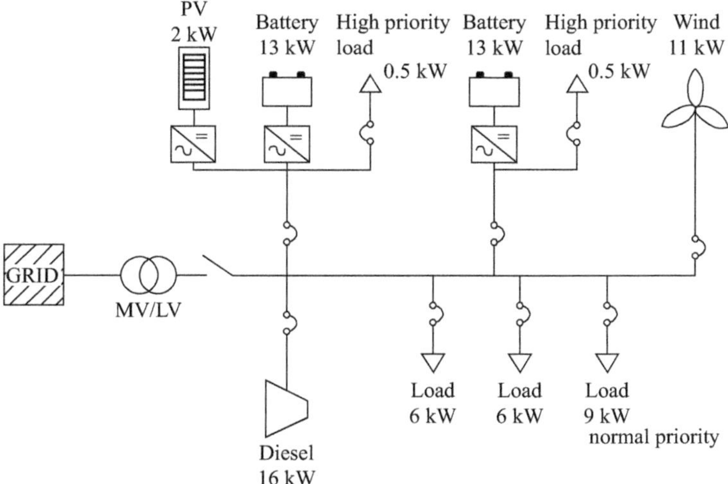

Figure 8.4   ISET microgrid in Kassel, Germany

8.4.6   Feldheim (Germany)

Feldheim is a town of 135 inhabitants in eastern Germany. Over the past two decades, it has witnessed a major transformation as it turned from a communist collective farm into a major RES hub. Today, Feldheim's grid has over 74 MW of installed wind turbines, a 400-kW CHP biogas unit powered by wood, pig manure and waste from corn plantations, over 300 kW of PV generation and a 10-MW battery storage system. The town only consumes under 1% of the power it generates.

Since 2010 Feldheim put in place not only a smart microgrid, but they also built a whole new grid, as they replaced the old utility's infrastructure with their own. Thus, the village has achieved complete self-sufficiency. Feldheim has turned into an example for other communities that also wish to implement RES in their regions and now offer RES training among other services, as plans to expand their current infrastructure and add new power sources, such as vehicle-to-grid systems, continue to develop [16].

8.4.7   Kythnos (Greece)

This project is one of the microgrids from the *Microgrids* project. It is a three-phase microgrid feeding 12 houses in Gaidouromantra, a small valley in the island of Kythnos in the Cyclades archipelago. The microgrid features 12-kWp PV generation, a 5-kVA diesel generator set and 85 kW h of battery energy storage. The site was used to test different control strategies and communication protocols between the various components of an islanded microgrid (Figure 8.5).

The microgrid has been in operation since 2005. It has been proposed to add wind generation to the microgrid and to connect the microgrid to the grid of the rest of the island [24,52].

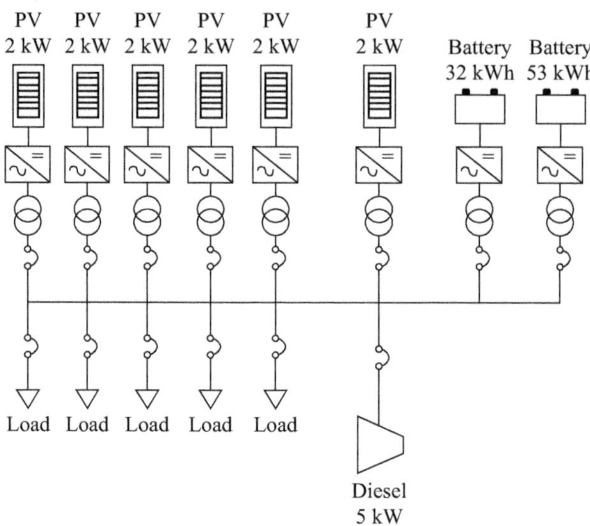

Figure 8.5   Kythnos microgrid in Greece

8.4.8   *ICCS-NTUA (Greece)*

The Institute of Communication and Computer Systems of the National Technological University of Athens (NCC-NTUA) is one of the partners of the *Microgrids Project*. This microgrid is located at their laboratories in Athens and comprises 1.2 kW of PV generation, 0.8 kW of wind generation and 5,500 Ah of battery energy storage. The microgrid is used to research communication between devices, control schemes and SCADA systems [46].

8.4.9   *CESI Ricerca DER test facility (Italy)*

This microgrid was also one of the pilot microgrids from the *More Microgrids* project. It is located in the facilities of the CESI (Italian Electrical and Technical Experimental Center) research centre for distributed energy resources in Milan, Italy. It is a low-voltage microgrid connected to the grid through an 800-kVA transformer. There are many different generators in this microgrid: 24 kWp of PV, a 7-kVA diesel generator, an 8-kVA wind generator, a 10-kW solar thermal plant with a parabolic dish and Sterling engine, a 10 kWe, 90 kWth CHP biomass unit, a 105 kWe, 170 kWth CHP gas microturbine (Figure 8.6).

The microgrid also features different electric energy storage systems, including different battery technologies (Vanadium, Lead acid, Zebra, for a total capacity of over 200 kW), and a flywheel (100 kW, 30 s). Finally, controllable loads, both resistive-inductive and capacitive, are available.

This microgrid is very flexible, with switching devices allowing many different configurations and controllable loads permitting simulation of very different conditions. That is, it is possible to dynamically change the topology and working

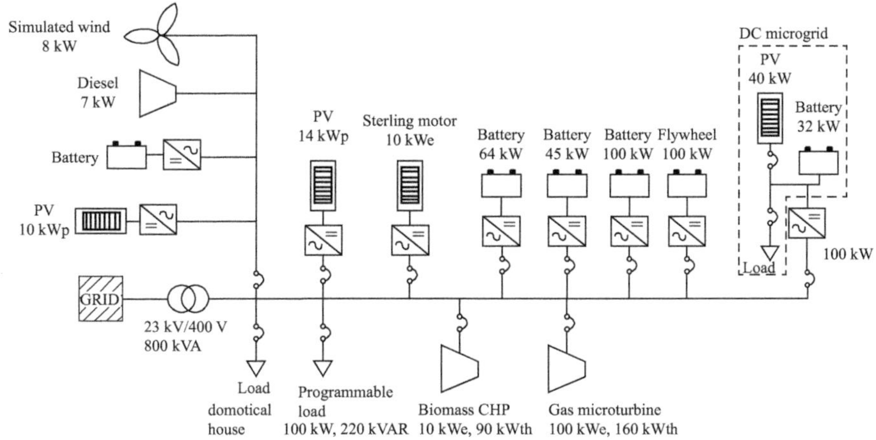

Figure 8.6   Test facility microgrid at RSE (previously CESI Ricerca)

scenarios of this microgrid. A number of communication protocols are also available for testing, thus making this microgrid a test bed for studying the behaviour of a real microgrid under many different circumstances.

After the end of the *More Microgrids* project, a Direct Current microgrid was added to the test facility. This DC microgrid consists of a 32 kW, 36 kW h battery bank, a 30 kW, six supercapacitor banks, 40 kW of emulated PV generation and 60 kW of controllable resistive load. The DC microgrid can be connected to the main AC microgrid through a 100-kW inverter.

After a transformation process of the research centre, this microgrid is now called *Laboratorio Test Facility di Generazione Distribuita in Bassa Tensione*, of the *Ricerca sul Sistema Energetico RSE* [53].

8.4.10   University of Genoa – Savona Campus (Italy)

The microgrid of the University of Genoa, at the Savona Campus, was developed by the state utility operator, *Enel*, in alliance with *Siemens*. It was developed with the goal of creating a concept of urban microgrid that would serve as reference for future developments. The microgrid has been in operation since 2014 and combines a number of RES and non-RESs, as well as energy storage systems.

The microgrid has three parabolic CSP systems generating both electric and thermal energy, PV generators with a capacity of 80 kWp and three 250 kWe, 300 kWth CHP gas microturbines. The microgrid also has many other features, such as absorption refrigerators, heat storage systems, a 100-kW h battery electric storage system and electric vehicle charging facilities.

The system is completed by a state-of-the-art control and monitoring system that allows managing energy consumption and generation in real time. Since beginning operation, the university campus has been able to cut its energy consumption costs by a third [54].

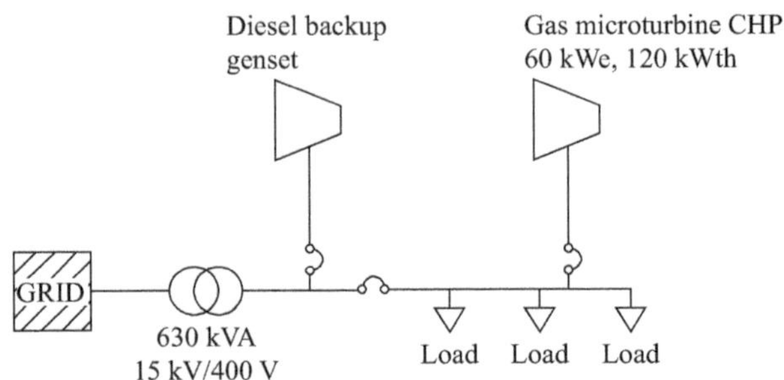

Figure 8.7   EDP swimming pool microgrid in Portugal

8.4.11   EDP swimming pool (Portugal)

This project was also one of the *More Microgrids* pilot projects. The microgrid is located in a public swimming pool in Ílhavo, Portugal. Before developing the project, the swimming pool had access to the utility grid through a 630-kVA transformer and had a backup diesel generator set. The project consisted of a 60 kWe, 120 kWth CHP gas microturbine and enabling both islanding capability and excess electrical energy export for the swimming pool installation. Islanding and parallel operation was tested successfully (Figure 8.7).

Thanks to this microgrids project, it can be shown that such microgrids for public swimming pools are feasible and attractive. It can also be determined that in order to operate such microgrids, some sort of load and generation control is necessary to maintain stability and power quality. After this project, the Portuguese utility operator, EDP, who developed the project, continued developing similar projects in the country [55].

8.4.12   Factory microgrid (Spain)

The factory microgrid is an industrial microgrid developed by CENER, the Spanish National Centre for Renewable Energy and Jofemar, a Spanish company specialized in the design and construction of vending machines. The microgrid was set in the company's facilities in Peralta, Spain, and was partially funded through the LIFE programme. This microgrid was created to demonstrate the advantages that such systems can have for industrial plants. Its objectives include the development of the microgrid itself as well as transfer of the knowledge acquired during the project to other interested factories.

The microgrid has a 120-kW wind turbine, 40 kWp of PV generation and 500 kW h of battery electrical energy storage. The project is ongoing; future development includes bidirectional and high-speed electric vehicle charging stations. Thanks to this project, the company has been able to diversify its portfolio, opening a new division specialized in energy storage systems.

When the project ends, in 2017, it is expected that over 160 MW h of the factory's electrical consumption will be generated by the microgrid, thus reducing its carbon footprint considerably [56].

8.4.13   Atenea (Spain)

This microgrid is located in the facilities of the Wind Generator Testing Laboratory of CENTER in Navarra, Spain. The microgrid has been operating since 2013 and was developed to provide energy for some of the laboratory equipment as well as lighting throughout the facility. This microgrid is made up of 25 kWp of PV generation, 20 kW of wind generation, a 55-kVA diesel generator set and over 6 h of 50 kW of electrical energy storage battery. The microgrid has a single connection point to the utility grid and can consume energy as well as provide energy to the grid through it [43].

8.4.14   Labein Experimental Centre (now Tecnalia) (Spain)

This microgrid was one of the microgrids of the *More Microgrids* project. It is located in Derio, Spain, in Tecnalia's (formerly Labein, a privately funded R&D centre) facilities. The microgrid is intended as a test bed for different control strategies. It has a total of 5.8 kW of PV generation, over 55 kW of wind turbine generation and two 55-kW diesel generators. It also includes a 15 s 250 kVA flywheel, over 2 MJ of ultra-capacitors energy storage and battery energy storage. The system is completed by controllable loads and flexible system controllers [39].

8.4.15   Centre for Alternative Technology (Wales)

The Centre for Alternative Technology (CAT) is an R&D Centre located in Wales and has been in operation since 2009. The microgrid consists of 20 kWp of PV generation, 7.5 kW hydro generation and 600 W of wind generation. Over 100 kW h of battery energy storage are available as well. Total generation is insufficient to deliver power to the whole centre, so power is consumed from the main grid through a single 30 kW point of connection.

The microgrid is capable of islanding in emergency situations, such as a fault in the main power lines outside the CAT. In that case, an electronic controller manages generation and demand and eventually shuts down part of the microgrid demand to maintain balance [57].

8.4.16   UK Microgrid (England)

This microgrid was developed in the laboratories of the University of Manchester between 2005 and 2008. The project aimed to develop a new control scheme that would not disconnect the microgrid from the utility grid during a fault. Instead, the microgrid should stay online and help the grid recover from its fault state, in a so-called model citizen behaviour (Figure 8.8).

The microgrid was actually only a simulated microgrid, since it had no real power sources. A 20-kW motor-generator set was available, but the motor was fed by an inverter connected to the utility grid. The microgrid also had a flywheel energy storage system. After the end of the project, this microgrid was decommissioned [40].

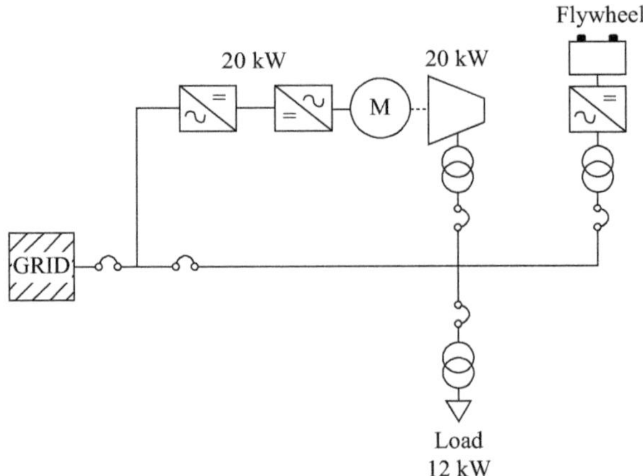

Figure 8.8   UK Microgrid, later decommissioned

8.4.17   Isle of Eigg (Scotland)

This microgrid powers the Isle of Eigg, a small island of 90 inhabitants off the west coast of Scotland. The microgrid started operating in 2008. It features 110 kW of hydropower, 24 kW of PV generation and 32 kW of wind turbines. A battery storage system is also present to ensure that power is available around the clock.

Thanks to this EU-funded project, the island went from a power system consisting of individual diesel generator sets for each household to a community-wide grid generating its power over 95% from RES.

Although, not described in detail here, numerous similar projects have been developed in other Scottish and British islands, such as the Island of Rum, Fair Isle, Isle of Muck, Isle of Foula and Horse Island [41].

8.4.18   Continuon (The Netherlands)

Also part of the *More Microgrids* project, the *Continuon* microgrid started operating in 2009 in Bronsbergen, The Netherlands. The project was developed in a holiday park where PV roofs were already installed, with a total capacity of 315 kWp. Peak consumption of the 109 cottages was 150 kW. The project consisted in installing a battery energy storage system (BESS), as well as a control system allowing the microgrid to manage generation and demand, and islanding. Another objective was reducing the harmonic components of the current at the connection point with the main grid [24,58] (Figure 8.9).

The project was successful and made this microgrid the first one in the Netherlands.

8.4.19   Power Matching City (The Netherlands)

Power Matching City was a two-phase project in Hoogkerk, the Netherlands. During the first phase, 25 houses in a small neighbourhood were equipped with

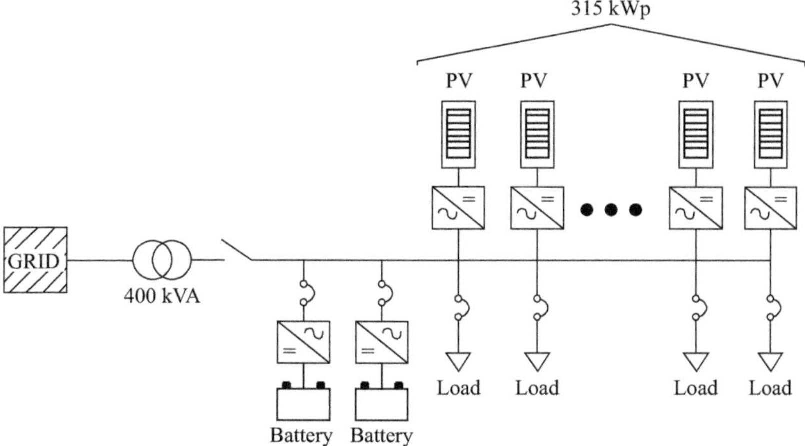

Figure 8.9   Continuon microgrid in the Netherlands

1.4 kW of PV generation, CHP generation, BESSs, electric vehicles and smart meters and household appliances. A 2-MW wind farm nearby also supplied the microgrid and also delivered power to the main grid. The second phase of the project focused on how the market should be managed to improve energy efficiency.

Both projects showed the feasibility of small community microgrids and smartgrids-oriented energy markets [44].

8.4.20   *Energie Kanton Zürich (EKZ) (Switzerland)*

The Energie Kanton Zürich (EKZ) microgrid is located in the EKZ office building in Zürich, Switzerland, and consists of a small PV generation installation, a diesel generator set and a 1-MW BESS. The load supplied is the office building and several electric vehicle charging stations, for an average consumption of 360 kW. An important feature of this microgrid is that all voltage and frequency control is done by the BESS, even when working on island mode and with the diesel generator set working in parallel. Other interesting features are being tested in this microgrid, such as using frequency deviation as means of communication [59].

8.4.21   *Nice Grid (France)*

The Nice Grid project, in Nice, in the French Riviera, is a demonstrator of a smart solar microgrid. The microgrid is located in a place with high availability of solar power, and far away from other energy sources (on the end of a long transmission line). Among the project's objectives are optimising the management of a distribution grid with high penetration of distributed PV generation, enabling the islanding of the microgrid and researching different models for involving the end user in the energy market. It is in part financed by the French government and the Grid4EU project. In total, the complete project includes 2.5 MW of PV generation, 3.5 MW of interruptible demand and energy storage systems with a capacity of

1.5 MW. All these elements are planned to be located at different voltage levels of the grid, and owned by different agents (grid operator, end users, etc.). The project is ongoing. To this date, islanding tests for part of the grid (eight industrial clients) have been completed successfully [60].

8.5 Microgrids perspectives in Europe

In this section, we analyse what could be expected of microgrids in the years to come in Europe. As it has been seen, numerous successful projects have been developed in the last decade across the whole continent. Most of the microgrids that have been developed, however, have purposes of R&D or demonstration. Furthermore, virtually all these projects have had some sort of financial aid through European or government funds. This means that although technical feasibility and environmental pertinence has been proven, there is still work to be done to increase economic attractiveness of microgrids for private project developers.

Over the next years, microgrids can be expected to begin their journey outside the R&D world and into the real electricity market. In this manner, we expect microgrids to continue their current development following four paths:

- The pilot projects developed so far have proven the advantages that microgrids have for isolated areas as means of replacing diesel generators and reducing operation costs of the local power grids. Microgrids can be expected to continue their growth in islands and other remote places that are currently served by fossil fuels.
- Smart grids will continue to have the lead on microgrids. Looking at the amount of current projects and investment on the continent, we can expect to find more and more European smart neighbourhoods, smart cities and virtual plants. In some cases, microgrids will be a part of these projects, especially in regions where the ability of islanding could be attractive, such as poorly interconnected zones.
- Some countries (e.g. Austria) might have little to no development of microgrids. As explained in the first section, the main reason for developing microgrids is replacing fossil fuels and nuclear plants. Some countries already have a low enough usage of those two energy sources, so making the investments necessary to develop microgrids might not be very attractive for them [61]. The *liberalization of the electricity market* (see the next item), and an increased will to improve the reliability of the power system, might change this.
- In many of the projects presented in this chapter, and other projects encountered during the research for writing it, some of the objectives include investigating how the electricity market should work to optimize energy efficiency and cost of operation of the electrical grid, taking into account participation of the end consumer as energy provider. In many countries, this so-called liberalization of the electricity market is yet to be regulated, waiting for the results of all these pilot projects. How the regulation is done in the end will determine how microgrids develop in the future [62].

References

[1]   ENTSO-E, "Electricity in Europe 2015," European Network of Transmission System Operators for Electricity (ENTSO-E), Brussels, Belgium, 2016.

[2]   M. Huber, D. Dimkova and T. Hamacher, "Integration of wind and solar power in Europe: Assessment of flexibility requirements," *Energy*, vol. 69, pp. 236–246, 2014.

[3]   J. Torriti, M. Hassan and M. Leach, "Demand response experience in Europe: Policies, programmes and implementation," *Energy*, vol. 35, no. 4, pp. 1575–1583, 2010.

[4]   A. Kander, P. Malanima and P. Warde, *Power to the People: Energy in Europe over the Last Five Centuries*, Princeton, NJ: Princeton University Press, 2013.

[5]   European Environment Agency, "Final Electricity Consumption by Sector, EU-27," 19 March 2013. [Online]. Available: http://www.eea.europa.eu/data-and-maps/figures/ds_resolveuid/7DTXTJ37B2 [Accessed 19 June 2016].

[6]   A. Omria, N. Mabrouka and A. Sassi-Tmarb, "Modeling the causal linkages between nuclear energy, renewable energy and economic growth in developed and developing countries," *Renewable and Sustainable Energy Reviews*, vol. 42, pp. 1012–1022, 2015.

[7]   P. Breeze, Power Generation Technologies, Waltham, MA: Elsevier, 2014.

[8]   dpa, "Kabinett beschließt Atomausstieg bis 2022," 2011. *Die Zeit*, [Online]. Available: http://www.zeit.de/politik/deutschland/2011-06/atomausstieg-energiewende-gesetzespaket [Accessed 6 June 2011].

[9]   Ministère de l'énergie, de l'environnement et de la mer, "La loi de transition énergétique est promulguée," Développement Durable, 24 August 2015. [Online]. Available: http://www.developpement-durable.gouv.fr/La-loi-de-transition-energetique,40895.html [Accessed 29 June 2016].

[10]   J. Merkisz, J. Pielecha and S. Radzimirs, *New Trends in Emission Control in the European Union*, London: Springer, 2014.

[11]   R. Montañés, M. Korpås, L. Nord and S. Jaehnert, "Identifying operational requirements for flexible CCS power plant in future energy systems," *Energy Procedia*, vol. 86, pp. 22–31, 2016.

[12]   Eurostat – Statistics Explained, "Energy Production and Imports," May 2015. [Online]. Available: http://ec.europa.eu/eurostat/statistics-explained/index.php/Energy_production_and_imports [Accessed 19 June 2016].

[13]   C. Rhodes, "The 2015 Paris climate change conference: COP21," *Science Progress*, vol. 99, pp. 97–104, 2016.

[14]   COP21, "Décryptage de l'accord," December 2015. [Online]. Available: http://www.cop21.gouv.fr/decryptage-de-laccord/ [Accessed 20 June 2016].

[15]   European Commission, "Strategic Energy Technology Plan," 2016. [Online]. Available: https://ec.europa.eu/energy/en/topics/technology-and-innovation/strategic-energy-technology-plan [Accessed 20 June 2016].

[16]   European Commission, "2020 Climate & Energy Package," 9 June 2016. [Online]. Available: http://ec.europa.eu/clima/policies/strategies/2020/ index_en.htm [Accessed 20 June 2016].

[17]   P. Hinesa, J. Aptb and S. Talukdarc, "Large blackouts in North America: Historical trends and policy implications," *Energy Policy*, vol. 37, no. 12, pp. 5249–5259, 2009.

[18]   I. Colaka, G. Fullia, S. Sagirogluc, M. Yesilbudakd and C.-F. Covriga, "Smart grid projects in Europe: Current status, maturity and future scenarios," *Applied Energy*, vol. 152, pp. 58–70, 2015.

[19]   Grid4EU, "Overview of Grid4EU," 2012. [Online]. Available: http://www. grid4eu.eu/overview.aspx [Accessed 30 June 2016].

[20]   L. Ardito, G. Procaccianti, G. Menga and M. Morisio, "Smart grid technologies in Europe: An overview," *Energies*, vol. 6, no. 1, pp. 251–281, 2013.

[21]   The Smart Cities Information System, "CONCERTO archive," 2015. [Online]. Available: http://smartcities-infosystem.eu/concerto/concerto-archive [Accessed 30 June 2016].

[22]   A. Immendöerfer, S. Volker, K.-R. Bräutigam and J. Jöerissen, "Existing building challenge and CONCERTO Project," *Journal of Civil Engineering and Architecture*, vol. 8, no. 10, pp. 1253–1259, 2014.

[23]   Eurelectric, "Innovation in Smartgrids," 2016. [Online]. Available: http://www.eurelectric.org/innovation/ [Accessed 30 June 2016].

[24]   The MICROGRIDS Project, "Microgrids and More Microgrids Projects," 6 August 2010. [Online]. Available: http://www.microgrids.eu/default.php [Accessed 30 June 2016].

[25]   N. Hatziargyriou, H. Asano, R. Iravani and C. Marnay, "Microgrids," *IEEE Power & Energy Magazine*, vol. 5, no. 4, pp. 78–94, 2007.

[26]   European Commission, "CORDIS," 20 September 2001. [Online]. Available: ftp://ftp.cordis.europa.eu/pub/eesd/docs/ev260901_poster_dispower. pdf [Accessed 10 July 2016].

[27]   D. Thomas, J. Schmid and P. Strauss, "DISPOWER – Distributed Generation with High Penetration of Renewable Energy Sources," Final public report, Kassel: Kassel, 2006.

[28]   European Commission, "The LIFE Programme," 8 June 2016. [Online]. Available: http://ec.europa.eu/environment/life/about/index.htm [Accessed 30 June 2016].

[29]   N. Balta-Ozkan, T. Watson, P. Connor, *et al.*, "Scenarios for the Development of Smart Grids in the UK," UK Energy Research Centre, London, 2014.

[30]   World Intellectual Property Organization (WIPO), "Patents," [Online]. Available: http://www.wipo.int/patents/en/ [Accessed 20 September 2016].

[31]   ABB, "Microgrids Solutions," [Online]. Available: http://new.abb.com/ microgrids [Accessed 20 September 2016].

[32]   EATON, 2016. [Online]. Available: http://www.eaton.com/Eaton/index.htm [Accessed 20 September 2016].

[33] Schneider Electric, 2016. [Online]. Available: http://www.schneider-electric.com/ww/en/ [Accessed 20 September 2016].

[34] Siemens, "Smart Grid," [Online]. Available: http://w3.siemens.com/smart-grid/global/en/pages/Default.aspx [Accessed 20 September 2016].

[35] ALSTOM, [Online]. Available: http://www.alstom.com/ [Accessed 20 September 2016].

[36] DNV GL, 2016. [Online]. Available: https://www.dnvgl.com/services/microgrids-17940 [Accessed 20 September 2016].

[37] M. Barnes, A. Dimeas, A. Engler, *et al.*, "MicroGrid Laboratory Facilities," in *2005 International Conference on Future Power Systems*, Amsterdam, 2005.

[38] AEG Power Solutions, "Smart Microgrid Campus Project Belecke," 2016. [Online]. Available: https://www.aegps.com/en/applications/storage-distribution/smart-micro-grid/ [Accessed 10 July 2016].

[39] M. Barnes, J. Kondoh, H. Asano, *et al.*, "Real-World Microgrids – An Overview," in *2007 IEEE International Conference on System of Systems Engineering*, San Antonio, TX, 2007.

[40] M. Barnes, A. Renfrew, J. Milanovic and N. Jenkins, "Universitas 21," University of Manchester, 2016. [Online]. Available: www.nottingham.ac.uk/esr21network/M%20Barnes%20UK%20microgrids.doc [Accessed 1 July 2016].

[41] Berkeley Lab, "Microgrids at Berkeley Lab – Isle of Eigg," 2016. [Online]. Available: https://building-microgrid.lbl.gov/isle-eigg [Accessed 1 July 2016].

[42] Berkeley Lab, "Microgrids at Berkeley Lab – Mannheim-Wallstadt," 2016. [Online]. Available: https://building-microgrid.lbl.gov/mannheim-wallstadt [Accessed 10 July 2016].

[43] Centro Nacional de Energías Renovables, "Microrred Atenea," [Online]. Available: http://www.cener.com/es/areas-de-investigacion/departamento-de-integracion-en-red-de-energias-renovables/infraestructuras-y-recursos-tecnicos/microrred-atenea/ [Accessed 1 July 2016].

[44] Commission de Régulation de L'Énergie, "Dossiers – Microgrids," [Online]. Available: http://www.smartgrids-cre.fr/index.php?rubrique=dossiers&srub=microgrids&action=imprimer [Accessed 1 July 2016].

[45] R. J. Schütt, "Future electrical energy supply for the Isle of Pellworm," *International Journal of Smart Grid and Clean Energy*, vol. 2, no. 3, pp. 376–382, 2013.

[46] Distributed Energy Resources Research Infrastructure, "Institute of Communication and Computer Systems – National Technical University of Athens (ICCS-NTUA)," [Online]. Available: http://www.der-ri.net/index.php?id=86 [Accessed 30 June 2016].

[47] V. Leite, Â. P. Ferreira and J. Batista, "On the implementation of a microgrid project with renewable distributed generation," in *Congreso Iberoamericano Sobre Microrredes con Generación Distribuida de Renovables*, Soria, Spain, 2013.

[48] L. Mariam, M. Basu and M. Conlon, "A review of existing microgrids architectures," *Journal of Engineering*, vol. 2013, pp. 1–8, 2013.

[49] T. Loix, "The residential micro grid of Am Steinweg in Stutensee, Germany," *Leonardo Energy, Distributed Generation*, pp. 1–5, 2009.

[50] S. Webel, "Mikronetz mit grossen Plänen," Siemens AG, 19 December 2014. [Online]. Available: http://www.siemens.com/innovation/de/home/pictures-of-the-future/energie-und-effizienz/smart-grids-und-energie-speicher-iren2.html [Accessed 10 July 2016].

[51] M. Vandenberg, A. Engler, R. Geipel, M. Landau and P. Strauss, "Interconnection Management in Microgrids," ISET, Kassel, 2006.

[52] G. Kariniotakis, A. Dimeas and F. Van Overbeeke, "Pilot Sites: Success Stories and Learnt Lessons," in *Microgrids: Architectures and Control*, N. Hatziargyriou, Ed., Chichester, UK: Wiley-IEEE Press, 2014, pp. 208–218.

[53] Ricerca Sistema Energetico (RSE), "Laboratorio Test Facility di Generazione Distribuita in Bassa Tensione," 2016. [Online]. Available: http://www.rse-web.it/laboratori/laboratorio/32 [Accessed 30 June 2016].

[54] Siemens, "Practice Makes Perfect: Smart Grids in Italy," 2015 February 2015. [Online]. Available: http://www.siemens.com/innovation/en/home/pictures-of-the-future/energy-and-efficiency/smart-grids-and-energy-storage-smart-grid-in-italy.html [Accessed 30 June 2016].

[55] N. Melo and F. Resende, "Field tests in the Ílhavo Municipal Swimming-Pool on transfer between Grid Connected and Islanding Modes," 31 December 2009. [Online]. Available: http://www.microgrids.eu/index.php?page=kythnos [Accessed 30 June 2016].

[56] Jofemar, "Factory Microgrid – Description," [Online]. Available: http://www.factorymicrogrid.com/es/el-proyecto/descripcion-del-proyecto.aspx [Accessed 1 July 2016].

[57] Xero Energy Limited, "Microgrids. A Guide to their Issues and Value," Highlands and Islands Enterprise, Glasgow, 2016.

[58] EMforce, Sunlight & Liander, "The first microgrid in the Netherlands," 2009. [Online]. Available: http://www.vsync.eu/fileadmin/vsync/user/docs/Workshop2/VanOverbeeke_MoreMicrogrids.pdf [Accessed 1 July 2016].

[59] M. Koller, J. Schmidli and B. Völlmin, "Frequency Regulation and Micro-grid Investigations with a 1 MW Battery Energy Storage System," in *CIRED Workshop*, Rome: CIRED, 2014.

[60] Nice Grid – ERDF, "Nice Grid," Tamaco, [Online]. Available: http://www.nicegrid.fr/l-architecture-du-reseau-8.htm [Accessed 10 July 2016].

[61] A. Einfalt, C. Leitinger, D. Tiefgraber and S. Ghaemi, "ADRES Concept – Micro Grids in Österreich," in *Internationale Energiewirtschaftstagung an der TU Wien*, Vienna, 2009.

[62] F. Dammeier and J. Rohrer, "Microgrids Chancen und Herausforderungen für Verteilnetzbetreiber," *Bulletin – Électrosuisse and VSE*, vol. 5/2013, pp. 40–43, 2016.

Chapter 9

Microgrids in the United States

Sergio Rivera and Miguel Leon

9.1 Introduction: USA electrical infrastructure

The United States of America (USA) is the third largest country in the world [1], covering 9,666,861 square kilometres. It comprises 48 states (the continental part), and Alaska and Hawaii. It is commonly divided into six regions: New England, Middle Atlantic, South, Middle West, Southwest and West. In 2015, the total energy consumption was around 4,500 TW h, with an installed power capacity of approximately 1,300 GW [1].

The USA power capacity can be grouped in ten areas according to the state location in order to supply electricity to the mentioned six regions [1]. These areas are [1,2]: (1) New England (Connecticut, Maine, Massachusetts, New Hampshire, Rhode Island, Vermont) with share of 3.1% of the total power capacity; (2) Middle Atlantic (New Jersey, New York, Pennsylvania) with a share of 9.6% of the total power capacity; (3) East North Central (Illinois, Indiana, Michigan, Ohio, Wisconsin) with share of 14.2% of the total power capacity; (4) West North Central (Iowa, Kansas, Minnesota, Missouri, Nebraska, North Dakota, South Dakota) with share of 8.2% of the total power capacity; (5) South Atlantic (Delaware, District of Columbia, Florida, Georgia, Maryland, North Carolina, South Carolina, Virginia, West Virginia) with share of 19.5% of the total power capacity; (6) East South Central (Alabama, Kentucky, Mississippi, Tennessee) with share of 8.4% of the total power capacity; (7) West South Central (Arkansas, Louisiana, Oklahoma, Texas) with share of 16.7% of the total power capacity; (8) Mountain (Arizona, Colorado, Idaho, Montana, Nevada, New Mexico, Utah, Wyoming) with share of 8.4% of the total power capacity; (9) Pacific Contiguous (California, Oregon, Washington) with share of 11.4% of the total power capacity; and (10) Pacific Non-contiguous (Alaska, Hawaii) with share of 0.4% of the total power capacity.

In order to meet US power needs, there were close to 20,000 electric generator facilities in 2015. A generator facility can have one or more generators, and some of those generators can use more than one type of fuel or primary energy source [2]. Fossil fuels such as coal, gas and liquid derivatives from oil represent 67% of the total energy generated in the USA; nuclear power, 19%; hydroelectric plants, 6%; and renewable energy, 8% [2]. The generation capacity also changes from state to state, and it depends on the availability of resources [3]. For instance, coal and gas plants are more common in the middle west and southwest of the country, whereas the west coast generates most of its energy from hydroelectric and gas plants [3].

The electric transmission grid of the USA consists of more than 258,000 km of lines that go across the whole country from east to west (98% AC lines and 2% DC lines) [4]. The North American transmission grid can be divided into the following interconnected electric grids [4]:

- The Eastern Interconnection, which covers two-thirds of the eastern region of the USA
- The Western Interconnection, which covers one-third of the area of the west of the USA and Alberta and British Columbia (Canadian provinces)
- The Texas Interconnection, which goes along all the state of Texas
- The Quebec Interconnection
- The Alaska Interconnection

Network systems in Hawaii and Alaska are not interconnected with the other 48 states. The distribution network in the USA is the final stage in the delivery of electricity. The primary objective is to supply electric energy to customers. It includes sub-transmission systems, distribution networks and transforming substations to lower voltage levels for use by commercial, industrial, institutional and residential customers. The largest distribution utilities in the USA are [5] Pacific Gas and Electric Company, Southern California Edison, Florida Power & Light, Commonwealth Edison and Georgia Power.

Residential, commercial and industrial customers account for two-thirds of the total energy consumption in the country [6]. The main customer characteristics are the following [6,7]:

- Residential customers: Residential sector represents more than one-third of the total electricity consumption in the USA [6]. The single major uses of electricity in this sector are heating and cooling (air conditioning), illumination, water heating, home appliances and consumer electronics. Load demand tends to be higher in summer afternoons due to air-conditioning use and in the evenings when lights are turned on in the residential sector.
- Commercial customers: Commercial sector includes governmental, public and private organisations among others. The primary uses of electricity in this sector are illumination, heating, ventilation and air conditioning. The usual electricity demand in this area tends to be higher at daytime and decreases significantly on nights and weekends.
- Industrial customers: Installations and equipment for industrial customers use electricity to process and produce different types of commodities, including mining, agriculture and building. The industrial sector uses less than one-third of the total electricity in the USA. Several US-wide surveys have shown that most of the energy in the industrial sector is used to feed motors, but other

applications are heating, cooling and electrochemical processes [6,7]. In many cases, electricity consumption in the industrial sector does not vary throughout the day or even the season [7].

- Transport: The transport sector consumes most of its energy by burning fossil fuels such as diesel, gasoline or jet fuel. However, some transportation systems use the electricity from the power grid [6]; among these are mass transportation systems such as some trains and automobiles types, and electric vehicles, which collect energy from the grid in energy storage systems [6,7]. Transport activities represent less than 10% of the total electricity consumption in the USA, but that percentage could increase if electric vehicles become more popular.

In order to meet and cooperate with the needs mentioned above of the US electric sector, smart grids and microgrids have been gradually introduced into the US power system in the last few years.

Smart grids refer to the activities of automation and programming intelligence into different devices of the power systems. Microgrids are small power system networks with high penetration of renewable energy sources and energy storage systems. These networks are able to handle controllable and non-controllable loads and must be robust against different disturbances [8,9]. In addition, they can be operated in islanded way or connected to the main power network. Microgrids have been mostly implemented in commercial buildings, such as school campuses, hotels, shopping malls, office buildings, hospitals, government buildings and special military installations [8]. Factors like climate change and environmental issues will result in an increased penetration of renewable generation and microgrids, but this generation is both variable and relatively unpredictable, compared to traditional fossil resources. This variability is an important feature to be considered in the penetration of microgrids in power systems. These factors create the need to study the current and future development of microgrids in the USA.

This chapter presents a summary of the current research, technology, standards and policy on microgrids in the USA. In addition, it shows some of the main microgrid projects in the USA and the envisaged developments in this area.

9.2   Microgrid research, technology, standards and policy in the USA

9.2.1   Microgrid research in the USA

Research on microgrids in the USA has been promoted by the need to improve and increase reliability, resilience and power quality of the US power systems [9]. Another reason for microgrids research is the usefulness of these facilities in the case of blackouts in the main grid, like those that occurred in 1965, 1971, 1977 and 2003 in the northeast of the USA, generating enormous economic, social and political issues [10]. Hurricane Andrew in 1992, which destroyed 17 mi of transmission lines; the ice storm of 1998, which damaged electricity supply in

New York due to the collapse of transmission towers; and Hurricane Katrina in 2005, which affected the states of Louisiana, Alabama, Kentucky, Tennessee and Florida, leaving approximately 1.6 million users without electricity, are some examples of the reason why microgrid research has been growing in the last decades in the USA [10].

The electric power system of the USA is in an evolution from a system of large-scale centralised generation to a decentralised one, in order to gain some advantages like improving reliability and security [8,11]. The US economy will depend more than ever on a robust electric energy system with high reliability to prevent collapses and blackouts like those mentioned above [11,12]. In order to develop and study new strategies to deal with the vulnerability of electric power networks, the Consortium for Electric Reliability Technology Solutions (CERTS) was formed in 1999. This was in response to a call from the US Congress to restart the federal programme for transmission and security of energy supply to meet the challenge of maximum reliability [12]. During the last few years, one of the significant findings by CERTS has been that it is possible to develop a robust energy supply system through microgrids. In this way, there have been several microgrid research initiatives in the USA [9]; two of the primary initiatives are described below: CERTS and Smart Power Infrastructure Demonstration for Energy Reliability and Security (SPIDERS).

CERTS is doing research on five different areas [12,13]:

- Real-time management of electric reliability
- Reliability and markets
- Energy storage
- Distributed resources and microgrids
- Evaluation of new needs and technologies for reliability

In 2003, CERTS defined a microgrid as an aggregation of loads and micro-sources operating as a unique system that provides power and heat [13]. Most micro-sources have power electronics interfaces to provide flexibility, operability and security to the microgrid [13]. A microgrid represents an entirely new focus on the integration of distributed resources; the new emphasis does not meet the requirements established in the IEEE standard 1547-2003 [13]. That standard focuses on ensuring that the interconnected generators will shut down automatically if there are problems on the grid. In addition, the new focus considers that microgrids must be designed to separate automatically from the main power grid and operate in islanded mode, trying to meet the maximum demand and reconnecting to the network once problems had been solved [14].

The main objective of the CERTS microgrid concept and the CERTS microgrid laboratory project [14–16] is to improve the integration of renewable and variable energy resources into microgrids. The control techniques, comprising the CERTS concept, are methods for making automatic reconfigurations between networked and islanded mode in order to maintain high service reliability [15]. Another objective of CERTS regarding microgrids is to achieve stability of voltage and frequency without high-speed communications between sources of generation [15,16]. These techniques were demonstrated at a bench-scale test microgrid built near Columbus, Ohio, and

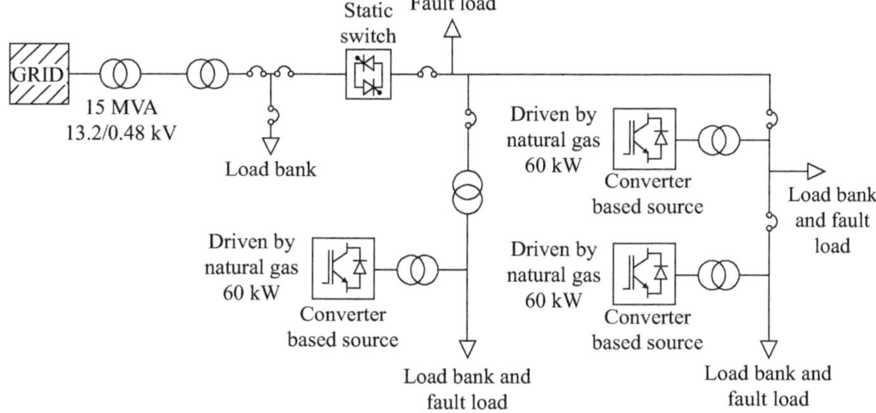

Figure 9.1   CERTS Microgrid in Ohio

operated by American Electric Power Company (Figure 9.1). The implementation of microgrids fully confirmed research conducted by CERTS through simulations of theoretical analysis and controls by testing different operating conditions. The CERTS microgrid testbed in Ohio is composed of five load test banks (with the possibility of being traditional loads or fault loads). These loads can be remotely controlled from 0–90 kW to 0–45 kVAR. Some of the load banks are able to act as remote fault loads, which range from bolted faults to high-impedance faults. Other loads correspond to induction motors with a capacity range of 0–20 HP [15,16].

The tested microgrid has a static switch which is the entry point for utility power between protected and unprotected parts of the microgrid testbed. The voltage controller for each source provides regional stability, preventing oscillations due to the high penetration of distributed generation [15]. The project has a master controller, and each source is connected point-to-point. The controllers increase system reliability compared to having a master-slave or centralised control scheme. It uses a central communication system to deal with set-points for each of the required points of generation, load or manoeuvre as necessary to improve overall system operation [15,16].

Another major microgrids research and demonstration programme is SPIDERS [8,17]. The objective of the SPIDERS programme is to address energy security and reliability by increasing the use of distributed energy during peak load periods, coordinate utility load shedding, and be a self-sustaining microgrid for medium or long periods. The following issues to be analysed in the project summarise the primary objectives of SPIDERS [17]:

- cyber security of electric grid, smart grid technologies (advanced metering infrastructure, substation and distribution automation, etc.),
- secure operation of microgrid in islanded and interconnected mode, integration of distributed and intermittent renewable sources (wind, solar, fuel cell, biofuels),

- demand-side management and
- applications of redundant backup storage systems such as batteries and vehicle to grid.

With the SPIDERS project, it is possible to handle some of the critical requirements needed to demonstrate enhanced electrical power reliability and robustness [17]:

- Protect priority assets from loss of power because of cyber-attack or electrical failures
- Integrate renewable and other distributed energy generation concepts to control critical assets in times of emergency
- Sustain critical operations during prolonged power outages

The SPIDERS programme is divided into three phases corresponding to three separate microgrid installations. Phase 1 is a developed project demonstration of a cyber-secure microgrid at Hickam Air Force Base in Hawaii. This microgrid has a single distribution feeder, two electrically isolated loads, two isolated diesel generators and a separate photovoltaic (PV) array. Phase 2 was developed in Fort Carson, Colorado. This microgrid consists of three distribution feeders, seven building loads, three diesel generators and a PV array. Phase 3 is the entire Camp Smith installation with multiple integrated microgrids of existing PV and wind energy, fuel efficiency during emergencies and the possibility of reducing risks of the energy system through cyber security.

As a final point in this section, it is important to note that the issue of patents on microgrids is a good indication of research in this area. Worldwide, there are more than 2,000 patents related with microgrids according to our search in the World Intellectual Property Organization [1,18]. The publication of patents in the last few years has been as follows: in 2006, 113 patents; in 2007, 117 patents; in 2008, 123 patents; in 2009, 137 patents; in 2010, 149 patents; in 2011, 186 patents; in 2012, 255 patents; in 2013, 235 patents; in 2014, 369 patents; in 2015, 369 patents; and half of 2016, 136 patents. The main applicants are State Grid Corporation of China, with 51 patents; Causam Energy Inc., with 31 patents; and General Electric Company, with 16 patents.

The patent office of the USA has processed around 1,003 patents related to branches of microgrid research. The percentages of these patents proposing new techniques and knowledge are control of microgrids, 92%; microgrids with controllable loads, 66%; microgrids with storage, 58%; direct current microgrids, 32%; communication within microgrids, 45%; cooperation within microgrids, 17%; microgrid optimisation, 61%; microgrid modelling, 55%; and clean energy microgrids 28%. The 5% of applicants for microgrids patents in the USA are from universities.

9.2.2 *Microgrids technology in the USA*

A key feature of distributed generation technologies is the interconnection to the grid through power electronics and communication links. With this technology, it is possible to improve response time and implement sophisticated control methodologies incorporating the dynamic behaviour of the microgrids. The technologies commonly implemented in US microgrids follow the CERTS directions. These are [19] as follows:

- Microturbines: The current capacity of these technologies is from 25 to 500 kW. The operation speed is between 50,000 and 100,000 rpm, frequently with aerodynamic bearings. They have power electronics interface with the loads to ensure higher controllability. Microturbines present different sources of primary energy; for instance, natural gas is promoted because is cleaner than liquid fuels.

- Fuel cells: They are electrochemical systems in which a chemical reaction is directly transformed into electricity. Unlike traditional batteries, fuel cells do not need recharging. A fuel cell consists of an anode in which fuel is injected, commonly hydrogen, ammoniac, or alcohol, and a cathode in which an oxidant is introduced, typically air and oxygen [20]. Fuel cells are appropriate for distributed generation and microgrids applications; they offer high efficiency and low emissions, but they are still very expensive [20].

- Diesel and gas low emission generators: This technology reduces the engine's carbon monoxide emission rate to the lowest possible level without negatively impacting power output [21].

- Renewable generation like solar panels or wind turbines that use sunlight and wind's force as primary energy source, respectively. These technologies are explained in detail in the first part of this book.

- Energy storage technologies: Energy storage systems are essential components within microgrids operation since they can be a backup to renewable energy source in certain periods [19]. The development of technology in this area has allowed mitigation of fluctuations in renewable generation [19]. Among the technologies for batteries, which produce electricity from chemical reactions, are lead–acid and lithium batteries. Supercapacitors store energy as electrostatic charges; these types of capacitors can handle more power than a battery and are less vulnerable to temperature changes. Other technologies are flywheels, which are essentially very heavy wheels that rotate at a high speed and store kinetic energy, and the Superconducting Magnetic Energy Storage, which stores energy in the form of a magnetic field created by the flow of current in a superconducting ring.

- Combined heat and power (CHP) units: CHP units allow generating electricity and thermal energy with a single integrated system, which makes them a valuable alternative for future residential supply [11]. CHP systems present high total efficiency; for instance, the new CHP technologies employ and recycle the heat produced in heating or refrigeration using a single fuel source like natural gas or biogas and capturing the heat produced in this process.

- Micro-hydroelectric units: These units are the most common and least expensive source of renewable electricity in the USA [22]. Hydropower technologies use flowing water to produce clean electricity. The classification of hydropower by type of installation is typically divided into five categories: impoundment (large system, uses dam to store water), diversion (river is diverted), run of river (uses natural flow of river) and pumped storage (when the demand is low, water is pumped back to a reservoir) [23]. Modern micro-hydroelectric power generation is usually integrated into the distribution network through power electronics at the interface with the grid providing a better performance through capability of the electricity generation and reactive power control or voltage regulation at the connection point of the distribution system.

- Electric vehicle to grid (V2G): Electric vehicles powered by batteries, fuel cells, or hybrid gasoline, have the ability to return the electric power to the grid. Electric vehicles must have capabilities to manage charging or support two-way interaction between vehicles and the grid. In order to do that, the vehicle can use profiles and behaviours of the battery life must be well understood, and grid interoperability standards must be developed [24].

These technologies can be grouped in five injected/consuming power categories according to the findings presented in [25]: non-controllable and controllable loads, energy storage units, stochastic and dispatchable generators. In addition, research in the USA in microgrids area has made developments in new technologies in the following activities: communication, control, protection, and simulations and modelling tools [9,13].

The top ten vendors of microgrid technology are [19–25]: General Electric, Pareto Energy, Power Analytics, Viridity Energy, ABB, Chevron Energy Solution, Echelon, Microgrid Solar, Siemens and Spirae Inc.

In order to develop the microgrid technology, specialized software is necessary. To conclude this subsection presents leading software in microgrids.

'ETAP Microgrid' is a microgrid software able to integrate power system simulation and planning, protection and real-time microgrid control. ETAP Microgrid solution combines distributed energy technologies with intelligent software in order to observe, forecast, manage and optimise energy supply and demand for microgrids (more information in http://etap.com/microgrid/etap-microgrid.htm).

The 'HOMER Pro Microgrid' software is the global standard for improving microgrid design in all segments, from island or isolated utilities to grid-connected hospitals, campuses and military bases (more information in [2,26]).

'RAPSim (Renewable Alternative Power Systems Simulation)' is a free and open source microgrid simulation framework to get the behaviour in smart microgrids with renewable sources. It is able to simulate grid-connected or standalone microgrids with renewable energy sources (more information in [3,27]).

'Paladin Microgrid Power Management System' is a software platform designed specifically for the online management and control of hybrid power systems incorporating both traditional utility power and renewable power generation (more information in [4,28]).

'DNV GL's Microgrid Optimization' tool is able to simulate the behaviour of customer premises with distributed energy resources such as distributed generation, electric and thermal storage, energy efficiency upgrades and building automation trough mathematical software modules (more information in [5,29]).

'UGE SET' is a proprietary microgrid optimisation software that designs microgrid systems. It is possible to find the optimal sizing and energy resources combination for the microgrid project (more information in [6,30]). 'SICAM Microgrid Manager', by Siemens, enables complete microgrid planning, monitoring and control (more information in [7,31]).

'GreenBus Microgrid Solution' by Green Energy Corp is an interoperability software platform that allows the implementation of developing smart grid technologies and integration with legacy power and communications infrastructures. With this tool, microgrid developers can design and implement an architecture that supports technology adoption over time, while realizing the business benefits (more information in [8,32]).

'Wave Microgrid Solution' by Spirae comes with power systems simulation capabilities such as Asset Monitoring and Control, Active and Reactive Power Import and Export Control, Scheduling and Dispatch, Islanding and Resynchronization, Voltage Control, Frequency Control and Spinning Reserves Management. In addition, it is possible to implement custom economic and optimisation logic to meet different customer and market requirements (more information in [9,33]).

The 'Distributed Energy Resources Customer Adoption Model (DER-CAM)' is an environmental model software that can be applied to microgrids. DER-CAM has microgrid specific applications. It is able to optimise the behaviour of generation and loads, either for individual customer sites or microgrids (more information in [10,34]).

The 'MicroGrid', from Concurrent Systems Architecture Group at the University of California, San Diego, is a software which provides the ability to emulate virtual grid infrastructures, enabling scientific study of grid resource management issues. The MicroGrid emulation tool allows controllable experimentation with dynamic resource management techniques (more information in [11,35]).

'GridOS', by Opus One Solutions, is an energy modeller and resource optimiser that realises the promises that microgrids bring (more information in [12,36]).

9.2.3   Clean energy standards and policy in the USA

The operation, interconnection, incentives, markets, reliability and sustainable development of microgrids have promoted policies and standards for integration and deployment of different projects throughout the USA [37]. For the US clean energy policy, it is a priority to increase penetration of renewable generation, reliability and security to the power users, using strong and robust competitive wholesale electricity markets [37]. The leading standards developed in the USA on clean energy and distributed generation are [37] as follows:

- Standard for Interconnecting Distributed Resources with Electric Power Systems (IEEE 1547) provides rules and procedure concerning the technical requirements (including design, construction, commissioning acceptance testing and maintenance/performance requirements) in order to interconnect dispatchable electric power sources with a capacity of more than 10 MV A [38].

- The Federal Energy Regulatory Commission (FERC) with Standardization of Small Generator Interconnection Agreements and Procedures established the national interconnection procedures applicable to generation projects that are 20 MW or less. They established procedures to apply to the retail interconnection and to control and operate generating facilities of the mentioned size [39].
- The state of California has been a leader in improving the interconnection of distributed generation systems. For instance, they developed a programme to streamline interconnection of small power plants called 'Rule 21' [40]. This programme details the interconnection, operating and metering requirements for generation facilities to connect to a utility distribution system. The programme estimates the impact of power quality due to the interconnection of distributed generation sources. In this way, it establishes the general requirements for Voltage Regulation Integration including the limits for direct current injection, voltage flicker and harmonics. In addition, it sets test design and metering of disturbance of voltage and frequency, as well as the evaluation of integration with interconnected bus conditions in islanded mode. Likewise, it reviews all protective interconnection functions.

In addition to clean energy standards, the clean energy policy has also evolved. For instance, the US Climate Action Plan was released during the administration of President Barack Obama. The objective was to cut the emissions that cause climate change and threaten public health. The plan has three pillars [41,42]:

- Cut carbon pollution in the USA
- Prepare the USA for the impacts of climate change
- Lead international efforts to combat global climate change and prepare for its impacts

Barack Obama's clean energy programme represents the primary policy of the government to fight global warming [41]. The programme is directed at the electricity sector, which is the largest emitter of greenhouse gasses in the USA (approximately 31% of them). This clean energy plan is a regulation proposed by the Environment Protection Agency, under the Clean Air Act, which aims to reduce carbon dioxide emissions from generating plants in the USA (at least 30% below 2005 levels by 2030) [41]. Each state has a different goal as well as a set of options to meet it. Among these options are increase in renewable sources like solar or wind, growth in efficiency of generating plants and transition from coal to natural gas in order to produce fewer emissions.

This new regulation includes the Clean Energy Incentive Program, which promotes an intensification in renewable and efficient energy due to the decline in gas resources [41]. It is worth mentioning that electricity generation from natural gas only produces half of the emissions in comparison with generation from coal. On the other hand, electricity generation from solar or wind produces no emissions. Some environmentalists argue that emissions from gas extraction outweigh the advantages of natural gas over coal; that is the main reason why most states started to use renewable energy in the USA [41,42].

All US clean energy standards and policies agree that paving the way for clean energy will be possible through the implementation of microgrids.

9.3   Cases of microgrid projects in the USA

In the penetration of smart grids in the USA for handling power systems, microgrids are essential components, since they can enhance reliability, power quality and efficiency [43]. The USA Energy Department defines a group of interconnected loads and distributed generation, which acts as a single controllable unit, as a microgrid [44]. They can be connected or disconnected to the main grid and operate in either islanded or grid-connected mode [44,45].

The different identified benefits that microgrids can provide to the US power grid are [43,46,47] as follows:

- Electric system modernisation and integration of several smart grid technologies
- Improvements in the integration of distributed resources and renewable energy
- Meeting of end-users' needs by guaranteeing energy supply for critical loads
- Power quality control and local reliability
- User participation through demand management (peak load reduction)
- Community involvement in electricity supply
- Losses reduction
- Incorporation of new micro sources into the system without modification of existing equipment
- System imbalances management within microgrids
- Meeting of the grid's dynamic load requirements by microgrids

Knowing these benefits, USA started a gradual penetration of microgrids. The first microgrid initiatives in the USA were launched in the states of Alaska and Hawaii, as they are isolated places with difficult access. The government encouraged the development of solutions for these places based on solid fuels. Other hybrid generation projects (solar and wind energy) were later developed in several states, especially in California, to provide power and emergency backup for contingencies.

The current US microgrid capacity is around 1,180 MW [48,49]. In this chapter, a list of some of the current microgrids in the USA is presented (see Table 9.1) [9,43,49–52], where the different kinds of microgrids are considered: isolated microgrids, islandable microgrids and non-synchronous microgrids. Data presented in Table 9.1 corresponds to the microgrids that handle 72% of the aggregate capacity of US microgrids (845.2 MW of 1,180 MW). Information about the first year of operation, whether the sources are renewable or not, the type of microgrid (Remote Community, Island, University Campus, City/Community, Public Institution, Military, Commercial, Utility, Interconnected) and total capacity is presented (Figures 9.2–9.4). The number of microgrids in operation is currently about 130 [48,49]. The states with the largest microgrid operational capacity are [49]: New York (221 MW from an Electric Power Industry Net Capacity of 39,852 MW), Georgia (168 MW from an Electric Power Industry Net Capacity of 36,351 MW),

Table 9.1   Current microgrids projects in the USA

Project	First year of operation	Conventional sources	Renewable sources	Storage	Type	Total capacity (kW)
Illinois Microgrid (Illinois)	2013	G	W, PV	Batt	UC	10,000
California University Microgrid (San Diego, California)	2007	S, G	W, PV, FC	Batt, Cap	UC	40,000
Santa Rita Jail Microgrid (Dublin)	2011	D, CHP	W, PV, FC	Batt	P	6,000
Manzanita Hybrid Power Plant (California)	2005		W, PV	Batt	U	15
Santa Cruz Island (California)	2005	D	PV	Batt	I	300
Woodstock Microgrid (Minnesota)	2001	–	W, PV	Batt	U	5
Dangling Rope Marina (Utah)	2001	–	PV	–	RC, P	160
Kotzebue Microgrid Plant (Alaska)	1997	D	W	–	RC	11,000
Wales Alaska Power Plant (Alaska)	2002	D	W	–	RC	500
St. Paul Power Plant (Alaska)	1999	D	W	–	RC	500
CERST Testbed (Ohio)	2009	G	–	Batt	UC	200
University of Wisconsin Madison Test-Bed (Wisconsin)	2008	D	PV	Batt	UC	20
University of Miami Testbed (Florida)	2007		PV, FC	Batt	UC	100
Sandia National Lab Testbed (Washington)	2012	D	PV, W	–	U	60
Texas University at Arlington Testbed (Texas)	2011	–	W, PV, FC	Batt	UC	7,300
FJU Testbed (Florida)	2008	M	W, PV, FC	Fly	UC	10
Laboratory Scale Microgrid Testbed (New Jersey)	2011	–	PV	Batt	UC, U	10
Texas University at Austin (Texas)	2010	D, G, M	–	Fly	UC, U	3,600
Microgrid Testbed in Albuquerque (New Mexico)	2010	CHP	W, PV	Batt	UC, U	2,500
Utility Microgrid in Los Alamos (New Mexico)	2011	–	PV	Batt	U	2,500
RIT Microgrid (New York)	2013	BG	W, PV, FC	–	UC, U	600
Lime Village Solar (Alaska)	2003	–	PV	Batt	RC	12
AVEC Kaltag Solar PV (Alaska)	2011	–	PV	–	RC	10
Kotzebue Microgrid (Alaska)	2000	D	W	–	RC, CC	1,100
Channel Islands (California)	1995	G	W, PV	–	I	95

North Manitou Island (Michigan)	1996	D	PV	Batt	I	70
Carol Springs Mountain (Arizona)	1996	D	PV	Batt	RC, P	85
Dangling Rope (Utah)	1996	D	PV	Batt	RC	380
Cherry Creek (Denver)	1996	–	PV	–	In	25
Pichtr (Hawaii)	1996	D	W, PV	Batt	RC, I	70
Pinnacles (California)	1996	–	PV	Batt	RC	10
Star Hybrid Test System (Arizona)	1997	D	PV	Batt	P	210
Kea (Alaska)	–	D	W	–	RC, CC	600
Volcanoes National Park (Hawaii)	1998	G	PV	Batt	RC, P	5
Joshua Tree (California)	1998	M	PV	Batt	RC	56
San Clement island (California)	1998	D	W	–	RC, I	3,350
Yuma (Arizona)	1998	D	PV	–	RC	330
Tanadgusix Corp (Alaska)	1999	D	W	–	RC, CM	525
Canyons National Park-Maze District (Alaska)	1999	D	PV	Batt	RC, P	67
Diamond Bar (California)	2000	–	PV	–	In	30
Fort Carson Microgrid (Colorado)	2014	D	PV	V2G	MI	5,000
Borrego Springs (California)	2010	D	PV	Batt	RC	4,000
South Oaks Hospital (New York)	1990	G, CHP	–	–	P, U	1,250
Greenwich Hospital (New York)	2008	D, G, CHP	–	–	P, U	1,250
Christian Health Care Center (New Jersey)	2008	CHP, G	–	–	P, U	260
Princeton University (Princeton)	1995	G, CHP, D	–	TH	UC, U	15,000
Salem Community College (New Jersey)	2009	CHP, M	–	–	UC, U	100
Co-op City (New York)	2011	CHP, G	–	–	CC, U	40,000
Twentynine Palms (California)	2003	CHP, M	–	–	MI, U	7,200
Louisiana State University (Louisiana)	2005	CHP, G	–	–	UC, U	2,370
Nassau Energy Corporation (New York)	1991	CHP, G	–	–	CM, U	57
Bergen County Utilities (New Jersey)	2008	CHP, G	–	–	CM. U	2,800
New York University (New York)	2010	CHP, G	–	–	UC, U	14,400
Sikorsky Aircraft Corporation (Connecticut)	2011	CHP, G	–	–	CM, U	10,700

(Continues)

Table 9.1   (Continued)

Project	First year of operation	Conventional sources	Renewable sources	Storage	Type	Total capacity (kW)
Niobrara Data Center Energy Park	2014	CHP, G	W, PV	Batt	CM, U	100,000
Idaho Power – US Air Force (Idaho)	2008	D	PV	–	MI	150
Mesa del Sol Microgrid (New Mexico)	2014	G	PV, FC	Batt	UC	500
Florida International University (Florida)	2013	D	PV	–	UC	60
APS – Greywolf Project (Arizona)	2006	D	PV	–	RC	90
The College of New Jersey (New Jersey)	1999	CHP, G	–	–	UC, U	5,200
Kodiak (Alaska)	1984	D, G, H	W	Batt	CC	39,000
California Valley Solar Ranch (California)	2013	–	PV	–	CC	250,000
Hawaii Hydrogen Power Park (Hawaii)	2012	–	W, PV, FC	Bat, Fly	U, CC	2,000
Mad River Park Microgrid (Vermont)	2005	D, G	PV	Batt	RC, P	500
Palmdale Water District Power (California)	2006	CHP	W	Cap	RC, P	4,000
Panoche Valley Solar Project (California)	2011	–	PV	–	CC	247,000
Other microgrids	–	–	–	–	–	334,460
Total microgrid capacity						1,179,757

Conventional sources: diesel (D), combined heat and power (CHP), stream (S), gas (G), hydraulic (H), motor driven gen (M), biogas (BG), cogeneration (Cog).
Renewable sources: wind (W), photovoltaic (PV), fuel cell (FC), biomass.
Storage: battery (Batt), flywheel (Fly), capacitor (Cap), thermal (TH), vehicle to grid (V2G), concentrated solar power (CSP).
Microgrid type: remote community (RC), island (I), university campus (UC), city/community (CC), public institution (P), military (MI), commercial (CM), utility (U), interconnected (In).

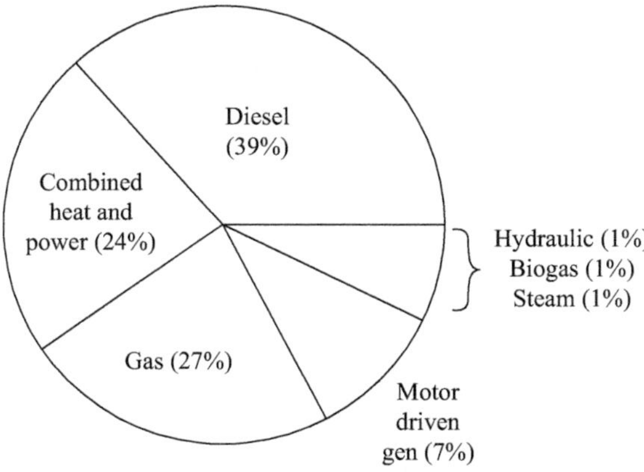

Figure 9.2   Microgrids with non-renewable energy sources

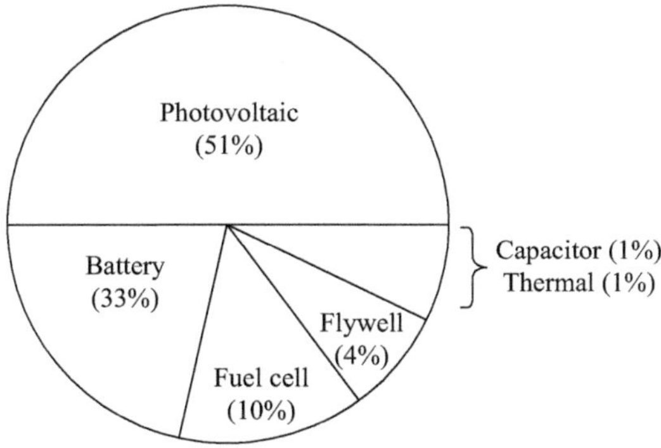

Figure 9.3   Microgrids with renewable energy

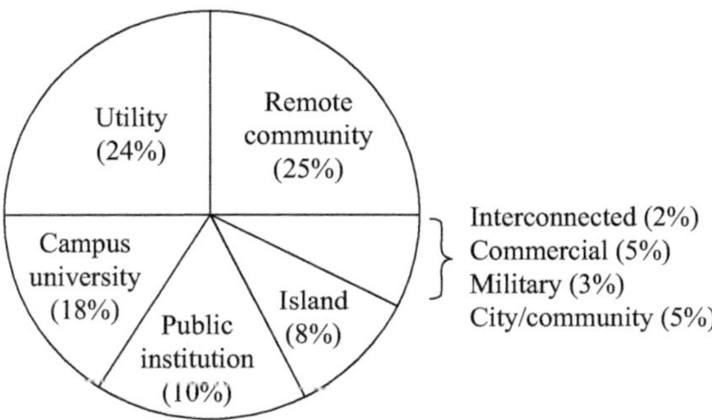

Figure 9.4   Types of microgrids in the USA

Texas (163 MW from an Electric Power Industry Net Capacity of 117,155 MW) and California (132 MW from an Electric Power Industry Net Capacity of 75,399 MW). Around 50% of microgrids are located in the Northeast and West Coast states [49].

The following subsections put forward details on the configuration of the main microgrids in the USA.

9.3.1   Microgrid in University of California

This microgrid supplies energy to a community of more than 30,000 people and covers several buildings and a wide range of distributed energy resources. The microgrid has a testbed connected to a step-down substation, which changes the voltage level from 66 to 12 kV and has a capacity of 15 MW [53]. It uses two transformers in order to cover ten circuits of 12 kV. This microgrid which includes more than 3 MW of solar energy is backed up by a combined cycle plant of 30 MW, includes a thermal plant (one of the largest thermal energy storage systems in the country with 4.5 million gallons (60,000 t-h)) and covers the main campus buildings with heating and cooling. The main features of the microgrid are 3-MW of PV modules on-campus and 1-MW off-campus, 2.8 MW of fuel cell, 28 kW of ultra-capacitors and 2.5 MW/5 MW h of energy storage (lithium-ion batteries) (Figure 9.5).

9.3.2   Fort Carson microgrid

This is one of the projects conducted in the USA following the clean energy policies, under the smart grid infrastructure for security and reliability [17]. This microgrid is a huge military base with nearly 14,000 people inside and covers 550

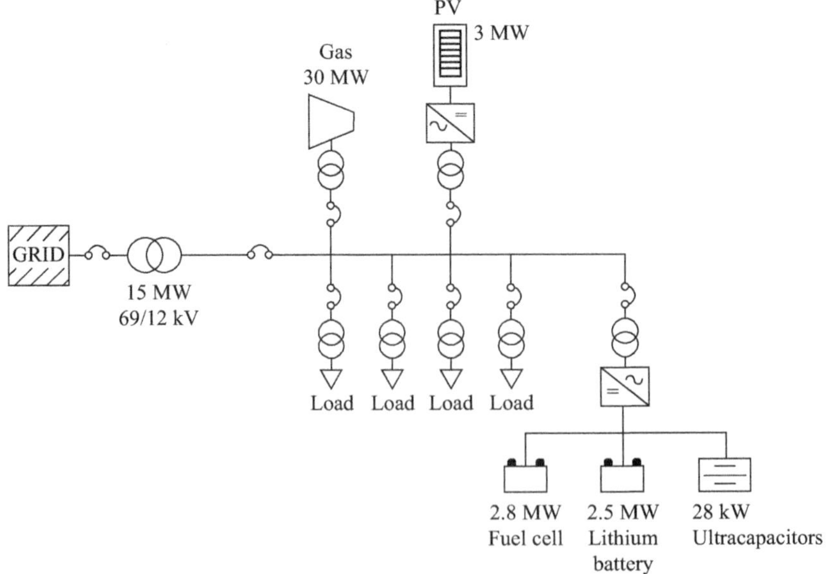

Figure 9.5   Wiring diagram of California University microgrid

square kilometres. The Fort Carson base is making an effort to become a zero-emission grid by mainly using PV resources (up to 100 MW in the future) as well as wind, geothermal energy, biomass and solar water heating. Currently, the grid is composed of 2 MW of PV generation [17]. The Fort Carson microgrid also includes advanced bi-directional charging electric vehicles in both microgrid and normal operations. Electric vehicles are used to support the microgrid as an energy storage resource. In addition, three diesel generators of 1,250, 1,000 and 900 kW are directly connected to the distribution grid using bypass breakers [54,55] (Figure 9.6).

9.3.3   Mesa del Sol microgrid

This project is an operating microgrid that has been created by NEDO (New Energy and Industrial Technology Development Organization), New Mexico State, New Mexico's Public Services Company and Sandia's national laboratory and some Japanese companies. The system is composed by 50 kW of PV power, a fuel cell of 80 kW, a diesel generator of 240 kW, a storage system with lead–acid batteries and an adsorption refrigerator [54,55] (Figure 9.7).

9.3.4   Santa Rita Jail microgrid

Santa Rita Jail is the third largest jail in California and the fifth largest in the USA. The prison houses up to 4,500 inmates and is located in Dublin, California, about 75 km east of San Francisco [56]. The objective of the microgrid project is to

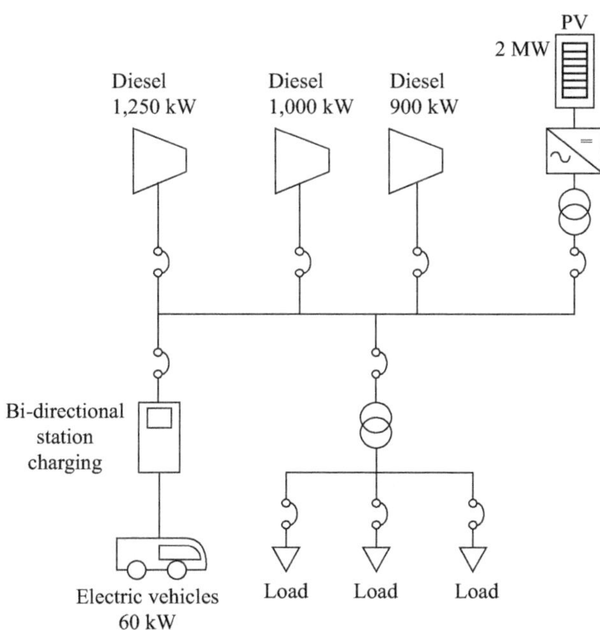

Figure 9.6   Fort Carson microgrid

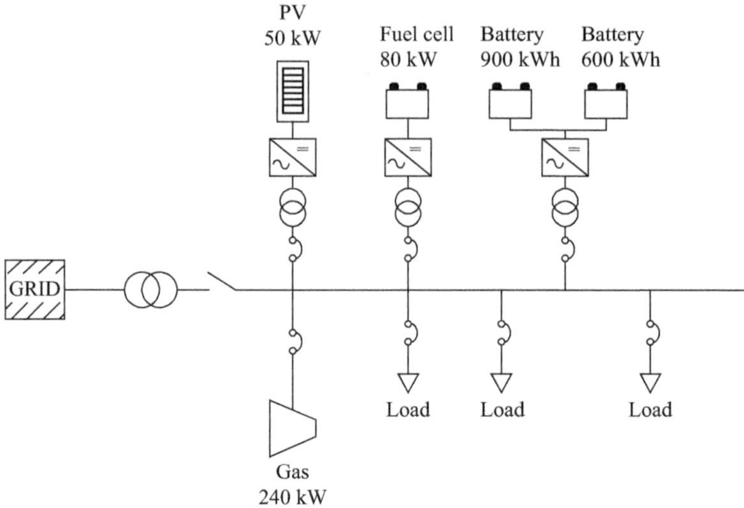

Figure 9.7   Mesa del Sol microgrid

demonstrate the implementation of the CERTS research technology combined with renewable energy, storage energy and fuel cell. Other objectives are to reduce peak load and monthly demand charges, to store overproduced renewable and fuel cell energy, to improve grid reliability and to reduce electrical overvoltage and sags [56].

In 2002, Santa Rita Jail installed a 1.2-MW PV array; in 2006, a 1-MW fuel cell with CHP (combined heat and power) capability. After that, a large (2 MW–4 MW h) lithium iron phosphate battery was installed. The static disconnect switch was installed between the utility and the microgrid for very fast islanding and autonomous operation of the microgrid [55–57] (Figure 9.8).

9.3.5   *Borrego Springs microgrid*

It covers 2,800 customers of the residential community in Borrego Spring. The total capacity of the microgrid is approximately 4 MW using two biodiesel generators of 1.8 MW, a battery system of 2,000 kW h and about 700 kW of PV power [54,55] (Figure 9.9).

9.3.6   *Illinois microgrid*

This microgrid aims to create maximum reliability and a security system with a smart grid that identifies and isolates failures, redirects the energy to meet changes in the load and generation, and reduces demand based on prices, meteorological forecasts and grid interruptions [54,55]. The maximum demand is about 10 MW; the sources include two combined cycle gas units of 4 MW and a small aerogenerator (8 kW). It is planned to add 200 kW of PV power as well as a 500 kW–1,000 kW h battery [56,57] (Figure 9.10).

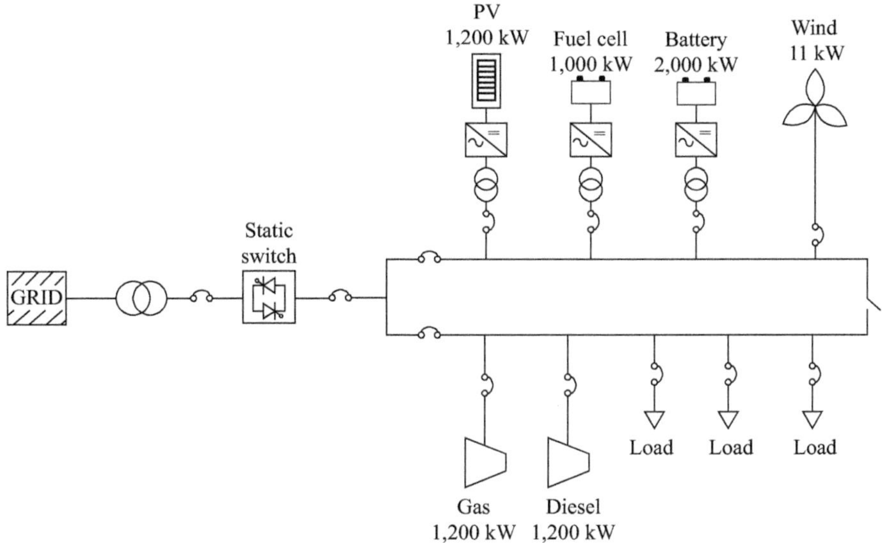

Figure 9.8   Santa Rita Jail microgrid

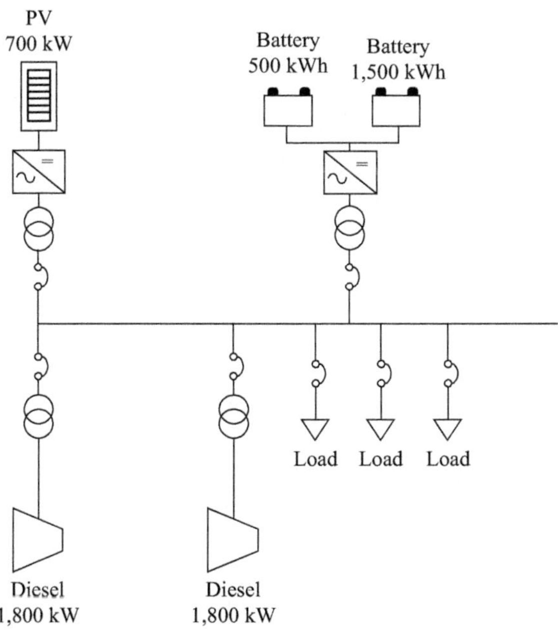

Figure 9.9   Borrego Springs microgrid

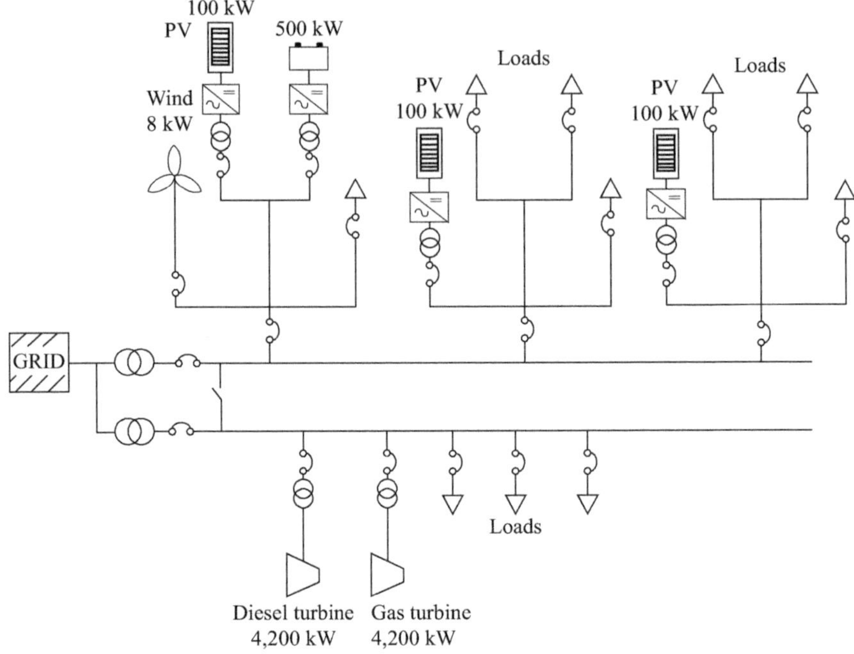

Figure 9.10   Illinois microgrid

9.3.7   Hawaii Hydrogen Power Park

The microgrid is under the direction of the Hawaii Natural Energy Institute (HNEI). It has a capacity of 24 kW. The HNEI is assessing the performance of integrated hydrogen energy systems by working on a real power supply system [54]. The Hawaii Hydrogen Power Park also grows public consciousness of the potential of hydrogen for a diversity of transportation applications [54] (Figure 9.11).

9.3.8   Kodiak microgrid

In the second largest island in the USA, the community of Kodiak, an Alaskan island, has been utilising hydro resources since the early 1980s [54]. The state board's vision drove the addition of multiple wind farms and a third hydro turbine. This microgrid consists of 35 MW of gas/diesel, hydroelectric (500 kW) and wind generation (9,000 kW), 2.5 MW/2 MW h of energy storage (lead–acid gel battery) that contributes to the stability of the system and recently it was added a 1-MW flywheel system [54] (Figure 9.12).

9.3.9   Microgrid in University of Wisconsin Madison

The microgrid of distributed generation in the University of Wisconsin Madison was designed and developed to control a diesel-based generator microgrid [43].

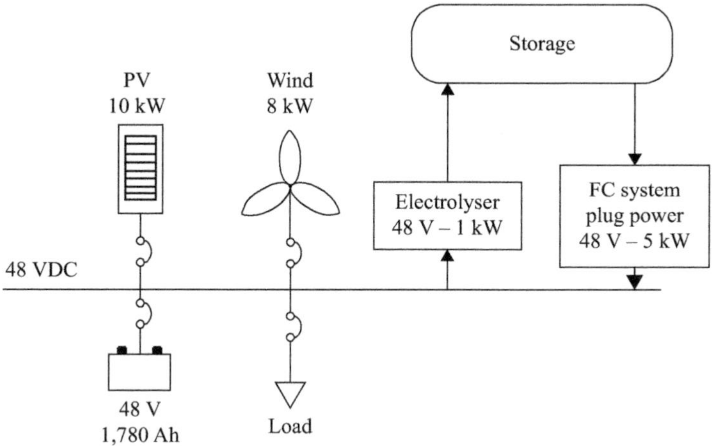

Figure 9.11   Hawaii Hydrogen Power Park

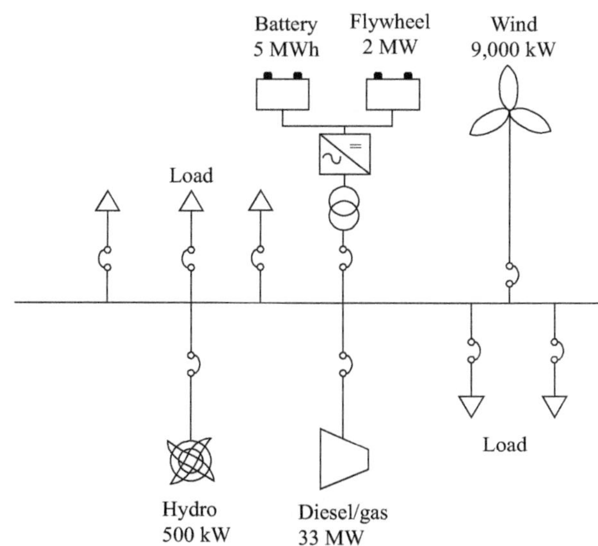

Figure 9.12   Kodiak microgrid

It consists of micro-sources which use converter-based sources including renewable sources such as wind and PV [43] (Figure 9.13).

9.3.10   Microgrid in University of Miami

The microgrid of Miami proposed a methodology utilising a scheme with various levels to implement the microgrid [43]. The interfaced DC microgrid integrates a

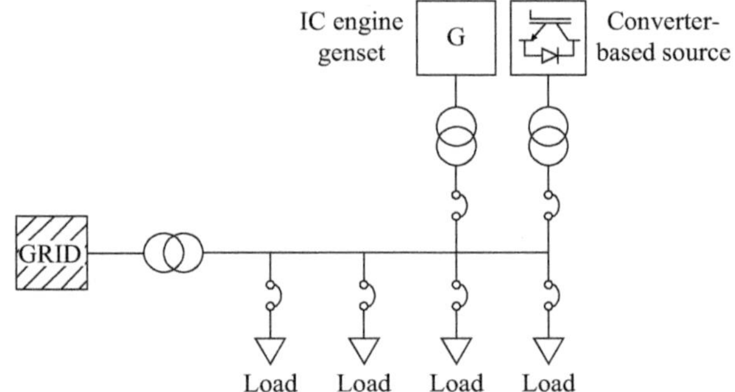

Figure 9.13   Microgrid in the University of Wisconsin Madison

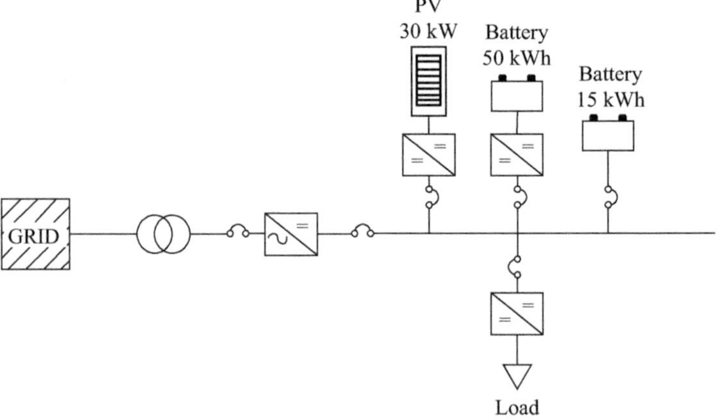

Figure 9.14   Microgrid in the University of Miami

PV system (30 kW) and an electric cell, coupled to the same DC voltage bus by suitable DC–DC power converters and controls [43]. The battery system has an energy storage capacity of 65 kW h. This system relies on the assumption that households will have PV systems or other energy sources, electricity storage and use of DC loads [43] (Figure 9.14).

9.3.11   *Microgrid in University of Texas at Arlington*

The University of Texas at Arlington's microgrid lab contains three independent representative microgrids working in interconnected system mode or islanded mode. Each representative network has a 24-VDC transport system. In standard configuration, energy controls contain lead–acid batteries, programmable loads and renewable energy sources such as PV and wind [43] (Figure 9.15).

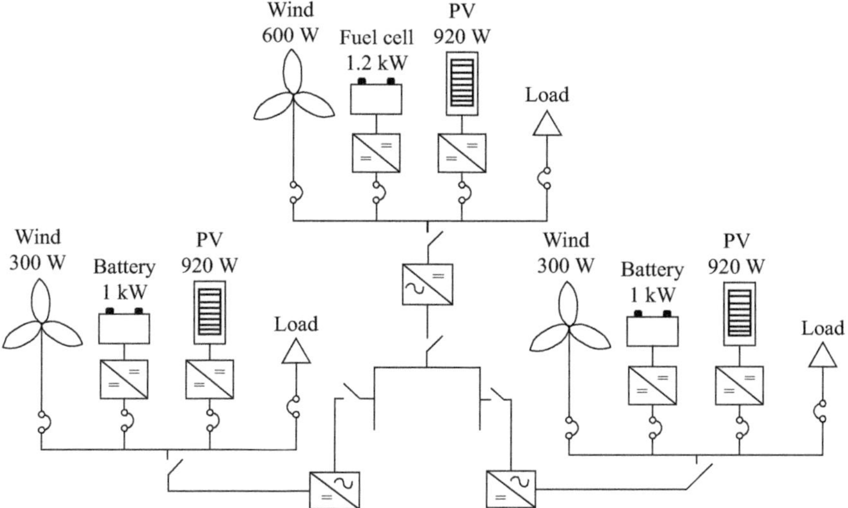

Figure 9.15   Microgrid test bed in the University of Texas at Arlington

9.3.12   Microgrid in Florida International University

The Florida International University microgrid is formed from four AC generators in a closed loop control topology with an adjustable load. Also, it has a DC bus with a solar panel of 4 kW and DC loads. The DC circuits are connected to the AC system through a bi-directional converter [43] (Figure 9.16).

9.3.13   Microgrid in University of Texas at Austin

Distributed generation has organised a DC bus that binds with numerous sources of energy and an assortment of variable and static loads. The microgrid transmission lines include more than 5 m between two laboratories in the UT-CEM (Center for Electro-Mechanics, University of Texas at Austin) offices.

The UT-CEM is a hybrid of microgrid that can provide power up to 3.6 MW to interact with the commercial power system and other distributed sources of electricity. It incorporates local primary engines driving generators and energy storage units [43] (Figure 9.17).

9.3.14   Microgrid in Albuquerque

This microgrid test consists of a gas engine generator (240 kW), a power module of fuel cells (80 kW), a battery system (60 kW/160 kW h) and a PV system (50 kW). The microgrid control has a system called Energy Management System able to manage the sources of electric power, heat supply and demand [43] (Figure 9.18).

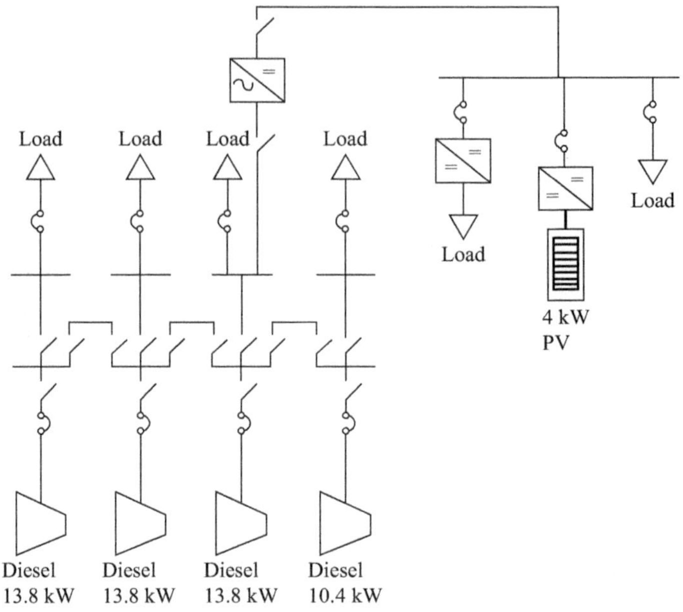

Figure 9.16   Florida International University Microgrid

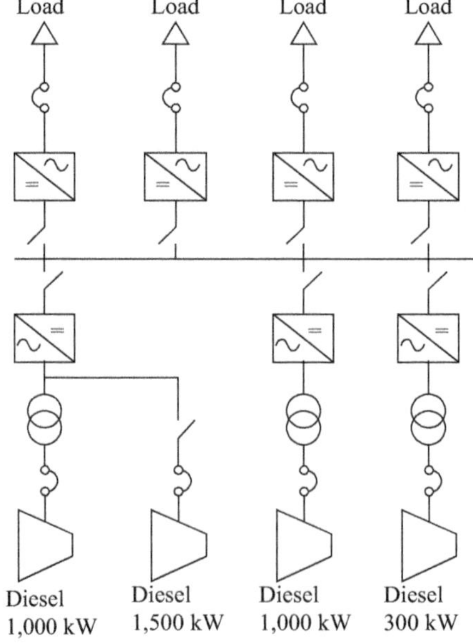

Figure 9.17   Microgrid in the University of Texas at Austin

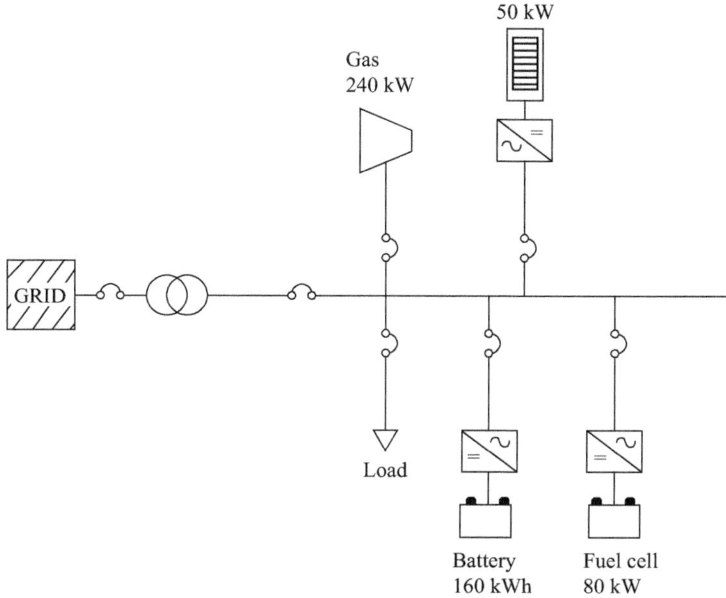

Figure 9.18   Microgrid at Albuquerque

9.4   Envisaged data of microgrids in the USA

In this book, several benefits of microgrids have been presented. For instance, they offer numerous potential benefits to customers, including improvement in reliability by providing portion of power to the grid, operating as an islanded microgrid of the electric power system during a utility outage and solving power quality problems like the reduction of total harmonic distortion in loads. There are also utility benefits such as solving overload problems by removing load from the system and by allowing a part of the system to island intentionally. The USA knows these advantages; therefore, microgrid implementation has evolved in the last few years.

For instance, this study has found that more than 82% of US microgrid capacity works with fossil fuels (natural gas and diesel); nevertheless, this percentage will decrease over the next three years to 71% [53]. Consequently, the USA will have a considerable number of clean microgrids in near future. Another important consideration is that increasing the use of renewable energy is possible through microgrids. Investment in renewable energy in the USA by 2016 is estimated at $46 billion, of which at least 16% will be through microgrids [58].

Another significant finding is that the USA had achieved 1.18 GW of installed capacity in microgrids by 2015 [58], about 0.1% of total USA installed electric generating capacity. Today, there are around 1,500 microgrids worldwide, with a capacity of 13 GW [46], which means the USA has around 42% of global microgrid capacity market share [46]. The US microgrid penetration has evolved in the following way: in 2013, there were 126 microgrids (81 in operation and 35 planned); in 2014, there were 149 microgrids (93 in operation and 52 planned); and in 2015, there were 216 microgrids (124 in operation and 92 planned) [58]. Worldwide, the microgrid share of the power market capacity for the USA has evolved from 66% in 2014 to 42% in 2015 [59]. The USA microgrids capacity is thus forecasted to reach approximately 1,843 MW by the end of 2017 and 2.8 GW by 2020 according to a 2015 forecast [58], or 3.7 GW by 2020 according to a 2016 forecast [58], and microgrids will become a dominant force in grid operations by 2030 and beyond [60]. That is to say, the future for microgrids is promising.

As a conclusion of this chapter, the following question arises: Will microgrids be the source of electricity supply across the USA as a way to move toward a clean energy future?

References

[1] "Electricity – U.S. Energy Information Administration (EIA)." [Online]. Available: http://www.eia.gov/electricity/ [Accessed 23-Mar-2016].

[2] "What is U.S. electricity generation by energy source? – FAQ – U.S. Energy Information Administration (EIA)." [Online]. Available: https://www.eia. gov/tools/faqs/faq.cfm?id=427&t=3 [Accessed 23-Mar-2016].

[3] "Electric Power Detailed State Data." 2016. [Online]. Available: https:// www.eia.gov/electricity/data/state/ [Accessed 29-Mar-2016].

[4] "Learn More About Interconnections | Department of Energy." [Online]. Available: http://energy.gov/oe/services/electricity-policy-coordination-and-implementation/transmission-planning/recovery-act-0 [Accessed 29-Mar-2016].

[5] "Factbox: Largest U.S. Electric Companies by Megawatts, Customers | Reuters." 2014. [Online]. Available: http://www.reuters.com/article/us-efh-bankruptcy-utilities-idUSBREA3S0P420140429 [Accessed 29-Mar-2016].

[6] Agency United States Environmental Protection (EPA), "End-Users of Electricity." [Online]. Available: https://www.epa.gov/energy/end-users-electricity [Accessed 7-Mar-2016].

[7] U.S. Department of Energy, "United States Electricity Industry Primer," 2015.

[8] D. T. Ton and M. A. Smith, "The U.S. Department of Energy's Microgrid Initiative," *Electr. J.*, vol. 25, no. 8, pp. 84–94, 2012.

[9] C. Marnay and N. Zhou, "Status of Overseas Microgrid Programs : Microgrid Research Activities in the U.S. (English version)," Japan Society of Energy and Resources, 2008.

[10] "The 12 Biggest Electrical Blackouts in History | Mental Floss." 2014. [Online]. Available: http://mentalfloss.com/article/57769/12-biggest-electrical-blackouts-history [Accessed 29-Mar-2016].

[11] A. Hampson, T. Bourgeois, G. Dillingham, and I. Panzarella, "Combined Heat and Power: Enabling Resilient Energy Infrastructure for Critical Facilities," ICF International, 2013.

[12] Lawrence Berkeley National Laboratory (LBNL), Oak Ridge National Laboratory (ORNL), Pacific Northwest National Laboratory (PNNL), Sandia

National Laboratories, National Science Foundation Power Systems Engineering Research Center (PSERC), and Electric Power Group (EPG), "Consortium For Electric Reliability Technology Solutions CERTS." 2016. [Online]. Available: https://certs.lbl.gov/ [Accessed 27-Nov-2015].

[13] W. Bower, D. Ton, R. Guttromson, *et al.*, "The Advanced Microgrid Integration and Interoperability," Sandia National Laboratories, 2014.

[14] R. Lasseter, A. Akhil, C. Marnay, *et al.*, "Integration of Distributed Energy Resources. The CERTS Microgrid Concept," California Energy Commission, 2002.

[15] R. H. Lasseter, J. H. Eto, B. Schenkman, *et al.*, "CERTS Microgrid Laboratory Test Bed," *IEEE Trans. Power Deliv.*, vol. 26, no. 1, pp. 325–332, 2011.

[16] J. Eto, R. H. Lasseter, B. Schenkman, *et al.*, "Overview of the CERTS Microgrid Laboratory Test Bed," *Integr. Wide-Scale Renew. Resource. Into Power Deliv. Syst. 2009 CIGRE/IEEE PES Jt. Symp.*, vol. 26, no. 1, pp. 325–332, 2009.

[17] Naval Facilities Engineering Command (NAVFAC), "SPIDERS Phase 2 Fort Carson Technology Transition Public Report," 2014.

[18] World Intellectual Property Organization (WIPO). [Online]. Available: http://www.wipo.int/patents/en/ [Accessed: 20-Sep-2016].

[19] C. Marnay, N. Zhou, M. Qu, and J. Romankiewicz, "Lessons Learned from Microgrid Demonstrations Worldwide," Berkeley National Laboratory, 2012.

[20] U.S. Department of Energy, "Fuel Cell Handbook," 2004.

[21] Consumer Product Safety Commission (CPSC), "Technology Demonstration of a Prototype Low Carbon Monoxide," 2012.

[22] S. Nababan, E. Muljadi, and F. Blaabjerg, "An Overview of Power Topologies for Micro-Hydro Turbines," *Proc. – 2012 Third IEEE Int. Symp. Power Electron. Distrib. Gener. Syst. PEDG 2012*, June 2012, pp. 737–744, 2012.

[23] M. Mohd and M. Abdul-Hakim, "Micro Hydro Energy Resource," in *Basic Design Aspects of Micro Hydropower Plant and Its Application*, University of Puerto Rico, pp. 1–35, 2009.

[24] W. Kramer, S. Chakraborty, B. Kroposki, and H. Thomas, "Advanced Power Electronic Interfaces for Distributed Energy Systems. Part 1: Systems and Topologies," National Renewable Energy Laboratory, 2008.

[25] S. Rivera, A. Farid, and K. Youcef-Toumi, "Chapter 15: A Multi-Agent System Coordination Approach for Resilient Self-Healing Operations in Multiple Microgrids," in *Book on Industrial Agents: Emerging Applications of Software Agents in Industry*, Elsevier, Amsterdam, 2015.

[26] HOMER Energy, 2014. [Online]. Available: http://www.homerenergy.com/HOMER_pro.html [Accessed 20-Sep-2016].

[27] Sourceforge, [Online]. Available: https://sourceforge.net/projects/rapsim/ [Accessed 20-Sep-2016].

[28] Power Analytics, 2016. [Online]. Available: http://www.poweranalytics.com/paladin-software/paladin-smartgrid/ [Accessed 20-Sep-2016].

[29] DNV GL, "Microgrids," 2016. [Online]. Available: https://www.dnvgl.com/services/microgrids-17929 [Accessed 20-Sep-2016].

[30] UGE, 2016. [Online]. Available: http://www.ugei.com/set [Accessed 20-Sep-2016].

[31] Siemens, "SICAM Microgrid Manager," [Online]. Available: http://w3.siemens.com/smartgrid/global/en/products-systems-solutions/distribution-grid-applications/microgrids/pages/sicam-microgrid-manager.aspx [Accessed 20-Sep-2016].

[32] Green Energy Corp, "GreenBus Microgrid Solution," 2016. [Online]. Available: http://www.greenenergycorp.com/solutions/green-bus-software-platform/ [Accessed 20-Sep-2016].

[33] Spirae, "Spirae Wave Microgrid," [Online]. Available: http://www.spirae.com/microgrid/about-microgrid [Accessed 20-Sep-2016].

[34] Microgrids at Berkeley Lab, "Distributed Energy Resources Customer Adoption Model (DER-CAM)," 2016. [Online]. Available: https://building-microgrid.lbl.gov/projects/der-cam [Accessed 20-Sep-2016].

[35] UC San Diego, "Microgrid," 2004. [Online]. Available: http://cseweb.ucsd.edu/groups/csag/html/projects/grid/ [Accessed 20-Sep-2016].

[36] Opus One Solution, "GridOS," [Online]. Available: http://www.opusone-solutions.com/solutions/ [Accessed 20-Sep-2016].

[37] M. Qu, C. Marnay, and N. Zhou, "Microgrid Policy Review of Selected Major Countries, Regions, and Organizations," National Laboratory, 2011.

[38] IEEE Standards Coordinating Committee 21, *IEEE Application Guide for IEEE Std 1547^{TM}, IEEE Standard for Interconnecting Distributed Resources with Electric Power Systems*, no. April 2009.

[39] Federal Energy Regulatory Commission (FERC), "Standardization of Small Generator Interconnection Agreements and Procedures," 2006.

[40] E. Prabhu and D. Michel, "Interconnecting Distributed Generation : The California Experience," *Int. J. Distrib. Energy Resour.*, vol. 1, no. 3, pp. 185–195, 2005.

[41] The White House president Barack Obama, "Energy, Climate Change, and Our Environment." [Online]. Available: https://www.whitehouse.gov/energy [Accessed 27-Nov-2015].

[42] "Fact Sheet: President Obama Announces New Actions to Bring Renewable Energy and Energy Efficiency to Households across the Country," *whitehouse.gov*, 2014. [Online]. Available: https://www.whitehouse.gov/the-press-office/2015/08/24/fact-sheet-president-obama-announces-new-actions-bring-renewable-energy [Accessed 27-Nov-2015].

[43] R. Bayindir, E. Hossain, E. Kabalci, and K. M. M. Billah, "Investigation on North American Microgrid Facility," *Int. J. Renew. Energy Res.*, vol. 5, no. 2, pp. 1–17, 2015.

[44] S. Parhizi, H. Lotfi, A. Khodaei, and S. Bahramirad, "State of the Art in Research on Microgrids: A Review," *IEEE Access*, vol. 3, no. 1, pp. 890–925, 2015.

[45] S. J. Ahn, J. W. Park, I. Y. Chung, S. Il Moon, S. H. Kang, and S. R. Nam, "Power-Sharing Method of Multiple Distributed Generators Considering Control Modes and Configurations of a Microgrid," *IEEE Trans. Power Deliv.*, vol. 25, no. 3, pp. 2007–2016, 2010.

[46] S. Rivera, A. M. Farid, and K. Youcef-Toumi, "Coordination and Control of Multiple Microgrids Using Multi-Agent Systems," *Energypath 2013 Our Glob. Sustain. Energy Futur.*, August 2013, pp. 1–5.

[47] S. Rivera, A. M. Farid, and K. Youcef-Toumi, "A Multi-Agent System Transient Stability Platform for Resilient Self-Healing Operation of Multiple Microgrids," *2014 IEEE PES Innov. Smart Grid Technol. Conf. ISGT 2014*, no. August 2014.

[48] GTM Research, "North American Microgrids 2014: The Evolution of Localized Energy Optimization," 2014. [Online]. Available: http://www.green techmedia.com/research/report/north-american-microgrids-2014 [Accessed 22-Mar-2016].

[49] GTM Research, "North American Microgrids 2015: Advancing Beyond Local Energy Optimization," 2015. [Online]. Available: https://www.green techmedia.com/research/report/north-american-microgrids-2015 [Accessed 23-Mar-2016].

[50] N. W. A. Lidula and A. D. Rajapakse, "Microgrids Research: A Review of Experimental Microgrids and Test Systems," *Renew. Sustain. Energy Rev.*, vol. 15, no. 1, pp. 186–202, 2011.

[51] L. Mariam, M. Basu, and M. F. Conlon, "A Review of Existing Microgrid Architectures," *Hindawi Publ. Corp. J. Eng.*, vol. 2013, pp. 1–8, 2013.

[52] Alliance for Rural Electrification, "Hybrid Power Systems Based on Renewable Energies: A Suitable and Cost-Competitive Solution for Rural Electrification," 2008.

[53] biocycle.net, "Symphony of the Microgrid at an Urban University, Bio-Cycle." 2014. [Online]. Available: http://www.biocycle.net/2014/07/16/symphony-of-the-microgrid-at-an-urban-university/ [Accessed 9-Feb-2016].

[54] "Microgrid Projects Map from Microgrid MediaMicrogrid Projects." [Online]. Available: http://microgridprojects.com/ [Accessed 22-Mar-2016].

[55] "Home | Building Microgrid." 2016. [Online]. Available: https://building-microgrid.lbl.gov/ [Accessed 22-Mar-2016].

[56] Chevron Energy Solutions Company, U.S. Department of Energy, and National Energy Technology Laboratory, "Certs Microgrid Demonstration with Large-Scale Energy Storage and Renewables at Santa Rita Jail," 2014.

[57] N. Deforest, J. Lai, M. Stadler, G. Mendes, C. Marnay, and J. Donadee, "Integration & Operation of a Microgrid at Santa Rita Jail," 2011.

[58] "US Microgrid Growth Beats Estimates: 2020 Capacity Forecast Now Exceeds 3.7 Gigawatts." [Online]. Available: http://www.greentechmedia. com/articles/read/u.s.-microgrid-growth-beats-analyst-estimates-revised-2020-capacity-project [Accessed 1-Jun-2016]

[59] Navigant Research, "Microgrid Deployment Tracker 4Q15", January 2016.

[60] A. M. Annaswamy and M. Amin, "Smart Grid Research: Control Systems – IEEE Vision for Smart Grid Controls: 2030 and Beyond," IEEE, 2013.

Chapter 10

Microgrids in developing countries

*Sergio Rivera, Wiston Ñustes, Miguel León
and Juan Rodríguez*

10.1 Developing countries and their electrical infrastructure

Developing countries are those whose economies are in the process of economic development from underdevelopment or a transitional economy [1]. These countries have not attained a high level of industrialization yet and have some problems related to weak infrastructures, inequality, and so on. Two-thirds of the world's population live in developing countries located mainly in Latin America, Africa, and Asia.

Some of the main social problems of developing countries are [1] poverty, understood as the lack of wealth to meet the needs of the inhabitants; high unemployment and corruption rates; economic inequality among their inhabitants; little or no funding for science and technology from the state; low per capita income; and technological dependence on other countries.

At the Fifth Ministerial Conference held in Cancun, Mexico, from September 10 to September 14, 2003, the World Trade Organization proposed the G20 (also known as G21, G22, G23, and G20+), a bloc of 24 developing nations established on August 20, 2003. The group comprises the following countries: Argentina, Bolivia, Brazil, Chile, China, Colombia, Cuba, Ecuador, Egypt, Philippines, Guatemala, India, Indonesia, Mexico, Nigeria, Pakistan, Paraguay, Peru, South Africa, Tanzania, Thailand, Uruguay, Venezuela, and Zimbabwe. Ten of these countries are in South America, three in Central America, seven in Asia, and four in Africa [1,2]. The characteristics of the electrical infrastructure in some developing countries are presented below. In addition, Table 10.1 provides data on installed capacity and type of generation in developing countries.

In Latin America, factors influencing power generation vary from one region to another due to the availability of natural resources, fuels, social impact, and regulatory framework of each country [3,4]. For instance, power generation in Bolivia comprises hydroelectric generation (15.21%) and natural-gas-fired thermoelectric generation (81.3%). Colombia' power generation includes hydroelectric plants (63.7%), thermal plants (gas and coal, 31.6%), minor plants (hydroelectric, thermal, and wind), and cogenerators (4.8%). Ecuador's system is based on

Table 10.1   Installed capacity and generation matrix in developing countries [1–17]

Region	Country	Capacity	Thermal (%)	Hydro-electrical (%)	Nuclear (%)	Wind (%)	Solar (%)
Latin America	Argentina	32,631 MW	67.00	27.00	5.60	0.36	0.04
	Bolivia	1,580 MW	81.30	15.21	–	–	3.49
	Brazil	96,294 MW	67.00	28.00	1.00	3.00	1.00
	Chile	19,393 MW	48.00	40.00	–	10.00	2.00
	Colombia	15,521 MW	31.60	63.70	–	3.80	1.00
	Cuba	3,500 MW	–	–	–	–	–
	Ecuador	5,675 MW	46.00	49.00	–	4.00	1.00
	Guatemala	3,630 MW	48.00	49.00	–	2.50	0.50
	Mexico	66,000 MW	82.00	13.60	4.10	0.10	0.20
	Paraguay	10,184 MW	–	–	–	–	–
	Peru	7,930 MW	47.00	46.00	–	4.00	3.00
	Uruguay	3,719 MW	45.60	41.36	–	12.93	0.11
	Venezuela	24,437 MW	29.00	69.00		1.60	0.40
Asia	China	1,565 GW	67.50	22.20	1.40	7.00	1.90
	India	303 GW	72.00	18.06	–	6.03	3.90
	Philippines	18,023 MW	87.20	11.60	–	0.80	0.40
	Indonesia	39,900 MW	39.00	48.00	–	9.00	4.00
	Pakistan	24,200 MW	58.00	32.00	4.00	4.50	1.50
	Tanzania	1,514 MW	62.00	37.00	–	0.83	0.04
	Thailand	16,376 MW	78.21	21.06	–	0.31	0.43
Africa	Egypt	26,010 MW	92.00	7.00	–	–	1.00
	Nigeria	12,806 MW	66.89	28.11	–	2.00	3.00
	South Africa	48,220 MW	81.00	10.00	5.00	3.20	0.80
	Zimbabwe	19,800 MW	89.00	9.50	–	0.96	0.54

nonrenewable energy sources (thermal, 46%) and renewable ones (water, wind, and solar, 54%). In Mexico, the sources of power generation are hydroelectric (13.6%), thermoelectric (73%), geothermal (2.3%), coal-fired (6.7%), wind (0.1%), nuclear (4.1%), and photovoltaic (PV) (0.2%) [5].

Argentina, Brazil, Chile, and Uruguay are also developing countries in Latin America. The Argentine electricity sector constitutes the third largest power market in Latin America; this sector includes thermal, water, nuclear, wind, and solar energy. Generating equipment installed in the Argentine Interconnected System (SADI (Sistema Argentino de Interconeccion, in Spanish)) is divided into four groups, according to the natural resource and the technology implemented: fossil fuels, nuclear, water, and renewable sources. This first one is classified into five subgroups according to the thermal cycling and fuels used to take advantage of the energy: steam turbines, gas turbines, combined cycles and diesel and biogas engines. The proportion of each generation source in SADI is fossil, 67%; water, 27%; nuclear, 5.6%; and other renewable sources, 0.4%. The capacity in the combined cycle technology is 47% of a thermal generation with 9.23 GW. Wind farms have a capacity of 187 MW, whereas PV technology has only 8 MW. Currently, Argentina has a net capacity of 32,631 MW [6].

Brazil is the third on the American continent and the fifth largest country in the world; it covers an area of 8,514,877 km^2. It has more than 200 million people, and it is the tenth largest consumer of energy in the world [4]. Electric power represents 18% of Brazil's total expenditure with an installed power capacity of 96,294 MW. Sixty-seven percent corresponds to the use of water resources; 28%, thermal resources; 4%, renewable sources such as wind and solar; and 1%, nuclear energy [4].

The sources of generation in Chile are as follows: hydropower (40%), thermal coal-fired power plants, natural gas and diesel (48%), and renewable resources such as solar, biomass, and wind (12%).

The installed power generation capacity of Uruguay was approximately 3,719 MW by December 2014; 1,538 MW correspond to hydroelectric power plants, 421 MW to biomass-fired power plants, 1,275 MW to fossil-fuel power plants, 481 MW to wind generators, and 4 MW to solar generators. Its interconnection with the Argentine Republic is approximately 2,000 MW, and the one with Brazil is 70 MW, being expanded to 500 MW. The solar PV installed power is 2,682.8 kW; the wind installed power is 27.5 kW [8].

China is the largest country in Asia continent and the third in the world, covering approximately 9.6 million km^2 [9]. China has a population of 1,369 million people. The installed power capacity in China is 1,565 GW; 67.5% is generated by thermal power plants, 22.2% by hydroelectric power plants, 7% by wind farms, 1.9% by solar PV power plants, and 1.4% by nuclear power plants. China's power system is divided mainly into six regions: *North, Northeast, Northwest, East, Central,* and *South.* Each region and its provinces have autonomy in making decisions on energy dispatch and planning. Eighty percent of solar and wind resources are located in the northern regions of China (North, Northwest, and Northeast), whereas 80% of hydroelectric plants are located in the South [10].

The Republic of Indonesia is an island country consisting of more than 17,000 islands. Java, Bali, and Sumatra are the most populated islands. The population in Indonesia was estimated at 250 million people in 2015 [11]. The traditional grid delivers electricity to around 80.51% of the population [11]. Since 2001, the installed capacity in Indonesia has increased from 23.7 to 39.9 GW by the end of 2011. In 2011, the renewable energy installed capacity in Indonesia amounted only to 13% of the total production [12]. Electric energy in Indonesia is generated mostly from coal and petroleum-derived fuels. Renewable energy is produced mainly by hydroelectric power plants. Between 2001 and 2011 in the field of renewable energy, the greatest increase in capacity occurred in hydroelectricity with 0.768 GW, followed by geothermal electricity that showed an increase in capacity of 0.424 GW [12,13].

Nonrenewable energy sources continue to be dominant because of their low costs of generation, taking into account that about 20% of the country is not connected to the traditional power grid [12]. Most of the islands that are not interconnected meet electricity demand by using mini hydroelectric plants and diesel generators. The demographic characteristics of Indonesia make the transportation of fuel for diesel generators a complex task. In addition, the oil price, which depends on the international market, has made the cost of power generation using this fuel comparable to the cost of renewable electricity generation [12,13].

The installed capacity in India is approximately 306 GW. Twenty-eight percent of power generation comes from renewable energy, and 72% from nonrenewable energy. The installed capacity of solar energy is approximately 6,763 MW and 26.74 GW of wind power [14].

In Egypt, most of the electricity supply comes from thermal and hydroelectric power stations: 92% and 7%, respectively. The main hydroelectric power stations currently operating in Egypt are Aswan, Esna, and Naga Hamady Barrages. Egypt has a high solar availability with a 4.5-MWp PV capacity [15].

The Republic of South Africa generated 20,800 GW h of energy in January 2016. Eighty-six percent of the population has access to electricity; since the population in South Africa is 54.7 million people, about 8 million inhabitants do not have access to electricity [16]. The electricity industry in the country is led by the state-owned company Eskom that generates approximately 95% of the country's electricity and approximately 45% of the total electricity consumed in Africa [16]. The fact that more than 90% of the energy is generated from coal has made South Africa one of the 15 countries with highest CO_2 emissions. The installed capacity in 2013 was 48,220 MW [17]. Regarding renewable energy, projects set a 1 GW increase in installed capacity for PVs, and a 1 GW capacity in wind power; finally, the country seeks to implement other projects in concentrated solar power with 200 MW capacity [16,17].

10.2   Microgrids research in developing countries

Renewable energy sources are widely used in the Latin American electricity sector and also in the Caribbean where there is great potential in hydroelectricity, geo-thermal, biomass, solar, and wind power which are, in general, limitedly used [18]. In Latin America, the leading countries in research on microgrids and renewable energy are Argentina, Brazil, and Mexico, where 0.58%, 1.15%, and 0.50% of the Gross Domestic Product (GDP), respectively, are invested in research and development [3,4]. In addition, Chile has also improved its research indicators since its investment policy reaches 0.36% of the GDP [3].

Argentina is currently carrying out some research on smart grids, civilian nuclear energy and renewable energy, with a particular focus on forecasts and integration of wind power into the grid. Likewise, there are some studies from Argentine universities about new technologies of distributed generation and energy storage. For example, models of batteries for microgrids, which provide support to the power quality and control on the demand peak, and control strategies, which regulate active and reactive power delivered by different renewable generation sources in a microgrid [6].

Recently, Brazil has conducted research on microgrids and distributed generation, mainly focused on the modeling of systems involving different renewable sources and control strategies applied to different scenarios. Their interest on the control and integration of hybrid microgrids with the interconnected system must be highlighted [19,20]. In the same way, the Federal University of Ceará researches into the modeling and simulation of a microgrid [21]; it has 10 kW of wind generation, 0.5 kW of fuel cell, 2 kW of PV system, and a 3-kW bank of batteries.

In Mexico, research is mainly conducted by universities and some specialized centers; the most important center for research in renewable energy is *Centro de Investigación en Energía* from *Universidad Nacional Autónoma de México*. This center offers departments of solar materials, energy systems and thermal sciences, as well as training, dissemination, and linkage [22]. Another leading institution is *Instituto de Investigaciones Eléctricas*, which conducts studies on fuel cells, PVs, tidal energy, solar thermal power systems, wind technology, small hydroelectric power stations, and geothermal resources [22].

In Chile, research on technologies related to microgrids is supported primarily by *Comité Nacional de Investigación Científica y Tecnológica*. This institution has created different funds to promote centers devoted to research and development of specialized topics with different universities; they aim at studying the viability of implementing geothermal energy projects and the integration of solar energy from four lines of research [23]. The first line is the application of solar energy in industry. They seek to strengthen the integration of solar sources and other systems, for example: PV converters into mining processes, hydrogen production from high-temperature solar converters, characterization of PV materials, modeling, design and state estimation in control systems for solar concentrators [23]. The second one is the integration into the interconnected system of large-scale solar sources [3]. The third one is a set of solutions of solar and wind energy for rural and urban communities, thus promoting microgrids as a solution to power supply in rural areas and strengthening the integration of PV sources into urban areas [3,24]. Finally, there are important developments of energy storage systems, with research on the photocatalytic activity of CdO/ZnO systems, characterization of tetrahydrate calcium chloride, development and design of lithium-ion batteries for PV systems storage, and use of ionic liquids for energy storage.

During the last few years, China became one of the largest producers of PV energy in the world with more than 40 GW of installed capacity [25]. From the need for further development of renewable energy sources and microgrids, different programs, platforms, and organizations have been created during the last years in China [26]. The China National Renewable Energy Centre, the National Centre for Wind Technology Research and the National Centre for Solar Energy can be mentioned among the most important organizations [27]. Among research and development programs is the National Research Program, implemented by the National Natural Science Foundation of China, which proposed to support research and implementation of important projects according to national needs [28]. Some of the National Research Program's fundamental research topics on distributed generation systems are the interaction between distribution companies and microgrids, the mechanisms to ensure stable grid operations, optimal microgrids operation and energy management, technologies of microgrid simulations, control and analysis of power quality in microgrids, control strategies to link the microgrids to the main grid, and the coordination of control on different renewable energy sources in a microgrid [26,29,30].

The State Intellectual Property Office (SIPO) of China has processed around 512 patents related on branches of microgrid research. The percentages of these patents proposing new techniques and knowledge in the different areas of microgrid research are control of microgrids, 55%; microgrids with controllable loads, 24%; microgrids with storage, 25%; Direct Current microgrids, 10%; communication within microgrids, 10%; cooperation within microgrids, 1%; microgrid optimization, 17%; microgrid modeling, 18%; and clean energy microgrids 2%. The change in publication date for these SIPO patents in the last few years has been: 2006, 2 patents; 2007, 5 patents; 2008, 7 patents; 2009, 11 patents; 2010, 16 patents; 2011, 35 patents; 2012, 64 patents; 2013, 44 patents; 2014, 141 patents; 2015, 154 patents; and half of 2016, 9 patents. A total of 31% of applicants for microgrid patents in SIPO are from universities [30].

India is the only country in the world with a ministry of renewable energy instructed directly by the president; it regulates and enforces research in renewable sources of energy, and this is achieving good results in the application of clean and sustainable energy models [31,32]. The Ministry has carried out different programs, covering practically all aspects of renewable energy [33]. These programs seek to replace conventional fossil-based energy sources gradually with PV and small hydroelectric plants, as well as to take the benefits from renewable energy to rural areas [33]. There are also urban, industrial, and commercial applications for these programs and the development of alternative fuels in transportation, thus supporting research and development of new technologies, products and services [34].

In South Africa, the Centre for Renewable and Sustainable Energy Studies is located at Stellenbosch University for the general field of renewable energy [35]. Some of the research projects are analysis of solar energy systems with different fluids for heat transfer, detailed analysis of components of solar energy plants (collector, receiver, storage, and refrigeration systems) and industrial applications of heat from solar energy.

In Egypt, an authority for the renewable energy field was established in 1986 to be the center for national coordination, development, and introduction of large-scale clean technologies in Egypt [15,36]. This ministry is in charge of planning and executing renewable energy programs through the evaluation of renewable power resources, research, development, demonstration, proof and assessment of the different renewable energy technologies focused on biomass, solar and wind energy, and the execution of renewable energy projects [36].

10.3 Microgrids technology in developing countries

Microgrids technology in developing countries can be grouped according to the technologies developed in Latin America and Asia. In Africa, technological developments in the field of microgrids cannot be found as traditional technologies are still used.

10.3.1   Latin America: Argentina and Brazil cases

10.3.1.1   Argentina

- *Wind turbines:* The great boom in wind resource use has led to the establishment of two leading local companies that develop high-power wind turbines for wind class 2; these companies are *Industrias Metalúrgicas Pescarmona* and *NRG Patagonia*. They built IWP70 and NRG1500 wind turbines with a 1.5-MW nominal power capacity installed in *El Tordillo*, a wind farm located in the province of Chubut [37,38]. The company *Investigación Aplicada (INVAP) Ingeniería* has also developed its own design for low and medium power wind turbines using different technologies; it aims at producing a prototype for high-power wind turbines with a range between 1.5 and 3 MW for winds class 1. Unlike the first two companies, INVAP does not develop wind turbines based on foreign patents but does the basic engineering designs and construction on its own [38].

- *Solar cells:* San Juan province has high-quality quartz and raw materials necessary for a less elaborate production of high-purity, solar-grade silicon [39]; this led to the foundation of the German company Schmid Branch in Argentina for the local manufacture of solar cells. Their goal is to produce 235,000 solar panels per year of 300 W each; the estimated equivalent is an annual production of 70 MW [40].

- *Hydrokinetic turbine:* This turbine or submerged generator is designed and manufactured by the company *INVAP Ingeniería*. Its turbines have different power ratings for a greater use of watercourses. Recent findings indicate that it is possible to have machines that exceed 1-MVA capacities and are suitable for tidal applications in the future [39]; this technology should also be coupled with power electronics so as to deliver AC 220 V/380 V – 50 Hz.

10.3.1.2   Brazil

The technological progress in Brazil has been proportional to the one developed for the integration of renewable sources in the country. As Brazilians have dabbled in large- and small-scale projects of distributed generation, the Brazilian industry has provided local technological resources mainly on the following elements: wind turbines, solar cells, turbogenerators, hydrogenerators, wind blades, and non-technical parts such as wind towers, casting parts, and electrical components. In that way, they have become one of the leaders providing technologies to the nonconventional sources of renewable energy.

The leading Brazilian developers of wind farms are CPFL Renováveis (CPFL: Companhia Paulista de Força e Luz, in Portuguese), Eletrobras, Renova Energia, and Casa dos Ventos; they carry out the different stages of a wind project: acquisition, design, construction, and operationalization [20]. WEG S.A., one of the most outstanding companies in the electricity sector worldwide, provides technology mainly in rotational machines such as motors, generators, hydraulic turbines, and hydromechanical equipment [41]. Tecsis produces approximately 1,500 MW/year from turbine blades, becoming a world leader in this technology [41]. Algolix and

PEVOUTO are specialists in supplying components like casting parts and general electrical devices for big wind turbine manufacturers [20]. Engebasa, Intecnial, Piratininga, and Tecnomaq manufacture wind towers and have great potential for local production. TECNOMETAL develops low-power turbines, equipment, hydro-mechanical plants, and PV panels (solar cells) [41].

They all contribute to the technological growth of the renewable energy generation in Brazil and all around the world.

The characteristics of some generation elements given by local manufacturers are mentioned below:

- *Hydrogenerators:* Technology developed for application in hydropower projects. There are different machines that have a broad range of power and rotational speed ranging from 500 kW to 150 MW and from 1,800 to 120 rpm. Supply voltages up to 13,800 V, 50 or 60 Hz. This type of generators can be applied to Kaplan, Francis or Pelton turbines, with different mechanical and electrical configurations [42].
- *Turbogenerators:* Technology applied to thermoelectric power plants that can be used in steam, gas, and other turbines. These systems have favorable characteristics and excellent electric and mechanical properties that provide robustness, efficiency, and operational reliability. The power ranges reach 150 MVA with rotations from 3,600 to 1,800 rpm [43].
- *Wind turbines:* This technology allows a variable rotation of the rotor and a direct control of the blade angles positioning the blades to always obtain a direction in the wind that takes significant advantage of this resource. Stages of power electronics that allow a controllable and safe connection to the microgrid are included. Cooling systems make it possible to install this equipment in corrosive environments as well as in environments near the sea. There are wind turbines with nominal power up to 2.1 MW and voltage of 34.5 kV [42].
- *Photovoltaic:* The production of this technology in Brazil is being currently developed. At this moment, TECNOMETAL has a plant to produce solar cells with a capacity of 25 MW/year. It is expected that in coming years the manufacture of PV modules increases in companies such as S4 Solar, SOLIKER and Globo Brasil by taking advantage of one of the main resources of this country: silicon, which is used in solar cells manufacturing. The company CSEM Brasil has contributed to research and advances in this technology by developing its third generation of solar cells [44].

10.3.2 Asia

Over the last few years, China has been a world power in the production of technology for renewable energy sources, mainly regarding PVs and wind power [25]. In this section, the advances in technological production of greater development in China are as follows:

- *Photovoltaics:* By the end of 2014, China had a PV installed capacity of 28 GW that accounted for approximately 17% of the global capacity; the country was ranked second after Germany [28]. By the end of 2015, this figure rose to 45 GW becoming, at that time, the country with greatest capacity in PV power generation with 20% of the global capacity [45]. China, being the leader in production since 2009, manufactures on average two-thirds of these modules globally. The main companies devoted to this technology are Trina, JinkoSolar, JA Solar, Yingli, SFCE (Formerly Suntech), and Rene-Solar [46].

- China has the largest global production of multi-crystalline silicon material, led by Jiangsu Zhongneng Polysilicon Technology Development [47]. In 2014, the production of this material reached 66,876 t representing 50% of China's total. In addition, there are some other companies devoted to the manufacture of this technology: Tbea Solar, China Silicon Corporation, DAQO New Energy Corp, Sichuan ReneSola Silicone Materials, Yichang CSG Polysilicon, ORISI Silicon, Asia Silicon, Shaxi Tianhong Silicon, DunAn Holding Group, among others [47].
- *Solar cells*: In 2014, Solar cell production was 51 GW worldwide (excluding thin-film batteries); China contributed with 33 GW, that is, 66% of the global production becoming the leading producer of this technology. Some new technologies have been implemented in the manufacture of solar cells, such as multi-printing, new grid line technology, passivated emitter rear cell, and black silicon technology [28].
- *Wind power*: Wind power generation installed capacity in China in 2015 reached 145 GW, the first in the world. The provinces that present a greater use of this renewable resource are Inner Mongolia (19%), Xiajiang (12.5%), Gansu (9.7%), and Hebei (7.9%) [48]. China has large companies devoted to wind power manufacture that develop technologies according to the wind resource needs [49]. In places where wind speed is low, special turbines have been developed to take full advantage of this resource [29]. Goldwind developed the GW115/2000 turbine, which is suitable for places with speeds of 5.2 m/s. In 2013 and 2014, United Power and Energy Envision developed ultra-low-speed turbines with rated power of 1.5–1.8 MW and rotor diameter of 97–106 m, respectively [48].

The Ministry of New and Renewable Energy, in India, is the authority in the government that addresses issues related to new technologies and renewable energy [50]. The ministry has classified renewable energy technologies as follows [51]:

- Connected to the main power grid.
- Off-grid power.
- Decentralized systems.
- New Technologies.

Renewable energy technologies are based on wind energy, biomass, small hydro, and solar technologies that are driven primarily by the private sector. It also takes energy from biomass generated from agricultural waste, plantations and urban and industrial waste. Due to the vastness of India and remoteness of most of the nonelectrified population, there is potential for renewable energy technologies to provide access to electricity for this population [51]. Some of the renewable energy technologies promoted in towns and rural areas are [51–53]:

- Biogas plants.
- Solar thermal systems.
- Solar flashlights and solar lighting systems for homes.
- Water heating solar systems.
- Solar cookers.
- Solar–biomass power generators.
- Wind pump.
- Micro-hydraulic plants.
- Hydrogen energy.
- Chemical energy sources (fuel cells).
- Geo thermal energy.
- Tidal energy.

To promote the development of these new technologies in India, research, development and demonstration projects are being promoted in various research, scientific and educational institutes, national laboratories, and universities [53].

Technology in Indonesia: Indonesia's electrical system is led by the state company *Perusahaan Listrik Negara* that dominated this service until 2002, when regulations on marketing of energy produced by private entities were introduced. Since then, participation of private entities has grown through the implementation of *Mini Hydro technology projects by hydroelectric power stations* [54]. On renewable energy, hydroelectric and geothermal plants are dominant. Solar energy is a source of power generation with more than 10 generators that have a capacity of 8 MW. The country is involved in a great project of technological development through the implementation of *mini hydro* plants; those mini plants have a generation capacity below 10 MW. In 2015, Indonesia had 67 mini hydropower stations. There are independent power producers that generate energy by means of about 313 hydro plants; 70% of them are located in the three most populated islands [54].

10.4   Clean energy standards and policy in developing countries

This section presents the main policies on clean energy in developing countries; countries were grouped according to three regions: Latin America, Asia, and Africa.

10.4.1   Latin America

Country	Policy
Argentina [55]	• National Energy Plan 2004–2008. This plan includes several mechanisms applied during this period • Emergency programs: installation of small mobile electricity units to support the main Argentine provinces during peak generation periods • Renewable Energy in the Rural Market Project (PERMER), which was expected to end in 2008, but was extended to 2011 • The intention is to reach 8% of total generation through renewable energy sources by 2016 is set by the National Law No. 26.190, declaring generation of electricity from wind, solar, and geothermal energy, waves, biomass, biogas and small hydro power plants a matter of national interest
Brazil [56]	• 2002: Renewable Energy Incentive Program, PROINFA (*Programa de Incentivo às Fontes Alternativas de Energía*) • 2004: Increase in the proportion of electricity produced by projects based on wind power, biomass and small hydroelectric power stations (SHPs) in the National Interconnected System • 2011: Renewable energy auction
Chile [23]	• 2008: Nonconventional Renewable Energies, NCRE (No. 20.257), which compels generators with capacities above 200 MW to generate a percentage of their sold electricity through NCRE. The established percentage is 5% from 2010 to 2014, and since 2015 the percentage is expected to increase by 0.5% annually until it reaches 10% by 2024
Colombia [57]	• 2001: Law 697 declares the rational and efficient use of energy as well as the interest in the promotion of nonconventional sources • 2003: Decree No. 3783 establishes mechanisms and incentives for research and funding of renewable and alternative energy sources • 2002: Tax reform that establishes 15-year tax exemptions for projects developed under the Clean Development Mechanism (CDM) • 2014: Law 1715 sets the legal framework and instruments necessary for the promotion and utilization of nonconventional energy sources, as well as for the promotion of investment, research and development of clean technologies for energy production, energy efficiency, and demand response
Ecuador [58]	• 2007: Electrical Sector Regime Law (LRSE), its regulations and standards aim to set guidelines for promotion and efficient and sustainable development of electrification
Cuba [57]	• 1997: National Environmental Strategy • 2006: Energy revolution with five main goals: energy efficiency, an increase in availability and reliability of the national grid, incorporation of more renewable energy technologies, growth in oil and local gas production and exploration and international cooperation • 2008: Introduction of renewable energies in the production and processing of agricultural products

(Continues)

(*Continued*)

Country	Policy
Mexico [59]	• Grid Interconnection Contract for Renewable Energy allows independent producers of electricity to inject surplus energy to be used later on. The value of electricity in any direction is calculated by using formulas set out in the contract terms • Law for the Use of Renewable Energy and Finance of the Energy Transition establishes conditions for electricity generation with renewable energy sources, wind energy particularly
Peru [57]	• 2008: Legislative Decree 1002 or RER Law and Supreme Decree No. 050-2008-EM (regulations), which establish incentives for the development of Non-Conventional Renewable Energy sources, NCRE, such as small-scale wind energy, geothermal, and hydro-electric power
Paraguay [60]	• 2006: Law 3009, allowing the generation by independent producers and electricity transportation for national or international consumption. It includes renewable energy generators except for hydroelectric power plants with capacities above 2 MW. This law also establishes transportation rates in which independent producers of electricity can sell surplus electricity at 70% of the price set by the national electricity administrator
Uruguay [60]	• 2006: Decree 77 introduces bidding processes for procurement of 60 MW from nonconventional renewable energy by the state electricity company. The outcome was based on contract formulation with biomass and wind generators • 2008: Law 18,362 offers an opportunity to wind power
Bolivia [57]	• 2002: Rural Electrification Plan • 2005: Rural electrification project with renewable energy through the popular participation law • 2006: Law for Universal Access to Electricity
Venezuela [61]	• The development plan of renewable energy sources (2007–2013) including policies, strategies, and incentives to promote renewable energy

10.4.2 Asia

Country	Policy
China [62]	First-level policies: They provide overall guidance and include some speeches by the state leaders regarding the development of renewable energy and some others explaining the point of view of the Chinese Government on global environment [63]: • 1995: Electrical Energy Law • 1996: Guideline for the Ninth Five-Year Plan and 2010: Long-Term Objectives on Economic and Social Development of China

Country	Policy
	• 1996: China's Technology and Energy Policy
	• 1997: Energy Saving Law
	• 2003: Renewable Energy Promotion Law

Second-level policies: they specify the objectives and development plans, which are focused on rural electrification and technologies of power generation based on renewable sources [63]:

- 1995: New and Renewable Energy Development Projects in Priority
- 1996: Ninth Five-Year Plan of Industrialization of New and Renewable Energy
- 1998: Incentive Policies for Renewable Energy Technology
- 2001: Tenth Five-Year Plan for New and Renewable Energy Commercialization Development

Third-level policies: They consist of specific incentives and support for the development and use of renewable energies [63]:

- 1997: Provisional Regulations on the Management of New Energy
- 1999: Circular on Further Supporting the Development of Renewable Energy
- 2001: Adjustment of Value-Added Tax for Some Resource Comprehensive Utilization Products by Ministry of Finance and State Tax Administration
- 2001: Electricity Facility Construction in Non-Electrification Townships in Western Provinces of China

Country	Policy
India [64]	• 2003: Renewable Energy Policies
	• 2005: Policies setting preferential tariffs for renewable energy
	• 2010: Jawaharlal Nehru National Solar Mission
	• 2012: Policies on grid-connected renewable energy
	• 2015: Telangana New Solar Energy Policy and Andhra Pradesh New Solar Energy Policy
Indonesia [65]	• 2007: Law No. 30 on Energy: The government should increase provision and utilization of new and renewable energy
	• 2008: Regulation of the management and development of geothermal energy sources for direct and indirect utilization
	• 2009: Law No. 30 that prioritizes the use of local primary resources for electricity generation
	• 2009: The State is compelled to purchase electricity from small- and medium-scale renewable energy power plants developed by cooperatives (community capacity ≤ 10 MW)
	• 2010: Tax facility for renewable energy in the form of income tax or import duty
Philippines [57]	• 2006: Biofuels Act of 2006
	• National Renewable Energy Program (2012–2030) aims to triple the capacity of renewable energy including hydropower (8,724 MW), geothermal (3,461 MW), and wind power (2,378 MW)

(Continues)

(*Continued*)

Country	Policy
Pakistan [57]	• 2002: Promotion to the use of local resources, including renewable energy [66] • 2006: Policy for Development of Renewable Energy for Power Generation Employing Small Hydro, Wind and Solar Technologies • 2010: National Energy Policy
Thailand [57]	• 2010: Power Development Plan that promotes energy efficiency and renewable energy • 2010: Development and Promotion of Renewable Energy Entrepreneurs to endorse the building of large-scale entrepreneurship in green technology

10.4.3   Africa

Country	Policy
Egypt [57]	• 1982: Energy conservation measures and renewable energy application • 1986: Renewable Energy Authority is established • 2008: The Renewable Energy Strategy aims at the contribution of renewable energy of 20% out of the total generated • 2013: Policy of net generation measurement to small interconnected generator • Egypt's National Plan (2012–2017), which includes the implementation of 100 MW of concentrated solar power and 20 PV MW interconnected to the transmission system
Nigeria [67]	• 2003: National Renewable and Sustainable Energy Development with the active participation of the private sector • 2005: Deregulation by the electricity sector reform • 2006: Renewable Energy Master Plan [68] • 2007: Biofuel policy and incentives to foster the use of E10 and B20 as automotive fuels
South Africa [57]	• 1998: White Paper on the Energy Policy of the country • 2004: National Strategy for Cleaner Production • 2007: Biofuels Industrial Strategy [69,70] • 2008: Energy law with energy planning, increase in power generation and renewable energy consumption • 2011: Integrated resource plan to help minimize greenhouse gas emissions
Zimbabwe [57]	• 2008: Energy Framework Project • 2016: The Ministry of Energy and Power Development is currently developing a renewable energy policy
Tanzania [57]	• 2003: National Energy Policy aimed to consider national energy needs, develop cost-effective national energy resources and improve the reliability, efficiency, and security of energy. • 2008: Rates scheme for small power producers (100 kW to 10 MW)

10.5   Current microgrids projects in developing countries

This section presents some projects of actual microgrids in developing countries. At the end of the section, a table summarizes the most relevant data of the main microgrids found throughout this work in Table 10.2.

10.5.1   Armstrong Microgrid (Argentina)

Microgrid projects in Argentina are few in comparison with the installed capacity of renewable generation; it means that the plants are directly connected to the SADI and cannot be isolated to meet certain amount of demand. Currently, there is only one pilot smart microgrid in Armstrong and several studies to imitate this project in Salta and San Martin [39] (Figure 10.1).

Armstrong is the first pilot smart microgrid with a generation capacity of approximately 520 kW [40]. The city has around 12,000 inhabitants out of which 4,486 are residential customers, 540 are commercial customers, 418 are rural, and 217 industrial customers, for a total of 8 MW power peak [39,40] This microgrid is connected to EPE Santa Fe through two three-phase circuits at 33 kV voltage level. The main substation has an installed capacity of 14 MW and two supply transformers for greater reliability. There is PV generation carried out using low voltage and a small hydroelectric plant that takes advantage of the water resources from Carcaraña River [40].

10.5.2   District Power Plant Microgrid (Brazil)

It is a small microgrid developed at the CERTI (The Foundation Centers of Reference in Innovative Technologies) Foundation, in order to feed their own loads from renewable sources such as wind, PVs, and diesel. The total size of the microgrid is 25.5 kW; it feeds loads of offices, refrigerators and AC outputs. It has autonomy in functioning together with the grid or separately through ATS (automatic transfer switch) operation. This project was developed through the adoption of the Net Metering Act of the National Electric Energy Agency (ANEEL) [41] (Figure 10.2).

10.5.3   Lençois Island Microgrid (Brazil)

It is an isolated microgrid that distributes electricity to the municipality of Cururupu, Maranhão, in the northeast of Brazil. It uses two renewable sources: wind (22.5 kW), solar PV (21 kW), and diesel generators (53 kW), which are connected to the DC and AC bus. The population consists of approximately 395 inhabitants and 90 houses. Energy demand exceeds 4,500 kW h per month, that is, the current capacity is sufficient to meet the load for six years if it has a 1.5% annual growth [19] (Figure 10.3).

10.5.4   Microgrid in sustainable building of the Federal University of Juiz de Fora (Brazil)

This microgrid was implemented in the building of the School of Engineering at the Federal University Juiz de Fora. It uses renewable sources: wind power, solar PV,

Table 10.2 Current microgrids projects in developing countries

Project	Country	First year of operation	Conventional sources	Renewable sources	Storage	Type	Total capacity (kW)
Armstrong Smart Microgrids (Santa Fe)	Argentina	2012	H	PV	–	CC	520
Diadema (Chubut)	Argentina	2010	–	W	–	In	6,300
Rawson I y II (Chubut)	Argentina	2012	–	W	–	In	77,400
Loma Blanca IV (Chubut)	Argentina	2013	–	W	–	In	51,000
El Jume (Santiago del Estero)	Argentina	2015	–	W	–	In	8,000
Xapuri (Acre)	Brazil	2005	–	PV	Batt	RC	0.3
San Antonio (Pará)	Brazil	2010	BG	PV	Batt	RC	56
Autazes	Brazil	2011	–	PV	Batt	RC	10.8
Barcelos	Brazil	2011	–	PV	Batt	RC	16.2
Beruri	Brazil	2011	–	PV	Batt	RC	10.8
Eirunepé	Brazil	2011	–	PV	Batt	RC	24.3
Maués	Brazil	2011		PV	Batt	RC	56.7
Novo Airao	Brazil	2011	–	PV	Batt	RC	43.2
Araras Pequena (Pará)	Brazil	2011	–	PV	Batt	RC	11.5
Araras Micro (Pará)	Brazil	2011	–	PV	Batt	RC	2.5
Araras Grande Sul (Pará)	Brazil	2011	–	PV	Batt	RC	9.4
Araras Grande Norte (Pará)	Brazil	2011	D	PV, W	Batt	RC	28.7
Cururupu-Maranhao	Brazil	2008	D	PV, W	Batt	I	96.5
Federal University of Juiz de Fora	Brazil	2012	–	PV, W, FC	Batt	UC	72
Condominium in Fortaleza	Brazil	2015	D	–	Batt	CM	25
District Power Plant	Brazil	2012	D	PV, W	Batt	I	25.5
Parintins (Amazonas)	Brazil	2013	D	PV	–	I	17,120
Paraná Smart Grid (Paraná)	Brazil	2012	–	PV	V2G	U	1.4
Lençois Island Microgrid	Brazil	2013	D	PV, W	Batt	I	312.5
Indústria sucroalcooleira	Brazil	2011	D, G	–	–	In	84,000
Coelba (Sao Paulo)	Brazil	2011	–	PV	Batt	U	5,066
Rio de Janeiro	Brazil	2006	H	–	–	In	10,000
Federal University of Ceara	Brazil	2012	–	W, PV, FC	Batt	UC	15

Goldwind Sci&Tech (Beijing)	China	2001	D, H	W, PV	Batt	In	4,000
Tsinghua University	China	2011	D	W, PV	Batt	UC	62
Shanghai ecological island	China	2013	–	W	Batt	I	156,000
Institute of Electrical Engineering, Chinese Academy of Sciences (IEE-CAS)	China	2014	–	W, PV	Batt	UC	70
Dongao Island	China	2002	D	W, PV	Batt	I	5,500
Hunlunber (Mongolia)	China	2012	–	W, PV	Batt	In	200
Tianjin University	China	–	–	W, PV	Batt	UC	3
Hefei University of Technology	China	–	S	W, PV, FC	Batt, Cap	UC	135
Zhongxin Tianjin Eco-city	China	–	–	W, PV	Batt	U	60
Nanlu Island	China	–	D	W,PV	Batt, V2G	I	3,165
Luxi Island	China	–	–	W, PV	Batt	I	1,860
Zhangbei	China	–		W, PV	Batt	U	710,000
Foshan City	China	–	CHP	–	–	U	600
Xi'an High Voltage Apparatus Research Institute	China	–	D	W, PV	Batt	UC	22
Mitsubishi Electric (Xinjiang)	China	2005	D	PV	Batt	In	250
Hangzhou Dianzi Technology University	China	2008	D	PV	Batt, Cap	UC	390
Zhairoushan Island (Zhejiang)	China	2014	–	W, PV	Batt, Cap	I	5,000
Yuquan Campus (Zhejiang)	China	2014	D	W, PV, FC	Batt, Cap	UC	94
Zhejiang Electric Power Test (Hangzhou)	China	2010	D	W, PV	Batt, Fly	In	680
Dongfushan Island	China	2011	D	W, PV	Batt	I	600
Nanjing Electric Power Company	China	2011	–	W, PV	Batt	In	95
Nandu Power Source Company	China	–	–	PV	Batt, Cap	In	505
Tianjin University	China	–	–	PV, W	Batt, Cap	UC	70
Henan College of Finance and Taxation	China	2010	–	PV	Batt	In	580
Shandong Power Institution	China	–	–	W, PV	Batt	In	60
China Electric Power Research Institution (Beijing)	China	–	D	PV	Batt	In	82
eco-city of Xinao (Langfang)	China	2011	G	PV, W	Batt	U	652
Singyes Solar Company (Zhuhai)	China	2012	G	PV, W	Batt	I	3,350

(Continues)

Table 10.2 (*Continued*)

Project	Country	First year of operation	Conventional sources	Renewable sources	Storage	Type	Total capacity (kW)
Chengde (Hebei)	China	2012	–	PV, W	Batt	In	190
Jiangsu Electric Power Research Institute	China	2013	D	PV, W	Batt	In	250
Xiamen University	China	–	–	PV	–	UC	150
Sino Danish (Shanghai)	China	–	–	PV, W	–	CC	200
CIGRE LV (Pathum Thani)	Thailand	–					
Naresuan University (Phitsanulok)	Thailand	2015	D, CHP	W, PV, FC	–	In	118
Mae Hong Son	Thailand	2009	D	PV	Batt	U	10,750
Provincial Electricity Authority (Dansai)	Thailand	2013	D, H	PV	–	In	10,750
Chiang Mai Rajabhat University (Chiang Mai)	Thailand	2013	D	PV	Batt	In	7,300
Mae-Sariang (Chiang Mai)	Thailand	2014	D, H	PV	Batt	UC	16,000
Cho Heng Fty	Thailand	2014	D, H	PV	Batt	In	16,000
National Park (Tarutao)	Thailand	2009	–	PV	Batt	In	200
National Park (Phu Kradung)	Thailand	2009	D	W, PV	Batt	I	184
National Park (Huay ka kang)	Thailand	2009	D	W, PV	Batt	CC	177
Kohjig (Chataburi)	Thailand	2009	D	PV	Batt	CC	174
Doi Intanon Royal (Chiang Mai)	Thailand	2008	D	PV, W	–	RC	82
Wat Chan Royal (Chiang Mai)	Thailand	2009	D	PV	–	RC	6
Kirimas (Sukothai)	Thailand	2009	D	PV	–	RC	6
Tha Takiab (Cha-Choeng Sau)	Thailand	2009	D	PV	–	RC	7
Huai Kha Khaeng (Uthai Thani)	Thailand	2008	D	W, PV	–	RC	5
Lan Island (Chonburi)	Thailand	2010	D	PV	–	RC	52
Institue of Technology Ladkrabang	Thailand	2010	D	W, PV	–	I	760
Toqui (Aysén)	Chile	2011	D, H	W	–	I	9,400
Ollagüe (Antofagasta)	Chile	2014	D	W, PV	–	I	235
Huatacondo (Tarapacá)	Chile	2012	D	W, PV	Batt	I	219
Cobija (Pardo)	Bolivia	–	D	PV	Batt	I	21,400
UPB Microgrid	Colombia	2012	–	W, PV	Batt, V2G	UC	130

Name	Country	Year					
Timatute	Colombia	2012	D	PV		RC	134
Santa Cruz del isolate	Colombia	2013	D	PV	Batt	I	193
Photovoltaic project in Baltra Island and Puerto Ayora	Ecuador	2015	–	W, PV	Batt	I	4,950
Hybrid Project Isabela	Ecuador	2014	D	PV	Batt	I	3,080
Isolated Microgrid Pucallpa (Peru)	Peru	2012	D, G	–	–	CC	24,800
Isolated Microgrid Iquitos (Peru)	Peru	2006	D, G, H	PV	–	CC	34,800
Isolated Microgrid Madre de Dios (Peru)	Peru	2008	D, G	–	–	CC	12,000
Isolated Microgrid Bagua-Jaén (Peru)	Peru	2005	D, G, H	PV	–	CC	46,000
RIAU&KEPRI (Tarempa Anambas island)	Indonesia	–	D	PV	–	I	200
Derawan Island	Indonesia	2011	D	PV	Batt	I	170
Morotai island	Indonesia	2012	D	PV	–	I	600
Tomia island	Indonesia	2011	D	PV	–	I	75
Miangas island	Indonesia	2011	D	PV	–	I	30
Bunaken island	Indonesia	2011	D	PV	Batt	I	335
Saonek Rampa Ampat islands	Indonesia	2010	D	PV	–	I	40
Banda Naira	Indonesia	2010	D	PV	–	I	100
Sumba, East Nusatenggara	Indonesia	2012	H	PV	Batt	I	1,300
Kalkeri Sangeet Vidyalaya DC Microgrid	India	2013	–	PV	Batt	RC	31
Baikampady Mangalore Microgrid	India	2014	–	PV	Batt	RC	2.8
Mendare Village Karnataka Microgrid	India	2015	–	PV	Batt	RC	1.8
Chief Ministers Official Residence Micro-grids, Rajbansi Nagar	India	–	–	PV	–	P	125
Dharnai Microgrid	India	–	–	PV	–	RC	100
Sundarbans Village Microgrid	India	–	–	PV	–	CC	120
Rampura Village Microgrid	India	–	–	PV	–	CC	8.7
KPCL Mandya Karnataka Microgrid	India	–	–	PV	–	RC	100
SYSYAN-Solar Desalination Microgrid (La Paz)	Mexico	–		PV	–	CM	100
Programa bandera blanca (Guásimas del Metate)	Mexico	2012	–	PV	Batt	RC	45.9

(Continues)

Table 10.2 (*Continued*)

Project	Country	First year of operation	Conventional sources	Renewable sources	Storage	Type	Total capacity (kW)
Programa bandera blanca (Tierra blanca del Picacho)	Mexico	2012	–	PV	Batt	RC	45.9
Ecopark Universidad Popular Autónoma el Estado de Puebla (Puebla)	Mexico	2014	–	PV	–	UC	174.1
eThekwini (Durban)	South Africa	2009	G	–	–	CM	7,500
Bokpoort (Groblershoop)	South Africa	2016	–	CSP	–	CM	50,000
KaXu Solar one (Poffader)	South Africa	2016	–	CSP	–	CM	100,000
Khi Solar one (Upington)	South Africa	2016	–	CSP	–	CM	50,000
Jasper (Postmasburg)	South Africa	2012	–	PV	–	U	90,000
Lesedi (Postmasburg)	South Africa	2014	–	PV	–	U	75,000
Letsatsi (Bloemfontain)	South Africa	2014	–	PV	–	U	75,000
Metrowind (Port Elizabeth)	South Africa	2014	–	W	–	U	9,000
Klipheuwel Wind Farm	South Africa	2003	–	W	–	U	3,200
Darling Wind Farm Company (DWP)	South Africa	2000	–	W	–	U	5,200
Hluleka hybrid mini-grid	South Africa	2002	D	W/PV	–	CC	13.1
Maasai Community Solar Microgrid	Tanzania	2014	–	PV	Batt	CC	5.6
Eyasi Lake Pilot Microgrid	Tanzania	2014	–	PV	–	CC	6.3
Northwestern Región microgrid	Tanzania	2016	–	PV	Batt	CC	16,000

Conventional sources: D, diesel; CHP, combined heat and power; S, stream; G, gas; H, hydraulic; M, motor driven gen; BG, biogas, Cog, cogeneration.
Renewable sources: W, wind; PV, photovoltaic; FC, fuel cell; biomass.
Storage: Batt, battery; Fly, flywheel; Cap, capacitor; TH, thermal; V2G, vehicle to grid; CSP, concentrated solar power.
Microgrid type: RC, remote community; I, Island; UC, university campus; CC, city/community; P, public institution; MI, military; CM, commercial; U, utility; In; interconnected.

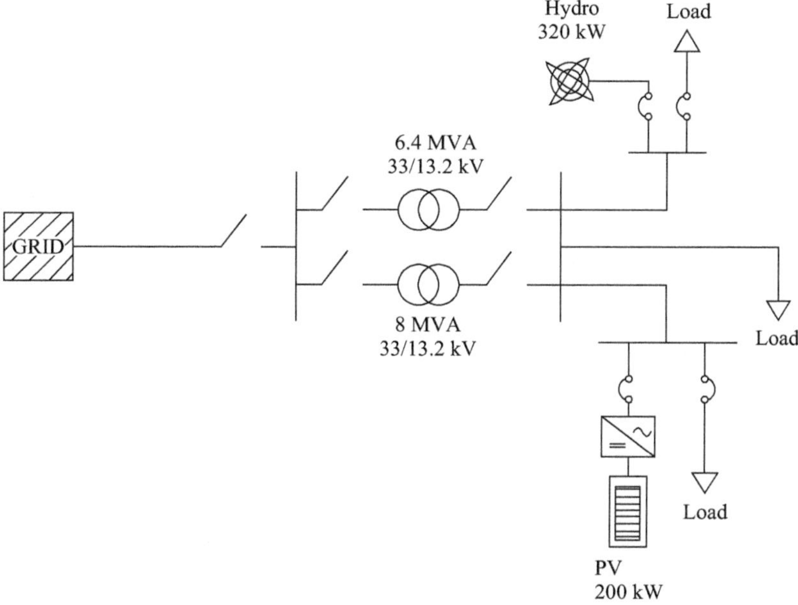

Figure 10.1   Armstrong smart microgrid

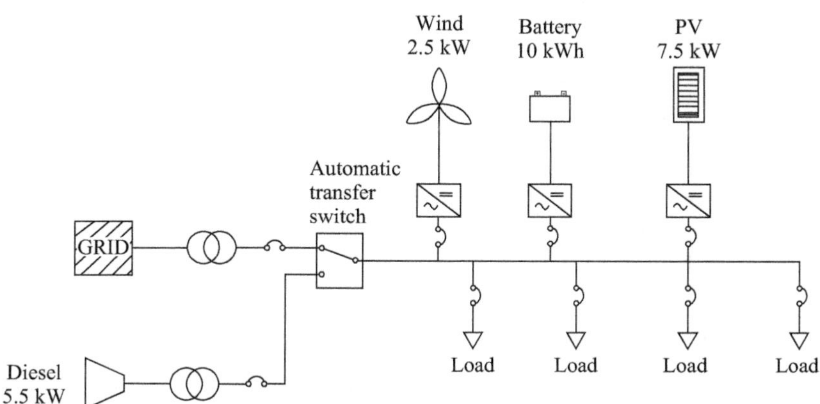

Figure 10.2   District Power Plant Microgrid

and fuel cells. In addition, it has a DC charging system for kart cross vehicles and an AC charging point compatible with SAE J1772 Level II. PV generation has a capacity of 15 kW from the Solar Photovoltaic Laboratory at the University, the third largest in Brazil [20] (Figure 10.4).

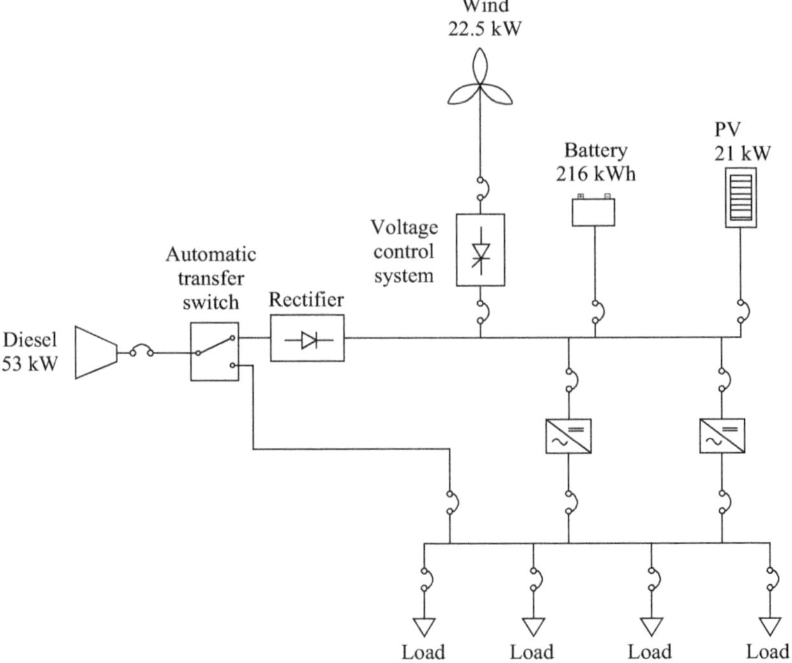

Figure 10.3   Lençois Island Microgrid

10.5.5   Huatacondo Microgrid (Chile)

This microgrid was implemented in the rural region of the Atacama Desert with an approximate population of 500 inhabitants. It operates separately to feed local loads in the area. It has three sources of generation: Solar PV, two wind turbines, a diesel generator, and an energy storage system. The total capacity of the system is 219 kW [71] (Figure 10.5).

10.5.6   UPB Microgrid (Colombia)

It is a microgrid implemented on the campus of the Pontifical Bolivarian University in Medellin for research, implementation of benchmarking and development of solutions according to the socioeconomic conditions of Colombia. It is made of solar cells distributed along different buildings with a capacity of 50–70 kWp, vertical axis wind turbines of 5–10 kW, generation scheme for non-interconnected areas such as an anaerobic bio-digester, production of bio-fertilizers, and methane for cooking. Electric vehicle management system with five charging stations and a modular charging station based on a solar cell for small vehicles [72] (Figure 10.6).

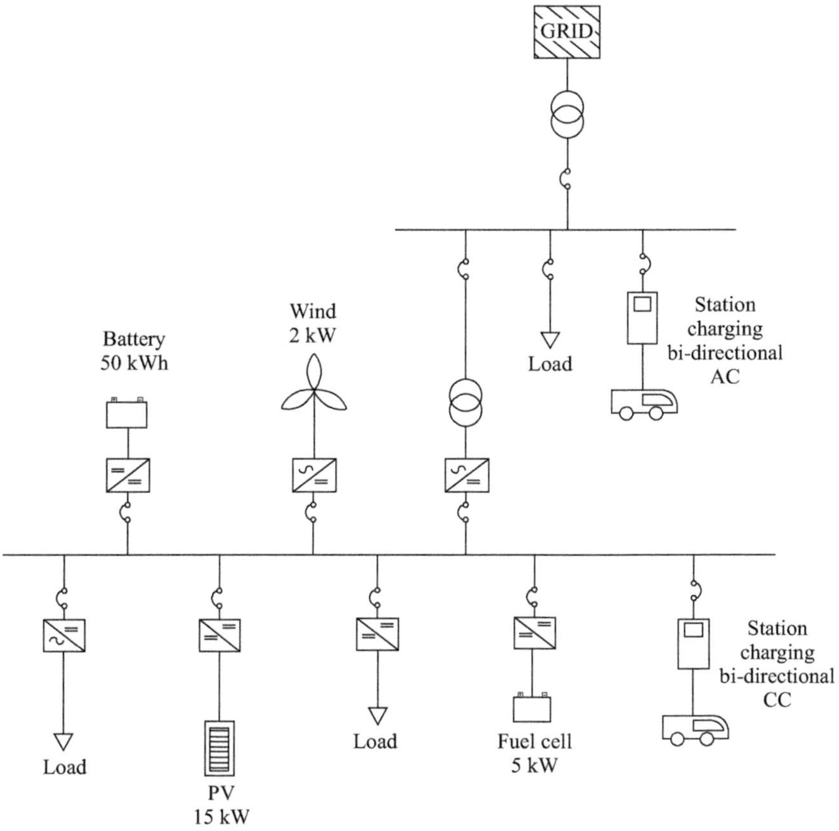

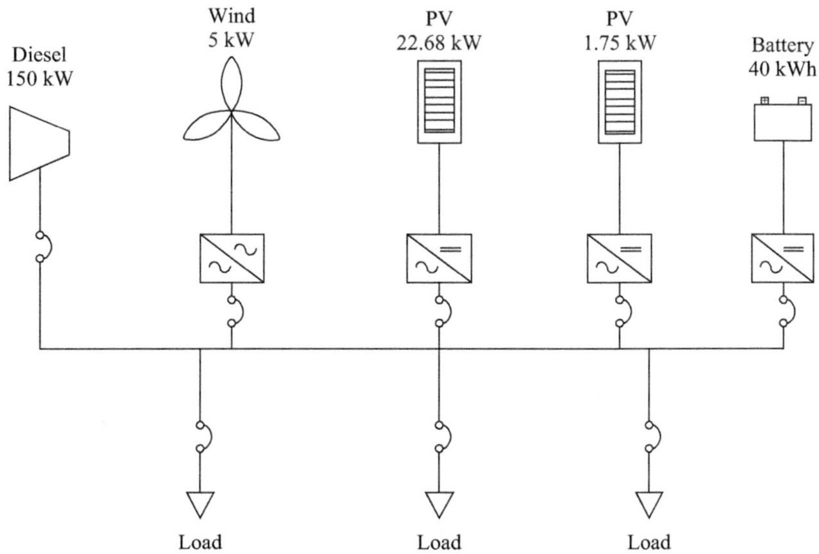

Figure 10.4   Microgrid in the building of the Federal University of Juiz de Fora

Figure 10.5   Huatacondo Microgrid

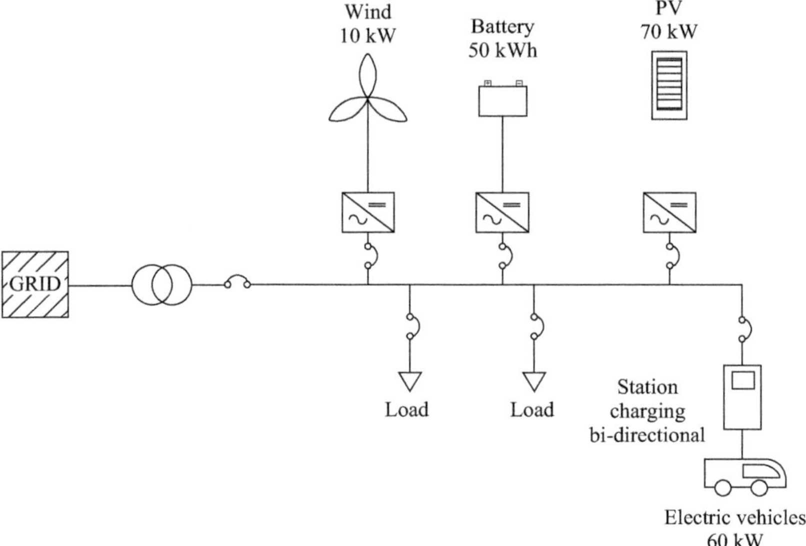

Figure 10.6   UPB Microgrid

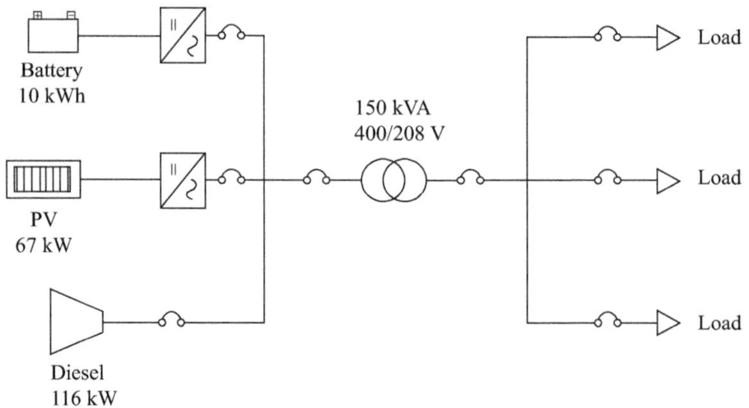

Figure 10.7   Santa Cruz del Islote Microgrid

10.5.7   *Santa Cruz del Islote Microgrid (Colombia)*

Santa Cruz del Islote belongs to the Archipelago of San Bernardo, located in the Gulf of Morrosquillo in the Colombian Caribbean Sea. Currently, it has 127 registered users. This town has a diesel–solar hybrid system with the following characteristics: A solar panel of 67 kWp, a bank of 144 batteries of 4,800 Ah 12 V, a diesel generation unit of 116 kW that feeds a distribution transformer from 150 to 208 kV A [73] (Figure 10.7).

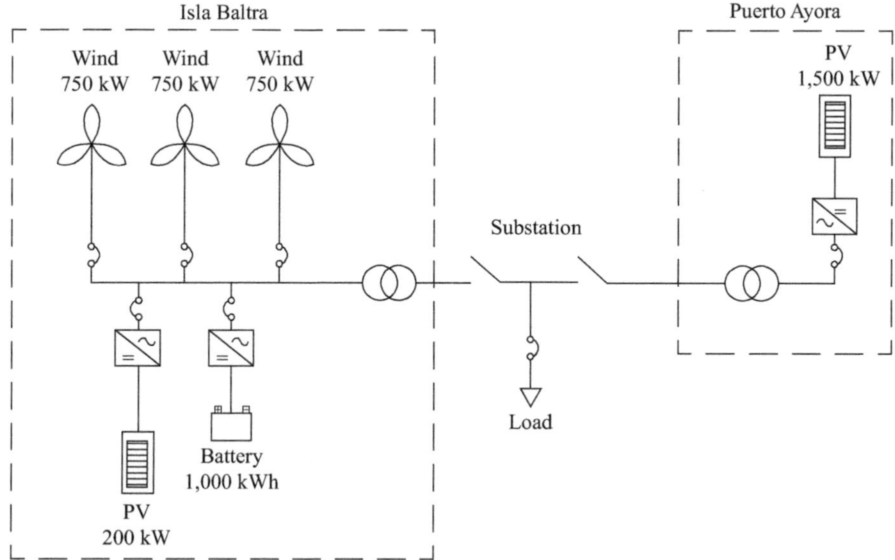

Figure 10.8　Photovoltaic project in Baltra Island and Puerto Ayora

10.5.8　Photovoltaic project in Baltra Island and Puerto Ayora (Ecuador)

The project aims to implement a 200-kWp PV system with a power storage system of 1,000 kW h in hybrid-type industrial batteries (lithium-ion and lead–acid). This project takes advantage of the solar resource of Baltra Island; its energy will be transported to the substation *ELECGALÁPAGOS* in Puerto Ayora through the transmission line implemented for the wind project [74] (Figure 10.8).

The Puerto Ayora PV project of 1.5 MWp will allow the coordination of penetration of the various renewable energy projects currently being developed in both Baltra Island and Santa Cruz [75].

10.5.9　Hybrid Project Isabela (Ecuador)

The project is developed on Isabela Island (see Figure 10.9) and covers [74,75]:

- Dual thermal plant (diesel/pinion pure oil) with an approximate capacity of 1.23 MW.
- PV plant of 1.1 MW.
- Energy storage system of 750 kW through lithium-ion batteries.
- Control system, operation, and monitoring of the entire plant.
- Fuel storage tanks: 2×60 cubic meters for pinion oil and 1×60 pinion cubic meters for diesel.

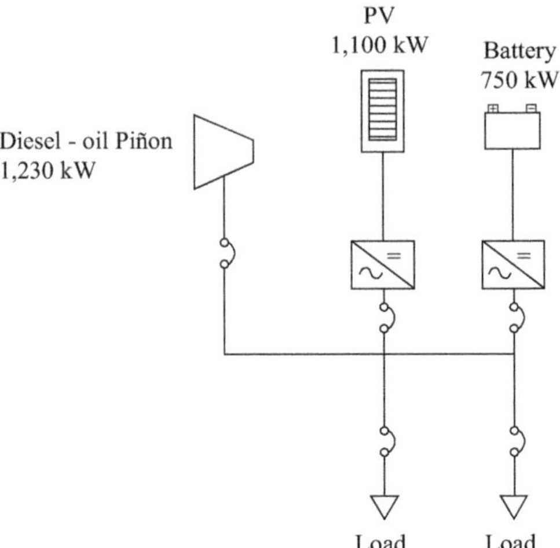

Figure 10.9   Hybrid Project Isabela

10.5.10   Microgrid in Hangzhou Dianzi Technology University (China)

In 2008, the microgrid of the University of Hangzhou Dianzi was built with the support of the New Energy and Industrial Technology Development Organization (NEDO), Development and Reform Commission of Zhejiang Province and National Development and Reform Commission (NDRC). It comprises 120 kW of diesel generation, 120 kW of PV generation using battery storage (50 kW h), and super-capacitors (100 kW). In addition, it has dynamic voltage regulators in order to improve the quality of the microgrid power [76] (Figure 10.10).

10.5.11   Microgrid in Langfang Eco-Smart City (China)

It was built by Xinao Energy Company in 2011 in order to present their own biomass, PV, and wind energy generation plants. This microgrid is integrated into a smart building in Langfang Eco-Smart City; it has 100 kW of PV power, a gas turbine functioning in combined cycle of 150 kW, a wind turbine of 2 kW, and energy storage through the use of 100 kW×4 h batteries [26,77] (Figure 10.11).

10.5.12   Microgrid in Chengde (China)

This microgrid was implemented in Chengde, Hebei, in 2012. It meets local residential demand and can supply electricity to the local utility. It has two arrays of PV solar cells with a capacity of 25 kW, two wind turbines of 30 kW, and an energy storage system of 80 kWh. In addition, the system has smart meters, energy management system, and active power filters to improve the power quality of the microgrid [78] (Figure 10.12).

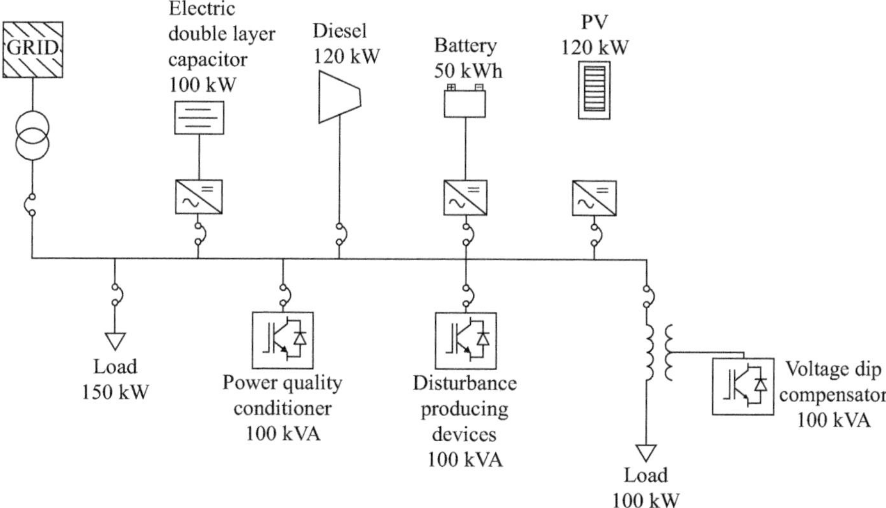

Figure 10.10   Microgrid at the Technological University of Hangzhou Dianzi

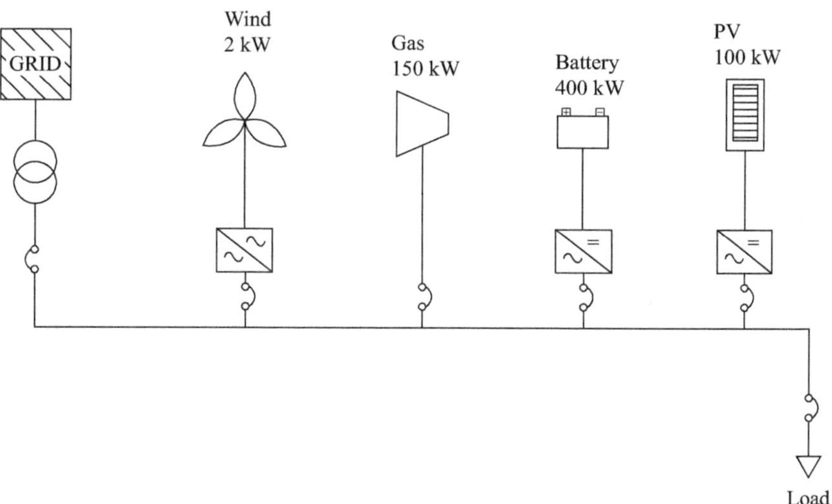

Figure 10.11   Microgrid in Langfang Eco-Smart City

10.5.13   Mae-Sariang (Chiang Mai) Microgrid (Thailand)

It is a microgrid implemented in the district of Mae Sariang (Chiang Mai), a region that is part of the Provincial Electricity Authority of Thailand. It can connect itself to the distribution network or operate separately. It has three local sources of generation: diesel, solar PV, and hydropower, as well as a battery storage system.

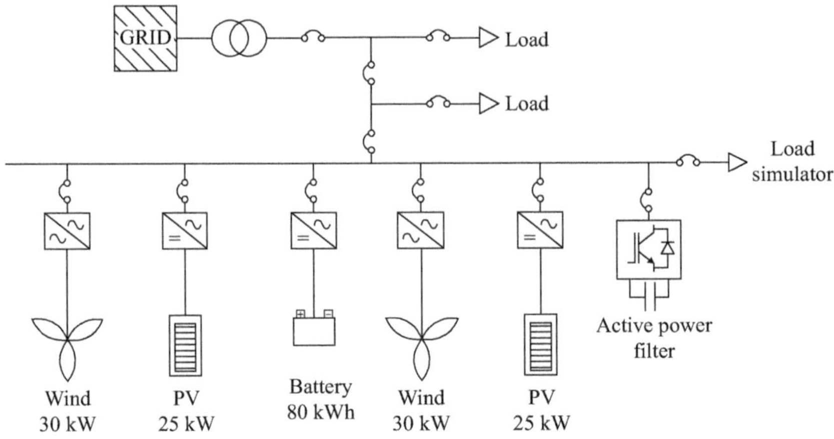

Figure 10.12   Microgrid in Chengde

The installed capacity of each one is 7.5, 4, 1.5, and 3 MW, respectively. The hydroelectric plant is located at a distance of 1.2 km from the substation, the solar system at 4 km, and the diesel generator and batteries at 0.2 km. The microgrid has an installed load of 5.1 MW, which is distributed in five distribution circuits with an average distance of 5 km [79] (Figure 10.13).

10.5.14   Mae Hong Son microgrid (Thailand)

Electricity Generating Authority of Thailand developed this microgrid as a smart grid pilot project in Mae Hong Son province [80,81]. It has two hydroelectric plants of total nominal capacity of 5.85 MW, a diesel generation facility (4.4 MW), and a solar PV system of 0.5 MW. The installed load on the system is 8 MW. It can operate connected to the distribution network or separately. The most typical system loads are Hospital, Administrative Center, Airport and Residential Zones [81] (Figure 10.14).

10.5.15   Community microgrids in India (Dharnai, Sundarbans Islands, Sagar Island)

The Republic of India is the second country with the world's largest population, but many of its regions face extreme poverty. One of these cities is Dharnai where all aspects of life are powered by solar energy [82]. Community microgrid of Dharnai integrates a 100-kW PV system, which is distributed to supply 450 housing units, 50 shops, three schools, a health center, and 30 kW for hydraulic pumps. The microgrids of the Sundarbans islands project are a pilot development to address energy needs and lighting with the implementation of 120 kW of PV panels on the roofs of houses [83]. The Sagar Island microgrid is composed of 250-kW PV panels and 400-kW diesel generation for approximately 1,500 residential users [83].

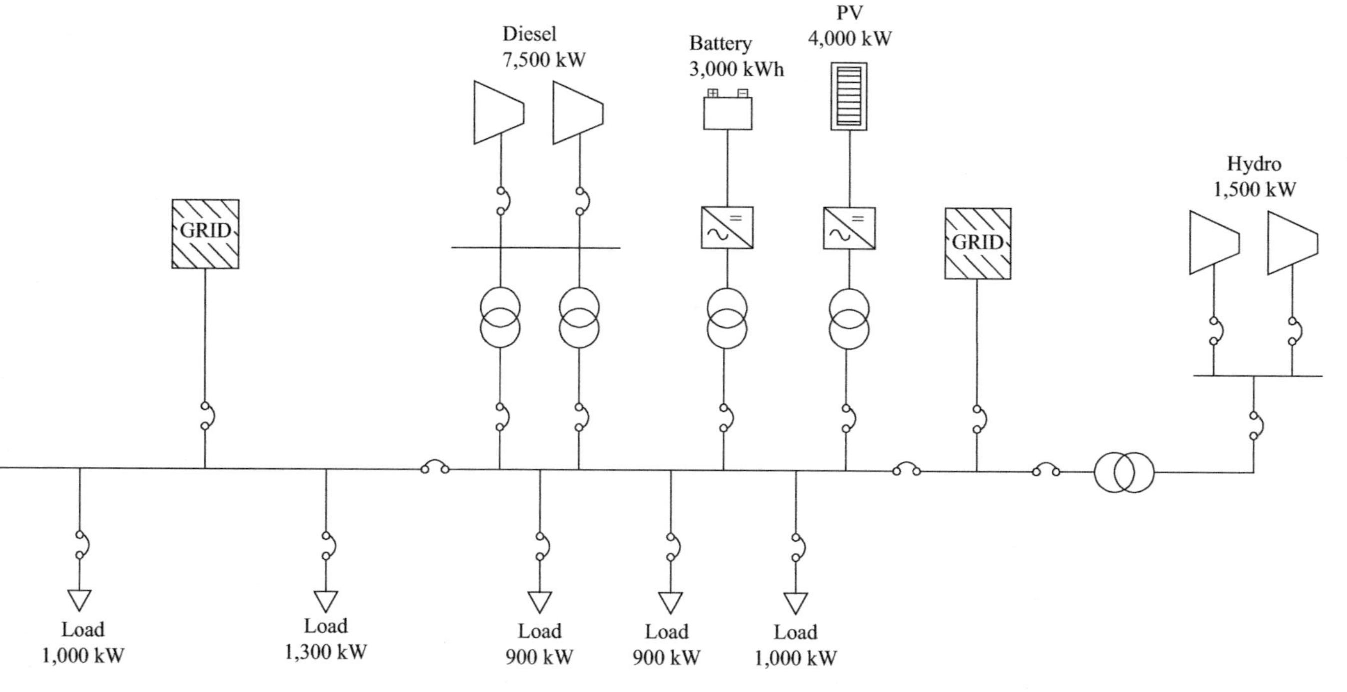

Figure 10.13   Mae-Sariang Microgrid (Chiang Mai)

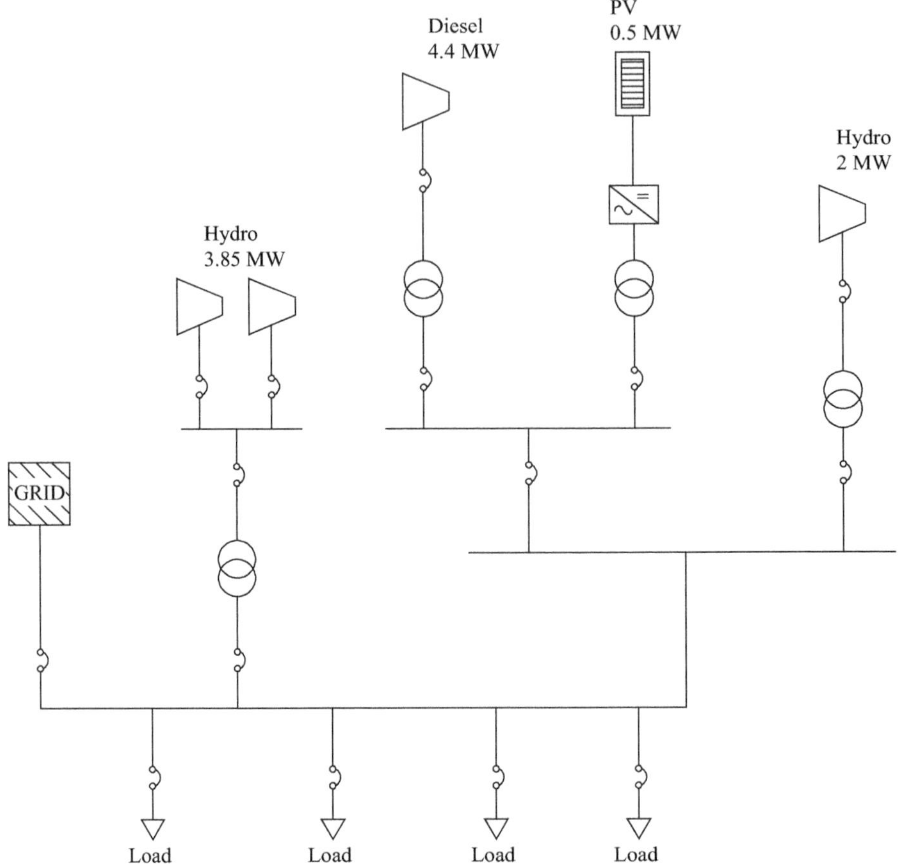

Figure 10.14   Mae Hong Son microgrid

References

[1]   World Trade Organization, "World Trade Organization", 2016 [Online]. Available: https://www.wto.org/.

[2]   P. Sadorsky, "Renewable energy consumption and income in emerging economies," *Energy Policy*, vol. 37, no. 10, pp. 4021–4028, 2009.

[3]   A. Vargas, O. Saavedra, M. Samper, S. Rivera, and R. Rodriguez, "Latin American energy markets, investment opportunities in nonconventional renewables," *IEEE Power Energy Mag.*, vol. 14, no. 5, pp. 38–47, 2016.

[4]   C. Flavin, M. Gonzalez, A. M. Majano, *et al.*, "Study on the Development of the Renewable Energy Market in Latin America and the Caribbean," Inter-American Development Bank, 2014.

[5] ABS Consulting Group, Diagnóstico de la Generación en América Latina y el Caribe: Bolivia, Latin American Energy Organization, 2013.

[6] Comisión Nacional de Energía Atómica (Argentinean National Atomic Energy Commission), "Sintesis del mercado electrico mayorista de la republica de Argentina," 2016.

[7] Ministry of Mines and Energy and Empresa de Pesquisa Energética (Energy Research Company), "Brazilian Energy Balance," 2015.

[8] Ministerio de Industria Energia y Mineria de Uruguay (Uruguayan Ministry of Industry, Energy and Mining), "Oferta de Energia. 50 años del Balance Energetico Nacional," 2015.

[9] K. Schneider, J. L. Turner, A. Jaffe, and N. Ivanova, "Choke point China: confronting water scarcity and energy demand in the world's largest country," *Vermont J. Environ. Law*, vol. 12, no. 3, pp. 713–733, 2010.

[10] The Climate Group, CDP, CREIA, and IRENA, "RE 100 China's Fast Track to a Renewable Future," 2015.

[11] Ministry of Energy and Mineral Resources (Republic of Indonesia), Handbook of Energy & Economic Statistic of Indonesia 2014, 2014.

[12] M. H. Hasan, T. M. I. Mahlia, and H. Nur, "A review on energy scenario and sustainable energy in Indonesia," *Renew. Sustain. Energy Rev.*, vol. 16, no. 4, pp. 2316–2328, 2012.

[13] S. Mujiyanto and G. Tiess, "Secure energy supply in 2025: Indonesia's need for an energy policy strategy," *Energy Policy*, vol. 61, pp. 31–41, 2013.

[14] D. Schnitzer, D. S. Lounsbury, J. P. Carvallo, R. Deshmukh, J. Apt, and D. M. Kammen, "Microgrids for Rural Electrification: A Critical Review of Best Practices Based on Seven Case Studies," United Nations Foundation, 2014.

[15] A. Ibrahim, "Renewable energy sources in the Egyptian electricity market: a review," *Renew. Sustain. Energy Rev.*, vol. 16, no. 1, pp. 216–230, 2012.

[16] Department of Energy (Republic of South Africa), "South African Energy Sector," 2013.

[17] Department of Energy (Republic of South Africa), "State of Renewable Energy in South Africa," 2015.

[18] A. S. Vazquez, "Participación de las Fuentes de energía renovables en América Latina y el Caribe," Centro de Gestión de la Información y Desarrollo de la Energía (Center of Information Management and Energy Development of Cuba), 2012.

[19] L. A. De Souza Ribeiro, O. R. Saavedra, S. L. De Lima, and J. G. De Matos, "Isolated micro-grids with renewable hybrid generation: the case of Lençois island," *IEEE Trans. Sustain. Energy*, vol. 2, no. 1, pp. 1–11, 2011.

[20] M. Rodriguez, A. Moura, L. Borges, and P. Almeida, "Microrrede híbrida cc/ca baseada em fontes de energia renovável aplicada a um edifício sustentável," *An. do XIX Congr. Bras. Autom. CBA* 2012, pp. 1–9, 2012.

[21] J. B. Almada, R. P. S. Leão, F. F. D. Montenegro, S. S. V Miranda, and R. F. Sampaio, "Modeling and simulation of a microgrid with multiple energy resources," in *IEEE EuroCon* 2013, July, pp. 1150–1157, 2013.

[22] F. Torres, E. Gomez, C. Alatorre, *et al.*, "Energías Renovables para el Desarrollo Sustentable en México," Secretaría de Energía (SENER) and Deutsche Gesellschaft für Technische Zusammenarbeit (GTZ) GmbH, 2006.

[23] Irena, "Renewable Energy Policy Brief: Chile," International Renewable Energy Agency (IRENA), June 2015.

[24] A. Halu, A. Scala, A. Khiyami, and M. C. González, "Data-driven modeling of solar-powered urban microgrids.," *Sci. Adv.*, vol. 2, no. 1, p. e1500700, 2016.

[25] K. Lo, "A critical review of China's rapidly developing renewable energy and energy efficiency policies," *Renew. Sustain. Energy Rev.*, vol. 29, pp. 508–516, 2014.

[26] X. Wu, X. Yin, Q. Wei, Y. Jia, and J. Wang, "Research on microgrid and its application in China," *Energy Power Eng.*, vol. 5, no. 4, pp. 171–176, 2013.

[27] C. Marnay, B. Kroposki, M. Mao, *et al.*, "The Tianjin 2014 Symposium on Microgrids: a meeting of the minds for international microgrid experts," *IEEE Electrif. Mag.*, vol. 3, no. 1 (Mar), pp. 79–85, 2015.

[28] B. T. Samad, E. Koch, and P. Stluka, "Automated demand response for smart buildings and microgrids: the state of the practice and research challenges," *Proc. IEEE*, vol. 104, no. 4, pp. 726–744, 2016.

[29] G. Kariniotakis, A. Dimeas, and F. Van, "Pilot Sites: Success Stories and Learnt Lessons," in *Microgrids: Architectures and Control*, Chichester: John Wiley & Sons Ltd., pp. 206–274, 2013.

[30] R. Zhao, L. Zhao, S. Deng, and N. Zheng, "Trends in patents for solar thermal utilization in China," *Renew. Sustain. Energy Rev.*, vol. 52, pp. 852–862, 2015.

[31] S. Mezzetti, "The Renewable Energy Sector in India: An Overview of Research and Activity," European Business and Technology Centre, 2011.

[32] V. S. K. M. Balijepalli, S. A. Khaparde, and C. V Dobariya, "Deployment of MicroGrids in India," in *IEEE PES General Meeting*, pp. 1–7, 2010.

[33] H. B. Dulal, K. U. Shah, C. Sapkota, G. Uma, and B. R. Kandel, "Renewable energy diffusion in Asia: can it happen without government support?," *Energy Policy*, vol. 59, pp. 301–311, 2013.

[34] O. Guillen Solis, "El ministerio de energías renovables de India, un caso de éxito," 2008. [Online]. Available: http://www.energiaadebate.com/Articu los/Mayo2008/OmarGuillenSolisMayo2008.htm.

[35] H. Patel and S. Chowdhury, "Review of technical and economic challenges for implementing rural microgrids in South Africa," in *2015 IEEE Eindhoven PowerTech*, pp. 1–6, 2015.

[36] M. Abdel-Salam, A. Ahmed, H. Ziedan, *et al.*, "Aggregation of microgrids for irrigation in Toshka area," in *2013 International Conference on Clean Electrical Power (ICCEP)*, pp. 496–502, 2013.

[37] R. De Dicco, "Diagnóstico y perspectivas del sector eólico en Argentina," Centro Latinoamericano de Investigaciones Cientificas y Tecnicas (Latin-American Center of Technical and Scientific Research), 2012.

[38] C. Giralt, "Energía eólica en Argentina: un análisis económico del derecho," *Rev. Let. Verdes*, vol. 9, no. 1, pp. 64–86, 2011.

[39] O. Medina, "Smart Grids The Development of Smart Grid Pilot Projects in Argentina," Secretaria de Energia, Ministerio de Planificacion Federal, Inversion Publica y Servicios (Energy Department, Ministry of Federal Planification, Public Investment and Services or Argentine), 2015.

[40] J. Sáenz, "Redes Inteligentes (Smart Grids) estado actual y su influencia en el aprovechamiento de las energías renovables. Ciudad de armstrong – 1ª Prueba Piloto En Argentina," Universidad Tecnologica Nacional (Argentinean National Technical University), 2013.

[41] F. Cassias, C. Pica, L. Gentilini, V. Maryama, and D. Makohin, "Design and Implementation of a Feasible Microgrid Model in Brazil," PCIM South America 2014 Conference, October, pp. 1–9, 2014.

[42] G. L. Susteras, D. S. Ramos, J. R. de A. Chaves, and A. C. V. J. Susteras, "Attracting wind generators to the wholesale market by mitigating individual exposure to intermittent outputs: an adaptation of the Brazilian experience with hydro generation," in *2011 Eighth International Conference on the European Energy Market (EEM)*, pp. 674–679, 2011.

[43] C. P. Salomon, W. C. Santana, E. L. Bonaldi, *et al.*, "A system for turbo-generator predictive maintenance based on Electrical Signature Analysis," in *2015 IEEE International Instrumentation and Measurement Technology Conference (I2MTC) Proceedings*, pp. 79–84, 2015.

[44] R. B. Heideier, A. L. V Gimenes, and M. E. M. Udaeta, "Photovoltaic Power: Review and Trends for Brazil," Business and Management Review (BMR), 2014.

[45] Z. Zeng, R. Zhao, H. Yang, and S. Tang, "Policies and demonstrations of micro-grids in China: a review," *Renew. Sustain. Energy Rev.*, vol. 29, pp. 701–718, 2014.

[46] T. Grau, M. Huo, and K. Neuhoff, "Survey of photovoltaic industry and policy in Germany and China," *Energy Policy*, vol. 51, pp. 20–37, 2012.

[47] L. Liu, S. Nakano, and K. Kakimoto, "Carbon concentration and particle precipitation during directional solidification of multicrystalline silicon for solar cells," *J. Cryst. Growth*, vol. 310, no. 7, pp. 2192–2197, 2008.

[48] W. Chen, X. Yin, and H. Zhang, "Towards low carbon development in China: a comparison of national and global models," *Clim. Change*, vol. 136, no. 1 (May), pp. 95–108, 2016.

[49] M. Yang, D. Patiño-Echeverri, and F. Yang, "Wind power generation in China: understanding the mismatch between capacity and generation," *Renew. Energy*, vol. 41, pp. 145–151, 2012.

[50] V. Khare, S. Nema, and P. Baredar, "Status of solar wind renewable energy in India," *Renew. Sustain. Energy Rev.*, vol. 27, pp. 1–10, 2013.

[51] A. Bhide and C. R. Monroy, "Energy poverty: a special focus on energy poverty in India and renewable energy technologies," *Renew. Sustain. Energy Rev.*, vol. 15, no. 2, pp. 1057–1066, 2011.

[52] R. Sen and S. C. Bhattacharyya, "Off-grid electricity generation with renewable energy technologies in India: an application of HOMER," *Renew. Energy*, vol. 62, pp. 388–398, 2014.

[53] S. Luthra, S. Kumar, D. Garg, and A. Haleem, "Barriers to renewable/ sustainable energy technologies adoption: Indian perspective," *Renew. Sustain. Energy Rev.*, vol. 41, pp. 762–776, 2015.

[54] N. U. Blum, R. Sryantoro Wakeling, and T. S. Schmidt, "Rural electrification through village grids: assessing the cost competitiveness of isolated renewable energy technologies in Indonesia," *Renew. Sustain. Energy Rev.*, vol. 22, pp. 482–496, 2013.

[55] IRENA, "Renewable Energy Policy Brief Argentina," International Renewable Energy Agency (IRENA), 2015.

[56] IRENA, "Renewable Energy Policy Brief Brazil," 2015.

[57] REN21, "Renewable Energy Policy Networks for the 21st Century," REN21. [Online]. Available: http://www.reegle.info/policy-and-regulatory-over views/.

[58] IRENA, "Renewable Energy Policy Brief: Ecuador," International Renewable Energy Agency (IRENA), June 2015.

[59] IRENA, "Renewable Energy Policy Brief Mexico," International Renewable Energy Agency (IRENA), June 2015.

[60] IRENA, "Renewable Energy Policy Brief Uruguay," International Renewable Energy Agency (IRENA), June 2015.

[61] IRENA, "Renewable Energy Policy Brief Venezuela," International Renewable Energy Agency (IRENA), June 2015.

[62] NREL, "Renewable Energy Policy in China: Overview," National Renewable Energy Laboratory (NREL), October 2004.

[63] L. Junfeng, Z. Li, H. Runqing, Z. Zhengmin, S. Jingli, and S. Yangin, "Policy analysis of the barriers to renewable energy development in the People's Republic of China," *Energy Sustain. Dev.*, vol. 6, no. 3 (Sep), pp. 11–20, 2002.

[64] P. Krithika and S. Mahajan, "Governance of Renewable Energy in India: Issues and Challenges," The Energy and Resources Institute, 2014.

[65] A. Senoaji, "Policy and Regulation of Renewable Energy in Indonesia," *GIZ Workshop*, pp. 1–19, 2011.

[66] M. Jabeen, M. Umar, M. Zahid, M. Ur Rehaman, R. Batool, and K. Zaman, "Socio-economic prospects of solar technology utilization in Abbottabad, Pakistan," *Renew. Sustain. Energy Rev.*, vol. 39, pp. 1164–1172, 2014.

[67] E. J. Bala, "Renewable energy activities and incentives in Nigeria," *Natl. Forum Renew. Energy, Energy Effic. Conserv.*, pp. 1–37, 2011.

[68] L. J. S. Baiyegunhi and M. B. Hassan, "Rural household fuel energy transition: evidence from Giwa LGA Kaduna State, Nigeria," *Energy Sustain. Dev.*, vol. 20, pp. 30–35, 2014.

[69] C. B. L. Jumbe, F. B. M. Msiska, and M. Madjera, "Biofuels development in Sub-Saharan Africa: are the policies conducive?," *Energy Policy*, vol. 37, no. 11, pp.4980–4986, 2009.

[70] A. Pradhan and C. Mbohwa, "Development of biofuels in South Africa: challenges and opportunities," *Renew. Sustain. Energy Rev.*, vol. 39, pp. 1089–1100, 2014.

[71] R. Palma-Behnke, C. Benavides, F. Lanas, *et al.*, "A microgrid energy management system based on the rolling horizon strategy," *IEEE Trans. Smart Grid*, vol. 4, no. 2, pp. 996–1006, 2013.

[72] Universidad Pontificia Bolivariana UPB, "Microred Inteligente UPB," Universidad Pontificia Bolivariana. [Online]. Available: https://microred.upb.edu.co.

[73] "Instituto de Planificacion y Promocion de Soluciones Energeticas para las Zonas No Interconectadas (IPSE) (Institute of Planification and Promotion of Energy Solutions for Non-Interconnected Areas, of Colombia)." [Online]. Available: http://www.ipse.gov.co/.

[74] Energias Renovables para Galapagos (ERGaL) (Renewable Energy for Galapagos), "Ayuda Memoria del proyecto de energias Renovables para las islas Galapagos," 2006.

[75] Empresa Eléctrica Provincial Galápagos (ELECGALAPAGOS) (Galapagos Provincial Electric Company), "Elecgalapagos." [Online]. Available: http://www.elecgalapagos.com.ec/proyectos.

[76] J. Zhang and Q. Zhang, "Microgrid Lab at the Hangzhou Dianzi University," Hangzhou Dianzi University, 2012. [Online]. Available: http://eetd.lbl.gov/news/events/2012/08/06/microgrid-lab-at-the-hangzhou-dianzi-university. [Accessed: 23 June 2016].

[77] W. Luan, "Non-Traditional Distribution Networks Microgrids for Rural Electrification," International Electrotechnical Commission (IEC), 2015.

[78] C. Kang, X. Chen, Q. Xu, *et al.*, "Balance of power: toward a more environmentally friendly, efficient, and effective integration of energy systems in China," *IEEE Power Energy Mag.*, vol. 11, no. 5 (Sep), pp. 56–64, 2013.

[79] P. Punjad, P. Yuthagovit, and S. Atchvsunthon, "Case study of micro power grid applications in remote rural area of Thailand," in *AORC Technical Meeting 2014*, pp. 1–6, 2014.

[80] W. Srithiam, S. Asadamonkol, and T. Sumranwanich, "Smart grid national pilot project in Mae Hong Son Province, Thailand," *Energy Environ.*, vol. 26, pp. 23–34, 2015.

[81] N. Hoonchareon, "Smart grid pilot project and study results," in *Technical Seminar "Future of Renewable Energy and Smart Grid Technologies"*, pp. 1–15, 2014.

[82] V. Vel, K. Meghana, R. Ramesh, H. Sekar, and V. Vijayaraghavan, "Flexible D-agent architecture for DER operation in a rural Indian microgrid," in *2015 IEEE Global Humanitarian Technology Conference (GHTC)*, pp. 26–32, 2015.

[83] K. Ravindra, B. Kannan, and N. Ramappa, "Microgrids: a value-based paradigm: the need for the redefinition of microgrids," *IEEE Electrif. Mag.*, vol. 2, no. 1 (Mar), pp. 20–29, 2014.

Index